LONDON MATHEMATICAL SOCIETY STUDENT TEXTS

Managing Editor: Professor D. Benson,
Department of Mathematics, University of Aberdeen, UK

London Mathematical Society Student Texts 88

Groups, Languages and Automata

DEREK F. HOLT
University of Warwick

SARAH REES
University of Newcastle upon Tyne

CLAAS E. RÖVER
National University of Ireland, Galway

CAMBRIDGE
UNIVERSITY PRESS

Shaftesbury Road, Cambridge CB2 8EA, United Kingdom

One Liberty Plaza, 20th Floor, New York, NY 10006, USA

477 Williamstown Road, Port Melbourne, VIC 3207, Australia

314–321, 3rd Floor, Plot 3, Splendor Forum, Jasola District Centre, New Delhi – 110025, India

103 Penang Road, #05–06/07, Visioncrest Commercial, Singapore 238467

Cambridge University Press is part of Cambridge University Press & Assessment,
a department of the University of Cambridge.

We share the University's mission to contribute to society through the pursuit of
education, learning and research at the highest international levels of excellence.

www.cambridge.org
Information on this title: www.cambridge.org/9781316606520

DOI: 10.1017/9781316588246

First published 2017

A catalogue record for this publication is available from the British Library

ISBN   978-1-107-15235-9   Hardback
ISBN   978-1-316-60652-0   Paperback

Contents

v

Preface

This book explores connections between group theory and automata theory. We were motivated to write it by our observations of a great diversity of such connections; we see automata used to encode complexity, to recognise aspects of underlying geometry, to provide efficient algorithms for practical computation, and more.

The book is pitched at beginning graduate students, and at professional academic mathematicians who are not familiar with all aspects of these interconnected fields. It provides background in automata theory sufficient for its applications to group theory, and then gives up-to-date accounts of these various applications. We assume that the reader already has a basic knowledge of group theory, as provided in a standard undergraduate course, but we do not assume any previous knowledge of automata theory.

The groups that we consider are all finitely generated. An element of a group G is represented as a product of powers of elements of the generating set X, and hence as a string of symbols from $A := X \cup X^{-1}$, also called words. Many different strings may represent the same element. The group may be defined by a presentation; that is, by its generating set X together with a set R of relations, from which all equations in the group between strings can be derived. Alternatively, as for instance in the case of automata groups, G might be defined as a group of functions generated by the elements of X.

Certain sets of strings, also called languages, over A are naturally of interest. We study the word problem of the group G, namely the set $\mathrm{WP}(G, A)$ of strings over A that represent the identity element. We define a language for G to be a language over A that maps onto G, and consider the language of all geodesics, and various languages that map bijectively to G. We also consider combings, defined to be group languages for which two words representing either the same element or elements that are adjacent in the Cayley graph fellow travel; that is, they are at a bounded distant apart throughout their length.

ix

We consider an automaton to be a device for defining a, typically infinite, set L of strings over a finite alphabet, called the language of the automaton. Any string over the finite alphabet may be input to the automaton, and is then either accepted if it is in L, or rejected if it is not. We consider automata of varying degrees of complexity, ranging from finite state automata, which define regular languages, through pushdown automata, defining context-free languages, to Turing machines, which set the boundaries for algorithmic recognition of a language. In other words, we consider the full Chomsky hierarchy of formal languages and the associated models of computation.

Finite state automata were used by Thurston in his definition of automatic groups after he realised that both the fellow traveller property and the finiteness of the set of cone types that Cannon had identified in the fundamental groups of compact hyperbolic manifolds could be expressed in terms of regular languages. For automatic groups a regular combing can be found; use of the finite state automaton that defines this together with other automata that encode fellow travelling allows in particular a quadratic time solution to the word problem. Word-hyperbolic groups, as defined by Gromov, can be characterised by their possession of automatic structures of a particular type, leading to linear time solutions to the word problem.

For some groups the set of all geodesic words over A is a regular language. This is true for word-hyperbolic groups and abelian groups, with respect to any generating set, and for many other groups, including Coxeter groups, virtually free groups and Garside groups, for certain generating sets. Many of these groups are in fact automatic.

The position of a language in the Chomsky hierarchy can be used as a measure of its complexity. For example, the problem of deciding whether an input word w over A represents the identity of G (which, like the set it recognises, is called the word problem) can be solved by a terminating algorithm if and only if the set $WP(G, A)$ and its complement can be recognised by a Turing machine; that is, if and only if $WP(G, A)$ is recursive. We also present a proof of the well-known fact that finitely presented groups exist for which the word problem is not soluble; the proof of this result encodes the existence of Turing machines with non-recursive languages.

When the word problem is soluble, some connections can be made between the position of the language $WP(G, A)$ in the Chomsky hierarchy and the algebraic properties of the group. It is elementary to see that $WP(G, A)$ is regular if and only if G is finite, while a highly non-trivial result of Muller and Schupp shows that a group has context-free word problem if and only if it has a free subgroup of finite index.

Attaching an output-tape to an automaton extends it from a device that defines a set of strings to a function from one set of strings to another. We call such a device a transducer, and show how transducers can be used to define groups. Among these groups are finitely generated infinite torsion groups, groups of intermediate growth, groups of non-uniform exponential growth, iterated monodromy groups of post-critically finite self-coverings of the Riemann sphere, counterexamples to the strong Atiyah conjecture, and many others. Our account is by no means complete, as it concentrates on introducing terminology, the exposition of some basic techniques and pointers to the literature.

There is a shorter book by Ian Chiswell [68] that covers some of the same material as we do, including groups with context-free word problem and an introduction to the theory of automatic groups. Our emphasis is on connections between group theory and formal language theory rather than computational complexity, but there is a significant overlap between these areas. We recommend also the article by Mark Sapir [226] for a survey of results concerning the time and space complexity of the fundamental decision problems in group theory.

Acknowledgements We are grateful to Professor Rick Thomas for numerous helpful and detailed discussions on the topics covered in this book. Some of the text in Section 1.2, and the statement and proofs of Propositions 3.4.9 and 3.4.10 were written originally by Rick.

We are grateful also to Professor Susan Hermiller and to an anonymous reviewer for a variety of helpful comments.

PART ONE

INTRODUCTION

1
Group theory

1.1 Introduction and basic notation

In this book we are assuming that the reader has studied group theory at undergraduate level, and is familiar with its fundamental results, including the basic theory of free groups and group presentations. However, in many of the interactions between group theory and formal language theory, it is convenient to consider group presentations as special cases of semigroup and monoid presentations, so we describe them from that aspect here.

We refer the reader to one of the standard textbooks on group theory, such as [223] or [221] for the definitions and basic properties of nilpotent, soluble (solvable) and polycyclic groups,

We also include some specific topics, mainly from combinatorial group theory, that will be required later. The normal form theorems for free products with amalgamation and HNN-extensions are used in the proofs of the insolubility of the word problem in groups, and we summarise their proofs. We introduce Cayley graphs and their metrical properties, and the idea of *quasi-isometry* between groups, which plays a central role in the area and throughout geometric group theory, and we define the small cancellation properties of presentations and describe related results.

The final section of the chapter is devoted to a brief introduction to some of the specific families of groups, such as Coxeter groups and braid groups, that arise frequently as examples throughout the book. The informed reader may prefer not to read this chapter in detail, but to refer back to it as necessary.

1.1.1 Some basic notation For g, h in a group, we define the conjugate of g by h, often written as g^h, to be hgh^{-1} and the commutator $[g, h]$ to be $ghg^{-1}h^{-1}$. But we note that some authors use the notations g^h and $[g, h]$ to mean $h^{-1}gh$ and $g^{-1}h^{-1}gh$, respectively.

3

We recall that a *semigroup* is a set with an associative binary operation, usually written as multiplication, a *monoid* is a semigroup with an identity element, and a *group* is a monoid G in which every element is invertible.

We extend the multiplication of elements of a semigroup S to its subsets, defining $TU = \{tu : t \in T, u \in U\}$ and we frequently shorten $\{t\}U$ to tU, as we do for cosets of subgroups of groups.

1.1.2 Strings and words Strings over a finite set are important for us, since they are used to represent elements of a finitely generated group.

Let A be a finite set: we often refer to A as an *alphabet*. We call the elements of A its *letters*, and we call a finite sequence $a_1 a_2 \cdots a_k$ of elements from A a *string* or *word* of length k over A. We use these two terms interchangeably. We denote by ε the string of length 0, and call this the *null string* or *empty word*. For a word w, we write $|w|$ for the length of w.

We denote by A^k the set of all strings of length k over A, by A^* the set (or monoid) of all strings over A, and by A^+ the set (or semigroup) of all nonempty strings over A; that is

$$A^* = \bigcup_{k=0}^{\infty} A^k, \quad A^+ = \bigcup_{k=1}^{\infty} A^k = A^* \setminus \{\varepsilon\}.$$

For $w = a_1 a_2 \cdots a_k$ and $i \in \mathbb{N}_0$, we write $w(i)$ for the prefix $a_1 a_2 \cdots a_i$ of w when $0 < i \le k$, $w(0) = \varepsilon$ and $w(i) = w$ for $i > k$.

In this book, A often denotes the set $X \cup X^{-1}$ of generators and their inverses for a group G; we abbreviate $X \cup X^{-1}$ as $X^{\pm}$. In this situation, we often refer to words in A^* as *words over X* even though they are really words over the alphabet A.

For $g \in G$, a word w over X of minimal length that represents g is called a *geodesic word* over X, and we denote the set of all such geodesic words by $\mathcal{G}(G, X)$. If w is an arbitrary word representing $g \in G$, then we write $|g|$ or $|w|_G$ (or $|g|_X$ or $|w|_{G,X}$ if X needs to be specified) for the length of a geodesic word over X that represents g. Similarly, we use $v = w$ to mean that the words v and w are identical as strings of symbols, and $v =_G w$ to mean that v and w represent the same element of the group.

We call a set of strings (i.e. a subset of A^*) a *language*; the study of languages is the topic of Chapter 2. It is convenient at this stage to introduce briefly the notation of a language for a group.

1.1.3 Languages for groups For a group G generated by X, we call a subset of $(X^{\pm})^*$ that contains at least one representative of each element in G a *language for G*; if the set contains precisely one representative of each element we call it a *normal form* for G. We shall be interested in finding *good* languages for a group G; clearly we shall need to decide what constitutes a good language. Typically we find good examples as the minimal representative words under a word order, such as word length or *shortlex*, $<_{\text{slex}}$, defined below in 1.1.4. The *shortlex normal form* for a group selects the least representative of each group element under the shortlex ordering as its normal form word. The set $\mathcal{G}(G, X)$ of all geodesic words provides a natural language that is not in general a normal form.

1.1.4 Shortlex orderings *Shortlex* orderings (also known as *lenlex* orderings) of A^* arise frequently in this book. They are defined as follows. We start with any total ordering $<_A$ of A. Then, for $u, v \in A^*$, we define $u <_{\text{slex}} v$ if either (i) $|u| < |v|$ or (ii) $|u| = |v|$ and u is less than v in the lexicographic (dictionary) ordering of strings induced by the chosen ordering $<_A$ of A.

More precisely, if $u = a_1 \cdots a_m$, $v = b_1 \cdots b_n$, then $u <_{\text{slex}} v$ if either (i) $m < n$ or (ii) $m = n$ and, for some k with $1 \le k \le m$, we have $a_i = b_i$ for $i < k$ and $a_k <_A b_k$.

Note that $<_{\text{slex}}$ is a well-ordering whenever $<_A$ is, which of course is the case when A is finite.

1.2 Generators, congruences and presentations

1.2.1 Generators If X is a subset of a semigroup S, monoid M or group G, then we define $\text{Sgp}\langle X \rangle$, $\text{Mon}\langle X \rangle$ or $\langle X \rangle$ to be the smallest subsemigroup, submonoid or subgroup of S, M or G that contains X. Then X is called a semigroup, monoid or group *generating set* if that substructure is equal to S, M or G respectively, and the elements of X are called *generators*.

We say that a semigroup, monoid or group is *finitely generated* if it possesses a finite generating set X.

1.2.2 Congruences If S is a semigroup and $\sim$ is an equivalence relation on S, then we say that $\sim$ is a *congruence* if

$$s_1 \sim s_2, \ t_1 \sim t_2 \implies s_1 t_1 \sim s_2 t_2.$$

We then define the semigroup $S/\sim$ to be the semigroup with elements the equivalence classes $[s] = \{t \in S : t \sim s\}$ of $\sim$, where $[s_1][s_2] = [s_1 s_2]$.

1.2.3 Presentations for semigroups, monoids and groups For a semigroup
S generated by a set X, let $\mathcal{R} = \{(\alpha_i, \beta_i) : i \in I\}$ be a set of pairs of words from
X^+ with $\alpha_i =_S \beta_i$ for each i. The elements of $\mathcal{R}$ are called *relations* of S. If $\sim$
is the smallest congruence on X^+ containing $\mathcal{R}$, and S is isomorphic to $X^+/\sim$,
then we say that $\mathcal{R}$ is a *set of defining relations* for S, and that $\mathrm{Sgp}\langle X \mid \mathcal{R}\rangle$ is a
presentation for S. In practice, we usually write $\alpha_i = \beta_i$ instead of (α_i, β_i). (This
is an abuse of notation but the context should make it clear that we do not mean
identity of words here.) Similarly the monoid presentation $\mathrm{Mon}\langle X \mid \mathcal{R}\rangle$ defines
the monoid $X^*/\approx$, for which $\approx$ is the smallest congruence on X^* containing $\mathcal{R}$.

For groups the situation is marginally more complicated. If G is a group
generated by a set X and $A = X^\pm$, then G is isomorphic to $A^*/\sim$, where $\sim$ is
some congruence on A^* containing $\{(aa^{-1}, \varepsilon), (a^{-1}a, \varepsilon) : a \in X\}$. We define a
relator of G to be a word $\alpha \in A^*$ with $\alpha =_G \varepsilon$. Let $R = \{\alpha_i : i \in I\}$ be a set of
relators of G. If $\sim$ is the smallest congruence on A^* containing

$$\{(\alpha, \varepsilon) : \alpha \in R\} \cup \{(aa^{-1}, \varepsilon) : a \in X\} \cup \{(a^{-1}a, \varepsilon) : a \in X\},$$

and if G is isomorphic to $A^*/\sim$, then we say that R is a *set of defining relators*
for G and that $\langle X \mid R\rangle$ is a *presentation* for G. Rather than specifying a relator
α, so that α represents the identity, we can specify a *relation* $\beta = \gamma$ (as in the
case of monoids or semigroups), which is equivalent to $\beta\gamma^{-1}$ being a relator.

We say that a semigroup, monoid or group is *finitely presented* (or, more
accurately, *finitely presentable*) if it has a presentation in which the sets of
generators and defining relations or relators are both finite.

1.2.4 Exercise Let $G = \langle X \mid R\rangle$ and let $A = X^\pm$. Show that

$$G \cong \mathrm{Mon}\langle A \mid \mathcal{I}_X \cup \mathcal{R}\rangle,$$

where $\mathcal{I}_X = \{(xx^{-1}, \varepsilon) : x \in X\} \cup \{(x^{-1}x, \varepsilon) : x \in X\}$ and $\mathcal{R} = \{(w, \varepsilon) : w \in R\}$.

1.2.5 Free semigroups, monoids and groups If S is a semigroup with pre-
sentation $\mathrm{Sgp}\langle X \mid \emptyset\rangle$ (which we usually write as $\mathrm{Sgp}\langle X \mid\rangle$), then we say that S
is the *free semigroup* on X; we see that S is isomorphic to X^+ in this case. Sim-
ilarly, if M is a monoid with presentation $\mathrm{Mon}\langle X \mid\rangle$, then we say that M is the
free monoid on X, and we see that M is then isomorphic to X^*. If $S = X^+$ and
$L \subseteq S$, then $\mathrm{Sgp}\langle L\rangle = L^+$; similarly, if $M = X^*$ and $L \subseteq M$, then $\mathrm{Mon}\langle L\rangle = L^*$.

If F is a group with a presentation $\langle X \mid\rangle$, then we say that F is the *free group*
on X; if $|X| = k$, then we say that F is the free group of *rank k* (any two free
groups of the same rank being isomorphic). We write $F(X)$ for the free group
on X and F_k to denote a free group of rank k.

1.2.6 Exercise Let $G = \langle X \mid R \rangle$ be a presentation of a group G. Show that the above definition of G, which is essentially as a monoid presentation, agrees with the more familiar definition $\langle X \mid R \rangle = F(X)/\langle R^{F(X)} \rangle$, where $\langle R^{F(X)} \rangle$ denotes the normal closure of R in $F(X)$.

1.2.7 Reduced and cyclically reduced words In $F(X)$, the free group on X, every element has a unique representation of the form $w = x_1^{\epsilon_1} x_2^{\epsilon_2} \ldots x_n^{\epsilon_n}$, where $n \geq 0$, $x_i \in X$ and $\epsilon_i \in \{1, -1\}$ for all i, and where we do not have both $x_i = x_{i+1}$ and $\epsilon_i = -\epsilon_{i+1}$ for any i; in this case, we say that the word w is *reduced*. Each word $v \in A^*$ is equal in $F(X)$ to a unique reduced word w.

If w is a reduced word and w is not of the form $x^{-1}vx$ or xvx^{-1} for some $x \in X$ and $v \in A^*$, then we say that w is *cyclically reduced*. Since replacing a defining relator by a conjugate in $F(X)$ does not change the group defined, we may (and often do) assume that all defining relators are cyclically reduced words.

1.3 Decision problems

In his two well-known papers in 1911 and 1912 [75, 76], Dehn defined and considered three decision problems in finitely generated groups, the word, conjugacy and isomorphism problems. While the word problem in groups is one of the main topics studied in this book, the other two will only be fleetingly considered. A good general reference on these and other decision problems in groups is the survey article by Miller [192].

1.3.1 The word problem A semigroup S is said to have *soluble word problem* if there exists an algorithm that, for any given words $\alpha, \beta \in X^+$, decides whether $\alpha =_S \beta$. The solubility of the word problem for a monoid or group generated by X is defined identically except that we consider words α, β in X^* or $(X^{\pm})^*$. For groups, the problem is equivalent to deciding whether an input word is equal to the identity element. The word problem for groups is discussed further in Chapter 3 and in Part Three of this book. Examples of finitely presented semigroups and groups with insoluble word problem are described in Theorems 2.9.7 and 10.1.1.

1.3.2 The conjugacy and isomorphism problems The conjugacy problem in a semigroup S is to decide, given two elements $x, y \in S$, whether there exists $z \in S$ with $zx = yz$. Note that this relation is not necessarily symmetric in x and *y*, but in a group G it is equivalent to deciding whether x and y are conjugate in G.

Since the word problem in a group is equivalent to deciding whether an element is conjugate to the identity, the conjugacy problem is at least as hard as the word problem, and there are examples of groups with soluble word problem but insoluble conjugacy problem. A number of such examples are described in the survey article by Miller [192], including Theorem 4.8 (an extension of one finitely generated free group by another), Theorem 4.11 (examples showing that having soluble conjugacy problem is not inherited by subgroups or over-groups of index 2), Theorem 5.4 (residually finite examples), Theorem 6.3 (a simple group), Theorem 7.7 (asynchronously automatic groups), and Theorem 7.8 (groups with finite complete rewriting systems) of that article.

The isomorphism problem is to decide whether two given groups, monoids or semigroups are isomorphic. Typically the input is defined by presentations, but could also be given in other ways, for example as groups of matrices. There are relatively few classes for which the isomorphism problem is known to be soluble. These classes include polycyclic and hyperbolic groups [232, 234, 72].

1.3.3 The generalised word problem Given a subgroup H of a group G, the generalised word problem is to decide, given $g \in G$, whether $g \in H$. So the word problem is the special case in which H is trivial. We shall encounter some situations in which this problem is soluble in Chapter 8. As for the conjugacy problem, the survey article [192] is an excellent source of examples (in particular in Theorems 5.4 and 7.8 of that article), in this case of groups with soluble word problem that have finitely generated subgroups with insoluble generalised word problem.

1.4 Subgroups and Schreier generators

Let H be a subgroup of a group $G = \langle X \rangle$, and let U be a right transversal of H in G. For $g \in G$, denote the unique element of $Hg \cap U$ by $\overline{g}$. Define

$$Z := \left\{ ux\overline{ux}^{-1} : u \in U, x \in X \right\}.$$

Then $Z \subseteq H$.

1.4.1 Theorem *With the above notation, we have $H = \langle Z \rangle$.*

Our proof needs the following result.

1.4.2 Lemma *Let $S = \{ux^{-1}\overline{ux^{-1}}^{-1} : u \in U, x \in X\}$. Then $Z^{-1} = S$.*

Proof Let $g \in Z^{-1}$, so $g = (ux\overline{ux}^{-1})^{-1} = \overline{ux}x^{-1}u^{-1}$. Let $v := \overline{ux} \in U$. Then, since the elements vx^{-1} and u are in the same coset of H, we have $\overline{vx^{-1}} = u$, and $g = vx^{-1}\overline{vx^{-1}}^{-1} \in S$.

Conversely, let $g = ux^{-1}\overline{ux^{-1}}^{-1} \in S$, so $g^{-1} = \overline{ux^{-1}}xu^{-1}$. Let $v := \overline{ux^{-1}}$. Then $\overline{vx} = u$, so $g^{-1} = vx\overline{vx}^{-1} \in Z$ and $g \in Z^{-1}$. □

Proof of Theorem 1.4.1 Let $U \cap H = \{u_0\}$. (We usually choose $u_0 = 1$, but this is not essential.) Let $h \in H$. Then we can write $u_0^{-1}hu_0 = a_1 \cdots a_l$ for some $a_i \in A := X^{\pm}$. For $1 \le i \le l$, let $u_i := \overline{a_1 \cdots a_l}$. Since $u_0^{-1}hu_0 \in H$, we have $u_l = u_0$. Then

$$h =_G (u_0 a_1 u_1^{-1})(u_1 a_2 u_2^{-1}) \cdots (u_{l-1} a_l u_l^{-1}).$$

Note that $u_{i+1} = \overline{a_1 \cdots a_{l+1}}$ is in the same coset of H as $u_i a_{i+1}$, so $\overline{u_i a_{i+1}} = u_{i+1}$, and

$$h =_G (u_0 a_1 \overline{u_0 a_1}^{-1})(u_1 a_2 \overline{u_1 a_2}^{-1}) \cdots (u_{l-1} a_l \overline{u_{l-1} a_l}^{-1}). \qquad (\dagger)$$

Each bracketed term is in Z if $a_i \in X$, and in Z^{-1} if $a_i \in X^{-1}$ by Lemma 1.4.2. So $H = \langle Z \rangle$. □

1.4.3 Corollary *A subgroup of finite index in a finitely generated group is finitely generated.*

1.4.4 Rewriting The process described in the above proof of calculating a word v over Z from a word w over X that represents an element of H is called *Reidemeister–Schreier rewriting*. We may clearly omit the identity element from the rewritten word, which results in a word over $Y = Z \setminus \{1\}$, which we denote by $\rho_{X,Y}(w)$. From the proof, we see immediately that:

1.4.5 Remark If $1 \in U$, then $|\rho_{X,Y}(w)| \le |w|$.

1.4.6 Schreier generators and transversals The above set Y of non-identity elements of Z is called the set of *Schreier generators* of H in G. Of course, this set depends on X and on U.

The set U is called a *Schreier transversal* if there is a set of words over X representing the elements of U that is closed under taking prefixes. Note that such a set must contain the empty word, and hence $1 \in U$. By choosing the least word in each coset under some *reduction ordering* of A^* (where $A = X^{\pm}$), it can be shown that Schreier transversals always exist. Reduction orderings are defined in 4.1.5. They include the shortlex orderings defined in 1.1.4.

It was proved by Schreier [228] that, if G is a free group and U is a Schreier transversal, then the Schreier generators freely generate H.

The following result, known as the *Reidemeister–Schreier Theorem*, which we shall not prove here, provides a method of computing a presentation of the subgroup H from a presentation of the group G. Note that it immediately implies the celebrated *Nielsen–Schreier Theorem*, that any subgroup of a free group is free. As with many of the results stated in this chapter, we refer the reader to the standard textbook on combinatorial group theory by Lyndon and Schupp [183] for the proof.

1.4.7 Theorem (Reidemeister–Schreier Theorem [183, Proposition II.4.1])
Let $G = \langle X \mid R \rangle = F/N$ be a group presentation, where $F = F(X)$ is the free group on X, and let $H = E/N \leq G$. Let U be a Schreier transversal of E in F and let Y be the associated set of Schreier generators. Then $\langle Y \mid S \rangle$ with $S = \left\{ \rho_{X,Y}(uru^{-1}) : u \in U, r \in R \right\}$ is a presentation of H.

1.4.8 Corollary *A subgroup of finite index in a finitely presented group is finitely presented.*

1.5 Combining groups

In this section we introduce various constructions that combine groups. We leave the details of many of the proofs of stated results to the reader, who is referred to [183, Chapter IV] for details.

1.5.1 Free products Informally, the *free product* $G * H$ of the groups G, H is the largest group that contains G and H as subgroups and is generated by G and H. Formally, it can be defined by its universal property:

(i) there are homomorphisms $\iota_G \colon G \to G * H$ and $\iota_H \colon H \to G * H$;
(ii) if K is any group and $\tau_G \colon G \to K$, $\tau_H \colon H \to K$ are homomorphisms, then there is a unique homomorphism $\alpha \colon G * H \to K$ with $\alpha \iota_G = \tau_G$ and $\alpha \iota_H = \tau_H$.

As is often the case with such definitions, it is straightforward to prove uniqueness, in the sense that any two free products of G and H are isomorphic, and it is not hard to show that $G * H$ is generated by $\iota_G(G)$ and $\iota_H(H)$. But the existence of the free product is not immediately clear.

To prove existence, let $G = \langle X \mid R \rangle$ and $H = \langle Y \mid S \rangle$ be presentations of G and H. Then we can take

$$G * H = \langle X \cup Y \mid R \cup S \rangle,$$

where ι_G and ι_H are the homomorphisms induced by the embeddings $X \to X \cup Y$ and $Y \to X \cup Y$; we tacitly assumed that X and Y are disjoint.

It is not completely obvious that ι_G and ι_H are monomorphisms. This follows from another equivalent description of $G * H$ as the set of alternating products of arbitrary length (including length 0) of non-trivial elements of G and H, with multiplication defined by concatenation and multiplications within G and H. With this description, ι_G and ι_H are the obvious embeddings, and G and H are visibly subgroups of $G * H$, known as the *free factors* of $G * H$. The equivalence of the two descriptions follows immediately in a more general context from Proposition 1.5.12.

The definition extends easily to the free product of any family of groups. The following result, which we shall not prove here, is used in the proof of the special case of the Muller–Schupp Theorem (Theorem 11.1.1) that torsion-free groups with context-free word problem are virtually free.

1.5.2 Theorem (Grushko's Theorem [183, IV.1.9]) *For a group G, let $d(G)$ denote the minimal number of generators of G. Then $d(G * H) = d(G) + d(H)$.*

1.5.3 Direct products The *direct product* $G \times H$ of two groups G, H is usually defined as the set $G \times H$ with component-wise multiplication. We generally identify G and H with the component subgroups, which commute with each other, and are called the *direct factors* of $G \times H$. Then each element has a unique representation as a product of elements of G and H. It can also be defined by a universal property:

(i) there are homomorphisms $\pi_G \colon G \times H \to G$ and $\pi_H \colon G \times H \to H$;
(ii) if K is any group and $\tau_G \colon K \to G$ and $\tau_H \colon K \to H$ are homomorphisms, then there is a unique homomorphism $\varphi \colon K \to G \times H$ with $\tau_G = \pi_G \circ \varphi$ and $\tau_H = \pi_H \circ \varphi$.

If $G = \langle X \mid R \rangle$ and $H = \langle Y \mid S \rangle$ are presentations, then $G \times H$ has the presentation

$$G \times H = \langle X \cup Y \mid R \cup S \cup \{[x, y] : x \in X, y \in Y\} \rangle.$$

We can extend this definition to direct products of families of groups as follows. Let $\{G_\omega : \omega \in \Omega\}$ be a family of groups. Then the *(full) direct product*, also known sometimes as the *Cartesian product*, $\prod_{\omega \in \Omega} G_\omega$ of the family consists of the set of functions $\beta \colon \Omega \to \cup_{\omega \in \Omega} G_\omega$ for which $\beta(\omega) \in G_\omega$ for all $\omega \in \Omega$, where the group operation is component-wise multiplication in each G_ω; that is, $\beta_1\beta_2(\omega) = \beta_1(\omega)\beta_2(\omega)$ for all $\omega \in \Omega$.

The elements of $\prod_{\omega \in \Omega} G_\omega$ consisting of the functions β with finite support (i.e. $\beta(\omega) = 1_G$ for all but finitely many $\omega \in \Omega$) form a normal subgroup of $\prod_{\omega \in \Omega} G_\omega$. We call this subgroup the *restricted direct product* of the family $\{G_\omega : \omega \in \Omega\}$. It is also sometimes called the *direct sum* of the family to distinguish it from the direct product.

1.5.4 Semidirect products Let N and H be groups, and let $\phi : H \to \text{Aut}(N)$ be a right action of H on N. We define the *semidirect product* of H and N, written $N \rtimes_\phi H$ or just $N \rtimes H$, to be the set $\{(n, h) : n \in N, h \in H\}$ equipped with the product

$$(n_1, h_1)(n_2, h_2) = (n_1 n_2^{\phi(h_1^{-1})}, h_1 h_2).$$

We leave it as an exercise to the reader to derive a presentation of $N \rtimes_\phi H$ from presentations of H and N and the action ϕ. We note that sometimes the notation $H \ltimes N$ is used for the same product. We identify the subgroups $\{(n, 1) : n \in N\}$ and $\{(1, h) : h \in H\}$ with N and H, and hence (n, h) with nh, so that the expression above reads

$$n_1 h_1 n_2 h_2 = n_1 n_2^{\phi(h_1^{-1})} h_1 h_2.$$

The direct product $N \times H$ is the special case when ϕ is the trivial action. The semidirect product is itself a special case of a *group extension*, which is a group G with normal subgroup N and $G/N \cong H$. Unfortunately roughly half of the set of mathematicians refer to this as an extension of N by H, and the other half call it an extension of H by N. An extension is isomorphic to a semidirect product if and only if N has a complement in G (that is, G has a subgroup K, with $N \cap K = \{e\}$, $G = NK$), in which case it is also called a *split extension*.

Note that we can also define a semidirect product of two groups N and H, from a left action of H on N.

1.5.5 Wreath products Let G and H be groups and suppose that we are given a right action $\phi : H \to \text{Sym}(\Omega)$ of H on the set Ω. We define the associated *(full) permutational wreath product* $G \wr H = G \wr_\phi H$ as follows.

Let $N = \prod_{\omega \in \Omega} G_\omega$, where the groups G_ω are all equal to the same group G. So the elements of N are functions $\gamma : \Omega \to G$. We define a right action $\psi : H \to \text{Aut}(N)$ by putting $\gamma^{\psi(h)}(\omega) = \gamma(\omega^{\phi(h^{-1})})$ for each $\gamma \in N$, $h \in H$, and $\omega \in \Omega$. We then define $G \wr_\phi H$ to be the semidirect product $N \rtimes_\psi H$. So the elements have the form (γ, h) with $\gamma \in N$ and $h \in H$. As in 1.5.4, we identify $\{(\gamma, 1) : \gamma \in N\}, \{(1, h) : h \in H\}$ with N and H, and hence (γ, h) with the product γh.

If we restrict elements of N to the functions $\gamma : \Omega \to G$ with finite support, then we get the *restricted wreath product*, which we shall write as $G \wr_R H$ or $G \wr_{R\phi} H$.

The special case in which ϕ is the right regular action of H (i.e. $\Omega = H$ and $h_1^{\phi(h_2)} = h_1 h_2$ for $h_1, h_2 \in H$) is known as the *standard* or *restricted standard* wreath product. This is the default meaning of $G \wr H$ or $G \wr_R H$ when the action ϕ is not specified.

Finally we mention that, if we are given a right action $\rho \colon G \to \text{Sym}(\Delta)$, then we can define an action $\psi \colon G \wr_\phi H \to \text{Sym}(\Delta \times \Omega)$ by setting $(\delta, \omega)^{\psi(\gamma, h)} = (\delta^{\rho \circ \gamma(\omega)}, \omega^{\phi(h)})$. This right action plays a central role in the study of imprimitive permutation groups, but it will not feature much in this book.

1.5.6 Exercise Show that the restricted standard or permutational wreath product $G \wr_R H$ is finitely generated if both G and H are finitely generated. Verify also that $G \wr H$ is not finitely generated unless H is finite and G is finitely generated.

1.5.7 Graph products Let Γ be a simple undirected graph with vertices labelled from a set I, and let G_i ($i \in I$) be groups. Then the *graph product* of the G_i with respect to Γ can be thought of as the largest group G generated by the G_i such that $[G_i, G_j] = 1$ whenever $\{i, j\}$ is in the set $E(\Gamma)$ of edges of Γ.

If $\langle X_i \mid R_i \rangle$ is a presentation of G_i for each i, then

$$\langle \cup_{i \in I} X_i \mid \cup_{i \in I} R_i \cup \{[x_i, x_j] : x_i \in X_i, x_j \in X_j, i, j \in I, \{i, j\} \in E(\Gamma)\} \rangle$$

is a presentation of the graph product.

Note that the right-angled Artin groups (see 1.10.4) can be described equivalently as graph products of copies of $\mathbb{Z}$.

1.5.8 Free products with amalgamation The amalgamated free product generalises the free product. Suppose that G and H are groups with subgroups $A \le G$, $B \le H$, and that there is an isomorphism $\phi \colon A \to B$.

Informally, the free product $G *_A H$ of G and H amalgamated over A (via ϕ) is the largest group P with $G, H \le P$, $\langle G, H \rangle = P$, and $a = \phi(a)$ for all $a \in A$.

1.5.9 Example Suppose that $\Gamma = \langle G, H \rangle$ and $G \cap H = A$, where A is a subgroup of both G and H with $|G : A| \ge 3$ and $|H : A| \ge 2$. Suppose also that Γ acts on the left on a set Ω and that Ω_1, Ω_2 are subsets of Ω with $\Omega_1 \nsubseteq \Omega_2$, such that

(1) $(G \setminus A)(\Omega_1) \subseteq \Omega_2$ and $(H \setminus A)(\Omega_2) \subseteq \Omega_1$;
(2) $A(\Omega_i) \subseteq \Omega_i$ for $i = 1, 2$.

Then $\Gamma \cong G *_A H$.

This result is often known as the *ping-pong lemma*. It is proved in [74, IIB.24] for the case $A = 1$ but essentially the same proof works for general A. The reader could attempt it as an exercise, using Corollary 1.5.13 below.

1.5.10 Exercise Let $\Gamma = \mathrm{SL}(2, \mathbb{Z}) = \langle x, y \rangle$ with

$$x = \begin{pmatrix} 0 & 1 \\ -1 & 0 \end{pmatrix} \quad \text{and} \quad y = \begin{pmatrix} 1 & -1 \\ 1 & 0 \end{pmatrix}.$$

Let $G = \langle y \rangle$, $H = \langle x \rangle$, and $A = G \cap H = \langle y^3 \rangle = \langle x^2 \rangle$. Show that $\Gamma \cong G *_A H$ by taking $\Omega = \mathbb{Z}^2$, $\Omega_1 = \{(x, y) \in \Omega : xy < 0\}$ and $\Omega_2 = \{(x, y) \in \Omega : xy > 0\}$.

1.5.11 Example As a consequence of the Seifert–van Kampen Theorem [189, Chapter 4, Theorem 2.1], we see that, for a topological space $X = Y \cup Z$ for which Y, Z and $Y \cap Z$ are open and path-connected, and the fundamental group $\pi_1(Y \cap Z)$ embeds naturally into $\pi_1(Y)$ and $\pi_1(Z)$, the fundamental group of X is isomorphic to the free product with amalgamation $\pi_1(Y) *_{\pi_1(Y \cap Z)} \pi_1(Z)$ (see [183, IV.2]).

Formally, $G *_A H$ can be defined by the following universal property:

(i) there are homomorphisms $\iota_G : G \to G *_A H$ and $\iota_H : H \to G *_A H$ with $\iota_G(a) = \iota_H(\phi(a))$ for all $a \in A$;

(ii) if K is any group and $\tau_G : G \to K$, $\tau_H : H \to K$ are homomorphisms with $\tau_G(a) = \tau_H(\phi(a))$ for all $a \in A$, then there is a unique homomorphism $\alpha : G *_A H \to K$ with $\alpha \iota_G = \tau_G$ and $\alpha \iota_H = \tau_H$.

The uniqueness of $G *_A H$ up to isomorphism follows easily, but not its existence, which is most conveniently established using presentations, as follows. Let $G = \langle X \mid R \rangle$ and $H = \langle Y \mid S \rangle$ be presentations of G and H. For each element of $a \in A$, let w_a and v_a be words over X and Y representing a and $\phi(a)$, respectively, and put $T := \{w_a = v_a : a \in A\}$. Then, as in [183, IV.2], we define

$$G *_A H := \langle X \cup Y \mid R \cup S \cup T \rangle,$$

and it is straightforward to show, using standard properties of group presentations, that $G *_A H$ has the above universal property, where ι_G and ι_H are defined to map words in G and in H to the same words in $G *_A H$.

Note that, in the definition of T, it would be sufficient to restrict a to the elements of a generating set of A so, if G and H are finitely presented and A is finitely generated, then $G *_A H$ is finitely presentable.

But we have still not proved that G and H are subgroups of $G *_A H$; that is, that ι_1 and ι_2 are embeddings. We do that by finding a normal form for the elements of $G *_A H$. Let U and V be left transversals of A in G and $B \in H$, respectively, with $1_G \in U$, $1_H \in V$. From now on, we shall suppress the maps ι_G, ι_H and just write g rather than $\iota_G(g)$.

1.5.12 Proposition *Every element of $G *_A H$ has a unique expression as $t_1 \cdots t_k a$ for some $k \geq 0$, where $a \in A$, $t_i \in (U \setminus \{1_G\}) \cup (V \setminus \{1_H\})$ for $1 \leq i \leq k$ and, for $i < k$, $t_i \in U \Leftrightarrow t_{i+1} \in V$.*

In particular, since distinct elements of G and of H give rise to distinct expressions of this form, G and H embed into $G *_A H$ as subgroups, and $G \cap H = A = \phi(A)$.

Proof By definition, each $f \in G *_A H$ can be written as an alternating product of elements of G and H, and working from the left and writing each such element as a product of a coset representative and an element of A (which has been identified with $\phi(A) = B$), we can write f in the specified normal form.

Let Ω be the set of all normal form words. We define a right action of $G *_A H$ on Ω, which corresponds to multiplication on the right by elements of $G *_A H$. To do this, it is sufficient to specify the actions of G and of H, which must of course agree on the amalgamated subgroup.

Let $\alpha = t_1 \cdots t_k a \in \Omega$ and $g \in G$. If $k = 0$ or $t_k \in V$, then we define $\alpha^g = t_1 \cdots t_k t_{k+1} a'$, where $t_{k+1} a' =_G ag$. Otherwise, $k > 0$ and $t_k \in U$, and we put $\alpha^g = t_1 \cdots t_{k-1} t_{k+1} a'$, where $t_{k+1} a' =_G t_k ag$. In both cases, $t_{k+1} \in U$, $a' \in A$ and we omit t_{k+1} if it is equal to 1. We define the action of H on Ω similarly.

It is easy to see that these definitions do indeed define actions of G and H on Ω that agree on the amalgamated subgroup A, so we can use them to define the required action of $G *_A H$ on Ω. This follows from the universal property of $G *_A H$. It is also clear from the definition that, taking $\alpha = \varepsilon \in \Omega$, and f to be the element of $G *_A H$ defined by the normal form word $t_1 \cdots t_{k-1} t_k a$, we have $\alpha^f = t_1 \cdots t_{k-1} t_k a$. So the elements of G represented by distinct normal form words have distinct actions on Ω, and hence they cannot represent the same element of $G *_A H$. $\qquad\qquad\square$

1.5.13 Corollary *Suppose that $f = f_1 f_2 \cdots f_k \in G *_A H$ with $k > 0$, where $f_i \in (G \setminus A) \cup (H \setminus B)$ for $1 \leq i \leq k$ and, for $i < k$, $f_i \in G \Leftrightarrow f_{i+1} \in H$. Then f is not equal to the identity in $G *_A H$.*

Conversely, suppose that the group F is generated by subgroups (isomorphic to) G and H with $G \cap H = A$, where $a =_F \phi(a)$ for all $a \in A$, and that $f \neq 1$ for every element $f = f_1 f_2 \cdots f_k \in F$ with $k > 0$, $f_i \in (G \setminus A) \cup (H \setminus B)$ for $1 \leq i \leq k$ and, for $i < k$, $f_i \in G \Leftrightarrow f_{i+1} \in H$. Then $F \cong G *_A H$.

Proof The assumptions ensure that, when we put f into normal form as described in the above proof, the resulting expression has the form $t_1 \cdots t_k a$ with the same k and, since we are assuming that $k > 0$, this is not the representative ε of the identity element.

For the converse, observe that the hypothesis implies that the normal form expressions for elements of F described in Proposition 1.5.12 represent distinct elements of F, and so the map $\alpha\colon G *_A H \to F$ specified by (ii) of the definition of the $G *_A H$ is an isomorphism.   □

A product $f = f_1 f_2 \cdots f_k$ as in the above corollary is called a *reduced form* for f. It is called *cyclically reduced* if all of its cyclic permutations are reduced forms. Every element of $G *_A H$ is conjugate to an element $u = f_1 \cdots f_n$ in cyclically reduced form, and every cyclically reduced conjugate of u can be obtained by cyclically permuting $f_1 \cdots f_n$ and then conjugating by an element of the amalgamated subgroup A [183, page 187].

If $f_1 f_2 \cdots f_k$ is a reduced form with $k > 1$ and $(f_1 f_2 \cdots f_k)^n$ is not a reduced form for some $n > 0$, then $f_k f_1$ cannot be reduced, and so $f_1 f_2 \cdots f_k$ cannot be cyclically reduced. This proves the following result.

1.5.14 Corollary [183, Proposition 12.4] *An element of finite order in $G *_A H$ is conjugate to an element of G or to an element of H.*

1.5.15 HNN-extensions Suppose now that A and B are both subgroups of the same group G, and that there is an isomorphism $\phi\colon A \to B$. The corresponding HNN-*extension*, (due to Higman, Neumann and Neumann [141]) with *stable letter t*, *base group G* and *associated subgroups A and B*, is roughly the largest group $G *_{A,t}$ that contains G as a subgroup, and is generated by G and an extra generator t such that $t^{-1}at = \phi(a)$ for all $a \in A$.

1.5.16 Example Analogously to Example 1.5.11, suppose that we have a path-connected topological space Y with two homeomorphic open subspaces U and V, of which the fundamental groups embed into that of Y, and suppose that we form a new space X by adding a handle that joins U to V using the homeomorphism between them. Then it is a consequence of the Seifert–van Kampen Theorem that the fundamental group $\pi_1(X)$ of X is isomorphic to the HNN-extension $\pi_1(Y) *_{\pi_1(U),t}$; see [183, IV.2].

We can also define an HNN-extension of $G = \langle X \mid R \rangle$ via a presentation. Again, for $a \in A$, we let w_a and v_a be words over X representing a and $\phi(a)$, respectively, and define

$$G *_{A,t} := \langle X, t \mid R \cup T \rangle, \quad \text{where} \quad T := \{t^{-1} w_a t = v_a : a \in A\}.$$

Again, we can restrict the elements a in T to a generating set of A, so $G *_{A,t}$ is finitely presentable if G is finitely presented and A is finitely generated.

There is a homomorphism $\iota\colon G \to G*_{A,t}$ that maps each word over X to the same word in $G*_{A,t}$. Once again, we can use a normal form to prove that ι embeds G into $G*_{A,t}$, and we shall henceforth suppress ι and write g rather than $\iota(g)$. Let U and V be left transversals of A and B in G, respectively, with $1_G \in U$, $1_H \in V$.

1.5.17 Proposition *Every element of $G*_{A,t}$ has a unique expression as $t^{j_0}g_1t^{j_1}g_2 \cdots g_kt^{j_k}g_{k+1}$ for some $k \geq 0$, where*

(i) *$g_i \in G$ for $1 \leq i \leq k+1$ and $g_i \neq 1$ for $1 \leq i \leq k$;*
(ii) *$j_i \in \mathbb{Z}$ for $0 \leq i \leq k$ and $j_i \neq 0$ for $1 \leq i \leq k$;*
(iii) *for $1 \leq i \leq k$, we have $g_i \in U$ if $j_i > 0$, and $g_i \in V$ if $j_i < 0$.*

In particular, since distinct elements of G give rise to distinct expressions of this form with $k = 0$ and $j_0 = 0$, G embeds as a subgroup of $G*_{A,t}$.

Proof Clearly each $f \in G*_{A,t}$ can be written as $t^{j_0}g_1t^{j_1}g_2 \cdots g_kt^{j_r}g_{k+1}$ for some $k \geq 0$ such that (i) and (ii) are satisfied. If $j_1 > 0$, then we write g_1 as $g_1'a$ with $g_1' \in U$ and $a \in A$ and, using the relation $t^{-1}at = \phi(a)$, replace at in the word by $t\phi(a)$. Similarly, if $j_1 < 0$, then we write g_1 as $g_1'b$ with $g_1' \in V$ and $b \in B$, and replace bt^{-1} in the word by $t^{-1}\phi^{-1}(b)$. By working through the word from left to right making these substitutions, we can bring f into the required normal form (i.e. satisfying (i), (ii) and (iii)).

Let Ω be the set of all normal form words. We define a right action of $G*_{A,t}$ on Ω, which corresponds to multiplication on the right by elements of $G*_{A,t}$. To do this, it is sufficient to specify the actions of G and of t provided that, for each $a \in A$ the action of a followed by that of t is the same as the action of t followed by that of $\phi(a)$. In fact it is more convenient to specify the actions of t and t^{-1} separately and then check that they define inverse mappings. Let $\alpha = t^{j_0}g_1t^{j_1}g_2 \cdots g_kt^{j_k}g_{k+1} \in \Omega$.

If $g \in G$, then we define $\alpha^g := t^{j_0}g_1t^{j_1}g_2 \cdots g_kt^{j_k}g_{k+1}'$, where $g_{k+1}' = g_{k+1}g$. We need to subdivide into three cases for the action of t.

(a) If $g_{k+1} \notin A$, then we write $g_{k+1} = g_{k+1}'a$ with $1 \neq g_{k+1}' \in U$ and $a \in A$, and
$$\alpha^t := t^{j_0}g_1t^{j_1}g_2 \cdots g_kt^{j_k}g_{k+1}'t\phi(a);$$
(b) If $g_{k+1} \in A$ and $j_k \neq -1$, then $\alpha^t := t^{j_0}g_1t^{j_1}g_2 \cdots g_kt^{j_k+1}\phi(g_{k+1})$;
(c) If $g_{k+1} \in A$ and $j_k = -1$, then $\alpha^t := t^{j_0}g_1t^{j_1}g_2 \cdots t^{j_{k-1}}g_k'$, where $g_k' = g_k\phi(g_{k+1})$ (and $g_k = 1$ if $k = 0$).

We have the corresponding three cases for the action of t^{-1}.

(a) If $g_{k+1} \notin B$, then we write $g_{k+1} = g_{k+1}'b$ with $1 \neq g_{k+1}' \in V$ and $b \in B$, and
$$\alpha^{t^{-1}} := t^{j_0}g_1t^{j_1}g_2 \cdots g_kt^{j_k}g_{k+1}'t^{-1}\phi^{-1}(b);$$

(b) If $g_{k+1} \in B$ and $j_k \neq 1$, then $\alpha^{t^{-1}} := t^{j_0}g_1 t^{j_1}g_2 \cdots g_k t^{j_k-1}\phi^{-1}(g_{k+1})$;

(c) If $g_{k+1} \in B$ and $j_k = 1$, then $\alpha^{t^{-1}} := t^{j_0}g_1 t^{j_1}g_2 \cdots t^{j_{k-1}}g'_k$, where $g'_k = g_k\phi^{-1}(g_{k+1})$ (and $g_k = 1$ if $k = 0$).

We leave it to the reader to verify that the action of $t^{\pm 1}$ followed by that of $t^{\mp 1}$ is the identity map on Ω and that, for each $a \in A$, the action of a followed by that of t is the same as the action of t followed by that of $\phi(a)$. This shows that we do indeed have an action of $G*_{A,t}$ on Ω.

As with the corresponding proof for free products with amalgamation, taking $\alpha = \varepsilon \in \Omega$, and f to be the element of $G*_{A,t}$ defined by the normal form word $t^{j_0}g_1 t^{j_1}g_2 \cdots g_k t^{j_k}g_{k+1}$, we have $\alpha^f = t^{j_0}g_1 t^{j_1}g_2 \cdots g_k t^{j_k}g_{k+1}$. So the elements of G represented by distinct normal form words have distinct actions on Ω, and hence they cannot represent the same element of $G*_{A,t}$.   □

The following corollary is known as *Britton's Lemma*, and is used in many arguments involving HNN-extensions. It will play a crucial role in the construction of groups with insoluble word problems in Chapter 10.

1.5.18 Corollary (Britton's Lemma (1963)) *Let $f = g_1 t^{j_1}g_2 \cdots g_k t^{j_r}g_{k+1}$ for some $k \geq 1$, where $g_i \in G$ for $1 \leq i \leq k+1$, $g_i \neq 1$ for $2 \leq i \leq k$, and each $j_i \in \mathbb{Z}\setminus\{0\}$. Suppose also that there is no subword in this expression of the form $t^{-1}g_i t$ with $g_i \in A$, or $tg_i t^{-1}$ with $g_i \in B$. Then f is not equal to the identity in $G*_{A,t}$.*

Proof The assumptions ensure that, when we put f into normal form as described in the above proof, there is no cancellation between t and t^{-1}, so the resulting expression is nonempty.   □

A product $f = g_1 t^{j_1}g_2 \cdots g_k t^{j_r}g_{k+1}$ as in the above corollary is called a *reduced form* for f [183, pages 181–186]. A subword of the form $t^{-1}g_i t$ with $g_i \in A$, or $tg_i t^{-1}$ with $g_i \in B$ is known as a *pinch*, so a reduced word contains no pinches.

It is easily seen that the presentation for the surface group

$$\langle a_1, b_1, \ldots, a_g, b_g \mid [a_1, b_1][a_2, b_2]\ldots[a_g, b_g]\rangle$$

can be rewritten as

$$\langle a_1, b_1, \ldots, a_g, b_g \mid a_1 b_1 a_1^{-1} = [b_g, a_g]\ldots[b_2, a_2]b_1\rangle,$$

and hence this group can be expressed as an HNN-extension of the free group $F(b_1, a_2, b_2, \ldots, a_g, b_g)$. We can find similar decompositions for the other surface groups.

One of the earliest applications of HNN-extensions was the following result of Higman, Neumann and Neumann [141].

1.5.19 Theorem *Every countable group can be embedded in a 2-generator group.*

Sketch of proof (see [223] for details) Let $G = \{1 = g_0, g_1, g_2, \ldots\}$ be a countable group and $H = G * F$ with F free on $\{x, y\}$. Then the subgroups $A = \langle x, g_i y^{-i} x y^i \, (i \geq 1) \rangle$ and $B = \langle y, x^{-i} y x^i \, (i \geq 1) \rangle$ of G are both freely generated by their generators, so there is an HNN-extension K of H in which the stable letter t conjugates x to y, and $g_i y^{-i} x y^i$ to $x^{-i} y x^i$ for all $i \geq 1$. The 2-generator subgroup $L = \langle t, x \rangle$ of K contains g_i for all i, which generate a subgroup isomorphic to G. □

Another application from [141] is the construction of torsion free groups in which all non-identity elements are conjugate.

1.5.20 Multiple HNN-extensions More generally, suppose that the group G has subgroups A_i and B_i with $i \in I$ for some indexing set I, with isomorphisms $\phi_i \colon A_i \to B_i$. Then we can define a corresponding multiple HNN-extension with stable letters t_i and relations $t_i^{-1} a_i t_i = \phi_i(a_i)$ for all $i \in I$, $a_i \in A_i$. We generally refer to this just as an HNN-extension, but with multiple stable letters. By ordering I, it can be regarded as an ascending chain of standard HNN-extensions. Groups of this type will arise in the the construction of groups with insoluble word problems in Chapter 10.

1.5.21 Graphs of groups The following construction, due to Bass and Serre, is a generalisation of both HNN-extensions and free products with amalgamation. Let Γ be a finite graph, possibly with loops at some vertices. For convenience we assign an orientation to each edge. Then to each vertex $v \in V(\Gamma)$ we assign a group G_v, and to each edge $e \in E(\Gamma)$ a group G_e, together with monomorphisms

$$\phi_{e,\iota(e)} \colon G_e \to G_{\iota(e)} \quad \text{and} \quad \phi_{e,\tau(e)} \colon G_e \to G_{\tau(e)}$$

of each edge group into the vertex groups corresponding to its initial and terminal vertices.

We call the system $\mathcal{G}$, consisting of the graph together with its associated groups and monomorphisms, a *graph of groups*.

Now suppose that T is a spanning tree of Γ. We define the *fundamental group* of $\mathcal{G}$ with respect to T, written $\pi_1(\mathcal{G}, T)$, to be the group generated by all of the vertex subgroups G_v ($v \in V(\Gamma)$) together with generators y_e, one for each oriented edge e, and subject to the conditions that

(1) $y_e \phi_{e,\iota(e)}(x) y_e^{-1} = \phi_{e,\tau(e)}(x)$ for each edge e and each $x \in G_e$,

(2) $y_e = 1$ for each edge e in T.

It turns out that the resulting group is independent of the choice of spanning tree, and hence we call it the *fundamental group of* $\mathcal{G}$, written $\pi_1(\mathcal{G})$. The groups G_v and G_e embed naturally as subgroups.

It can be seen from the definition that $\pi_1(\mathcal{G})$ can be constructed by first taking free products of the groups G_v with subgroups amalgamated along the edges of the spanning tree, and then taking HNN-extensions of the result, with one extra generator for each remaining edge. In particular, when Γ is a graph with 2 vertices v, w and a single edge e, we get the free product with amalgamation $G_v *_{G_e} G_w$; and when Γ is a graph with a single vertex v and a loop e on v, we get the HNN-extension $G_v*_{G_e,t}$.

We can construct an infinite graph on which the fundamental group acts. The vertex set consists of the union of all of the sets of left cosets in $\pi(\mathcal{G})$ of the subgroups G_v for $v \in V(\Gamma)$, while the edge set consists of the union of all of the sets of left cosets in $\pi(\mathcal{G})$ of the subgroups G_e, and incidence between vertices and edges is defined by inclusion. This graph turns out to be a tree, which is called the *Bass–Serre tree*. The fundamental group $\pi_1(\mathcal{G})$ acts naturally as a group of automorphisms on the tree, and the vertices of the original graph Γ form a fundamental domain.

1.6 Cayley graphs

Here we define the Cayley graph and collect some basic results. A metric space (Γ, d) is called *geodesic* if for all $x, y \in \Gamma$ there is an isometry (distance preserving bijection) $f : [0, l] \to \Gamma$ with $f(0) = x$ and $f(l) = y$, where $l = d(x, y)$. The image of such a map f is called a *geodesic path* from x to y, often written $[x, y]$ (and hopefully never confused with a commutator). A familiar example is $\mathbb{R}^n$ with the Euclidean metric.

1.6.1 Cayley graphs Let G be a group with finite generating set X. Then we define the Cayley graph $\Gamma(G, X)$ of G over X to be the graph with vertex set G and with a directed edge labelled by x leading from g to gx, for each $g \in G$ and each $x \in X^{\pm}$. The vertex labelled 1 is often referred to as the *base point* or the *origin* of $\Gamma(G, X)$. It is customary (and convenient, particularly when drawing pictures of the Cayley graph), to regard the edges labelled x and x^{-1} that connect the same two vertices as a single edge with two orientations. If x has order 2 then the edge is shown undirected. More generally, one can define the Cayley graph $\Gamma(G, A)$ for an arbitrary subset A of G that generates G as a

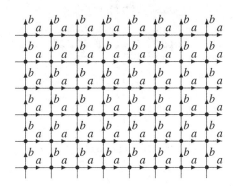

Figure 1.1 Cayley graph of $\mathbb{Z}^2 = \langle a, b \mid ab = ba \rangle$

monoid, but we will restrict to inverse-closed monoid generating sets in this book. Some examples of Cayley graphs are depicted in Figures 1.1–1.3.

1.6.2 Paths and geodesics The graph metric in which each edge has length 1 makes $\Gamma(G, X)$ into a geodesic metric space, on which G acts vertex transitively on the left as a group of isometries.

Each word w over X labels a distinct path of length $|w|$ in Γ from each vertex g to gw. We denote this path by $_gw$, or sometimes just by w when $g = 1$. A word w is geodesic if and only if it labels a geodesic path in Γ. In particular, the shortest paths in Γ from 1 to g have length $|g|$ and, for $g, h \in G$, we have $d(g, h) = |g^{-1}h|$.

1.6.3 Schreier graphs The Cayley graph generalises to the *Schreier graph* $\Gamma(G, H, X)$, which is defined for a subgroup $H \leq G$. Here the vertices are labelled by the right cosets Hg of H in G, with an edge labelled x from Hg to Hgx.

1.6.4 Embeddings of Cayley graphs in spaces The facts that the Cayley graphs for $\mathbb{Z}^2$ and the Coxeter group of type $\tilde{A}_2$ (see 1.10.2) embed naturally in $\mathbb{R}^2$, and those for F_2 and the 237-Coxeter group in the Poincaré disk are consequences of the natural cocompact actions of the groups on those spaces as groups of isometries. An embedding of the graph in the appropriate space is found by selecting a point v of the space to represent the vertex 1, and then setting the image of v under g in the space to represent the vertex g for each

Group theory

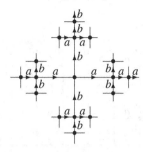

Figure 1.2 Cayley graph of $F_2 = \langle a, b \mid \rangle$

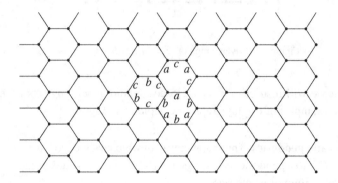

Figure 1.3 Cayley graph of $\tilde{A}_2 = \langle a, b, c \mid a^2 = b^2 = c^2 = (ab)^3 = (bc)^3 = (ca)^3 = 1 \rangle$

$g \in G$. In each of these cases the graph metric d_Γ on Γ and the natural metric d_X on the space X in which the graph embeds are related by inequalities

$$d_\Gamma(g, h) \leq \lambda d_X(g, h) + \epsilon, \quad d_X(g, h) \leq \lambda d_\Gamma(g, h) + \epsilon,$$

for fixed λ, ϵ and all $g, h \in G$.

An embedding satisfying these conditions is called a *quasi-isometric embedding* (formally defined in 1.7). We shall discuss its relevance to the theory of hyperbolic groups in Section 6.2.

1.6.5 Growth of groups For a group $G = \langle X \rangle$ with X finite, we define the growth function $\gamma_{G,X} \colon \mathbb{N}_0 \to \mathbb{N}$ by

$$\gamma_{G,X}(n) = |\{g \in G : |g|_X \leq n\}|.$$

Equivalently, $\gamma_{G,X}(n)$ is the number of vertices of $\Gamma(G, X)$ in the closed ball of radius n centred at the origin.

It is straightforward to show that, for two finite generating sets X and Y of G, there is a constant $C \geq 1$ such that

$$\gamma_{G,Y}(n)/C \leq \gamma_{G,X}(n) \leq C\gamma_{G,Y}(n)$$

for all $n \in \mathbb{N}_0$. Hence we can meaningfully refer to groups as having, for example, linear or polynomial growth without reference to the (finite) generating set.

It is also not difficult to see that, if $H = \langle Y \rangle \leq G$ with $|G : H|$ finite, then

$$\gamma_{H,Y}(n)/C \leq \gamma_{G,X}(n) \leq C\gamma_{H,Y}(n),$$

for some constant $C \geq 1$ and all $n \in \mathbb{N}_0$.

By a deep theorem of Gromov [124], a group has polynomial growth if and only if it is virtually nilpotent. We shall use an argument based on growth in our proof that groups with 1-counter word problem are virtually cyclic (Theorem 11.2.1), but for that we only need the more elementary result that groups with linear growth are virtually cyclic.

The *growth series* of the group G with respect to the generating set X is defined to be the power series $\sum_{n=0}^{\infty} \gamma_{G,X}(n)t^n$.

1.7 Quasi-isometries

Let (X_1, d_1) and (X_2, d_2) be metric spaces. A map $f : X_1 \to X_2$ is called a *quasi-isometric embedding* if there exist real constants $\lambda \geq 1$, $\epsilon \geq 0$ such that, for all $x, y \in X_1$,

$$\frac{d_1(x, y)}{\lambda} - \epsilon \leq d_2(f(x), f(y)) \leq \lambda d_1(x, y) + \epsilon.$$

The map f is called a *quasi-isometry* if, in addition, there exists $\mu > 0$ such that for all $y \in X_2$ there exists $x \in X_1$ with $d_2(f(x), y) \leq \mu$. The Cayley graph embeddings mentioned in 1.6.4 above are examples of quasi-isometries.

Note that an isometry is a quasi-isometry with $\lambda = 1$ and $\epsilon = \mu = 0$. Note also that, despite the names, a quasi-isometric embedding is not necessarily injective, and a quasi-isometry is not necessarily bijective.

We say that the spaces (X_1, d_1) and (X_2, d_2) are *quasi-isometric* if there is a quasi-isometry between them. So, for example, any space with finite diameter is quasi-isometric to the one-point space.

1.7.1 Exercise Show that being quasi-isometric is an equivalence relation on metric spaces.

1.7.2 Quasi-isometric groups We say that the two finitely generated groups $G = \langle X \rangle$ and $H = \langle Y \rangle$ are *quasi-isometric* if (the 1-skeletons of) their associated Cayley graphs $\Gamma(G,X)$ and $\Gamma(H,Y)$ are. The next result implies that this notion does not depend on the choice of finite generating sets of G and H, and so we can henceforth talk about two finitely generated groups being quasi-isometric.

1.7.3 Proposition *Let X and Y be finite generating sets of the group G. Then $\Gamma(G,X)$ and $\Gamma(G,Y)$ are quasi-isometric with $\epsilon = 0$ and $\mu = 1/2$.*

Proof For each $x \in X$ let w_x^Y be a word over Y with $x =_G w_x^Y$, and define w_y^X similarly. We define $f \colon \Gamma(G,X) \to \Gamma(G,Y)$ to map the vertex labelled g in $\Gamma(G,X)$ to the vertex with the same label in $\Gamma(G,Y)$, and an edge labelled $x^{\pm 1}$ continuously to the corresponding path labelled $(w_x^Y)^{\pm 1}$ in $\Gamma(G,Y)$. It is straightforward to verify that f is a quasi-isometry with

$$\lambda = \max(\{|w_x^Y| : x \in X\} \cup \{|w_y^X| : y \in Y\}),$$

$\epsilon = 0$, and $\mu = 1/2$. (Note that the edges of $\Gamma(G,Y)$ are not necessarily in the image of f.) □

1.7.4 Proposition *Let $H \leq G$ be finitely generated groups with $|G : H|$ finite. Then H and G are quasi-isometric.*

Proof Let $G = \langle X \rangle$, and let Y be the associated Schreier generators of H (see Section 1.4) for some right transversal T of H in G. So each $y \in Y$ is equal in G to a word w_y over X, and we define $f \colon \Gamma(H,Y) \to \Gamma(G,X)$ by mapping each vertex of $\Gamma(H,Y)$ to the vertex of $\Gamma(G,X)$ that represents the same group element, and mapping each edge of $\Gamma(H,Y)$ labelled y continuously to the path with the same initial and final vertices labelled w_y in $\Gamma(G,X)$.

Since, by Remark 1.4.5, any word w over X that represents an element of H can be rewritten as a word v over Y with $v =_H w$ and $|v| \leq |w|$, we see that f is a quasi-isometry with $\lambda = \max\{|w_y| : y \in Y\}$, $\epsilon = 0$ and $\mu = \ell + 1/2$, where ℓ is the maximum X-length of an element of the transversal T. □

1.7.5 Quasigeodesics Let p be a rectifiable path in a metric space. For real numbers $\lambda \geq 1$ and $\epsilon \geq 0$, we say that p is a (λ, ϵ)-*quasigeodesic* if $d_p(x,y) \leq \lambda d(x,y) + \epsilon$ for all $x,y \in p$. The path is a *quasigeodesic* if it is a (λ, ϵ)-quasigeodesic for some λ and ϵ. In this book we shall be only concerned with quasigeodesics in the Cayley graphs of groups. A word w over X that labels a quasigeodesic path in $\Gamma(G,X)$ is called a *quasigeodesic word*.

We should like to be able to say that the image of a geodesic path under a quasi-isometry is a quasigeodesic, but the lack of any continuity requirements on quasi-isometries prevents this being true. There is however a remedy that is sufficient for our applications. The arguments here are based on Bowditch's notes [36, Section 6.9].

The *Hausdorff distance* between two subsets X, Y of a metric space is defined to be the infimum of $r \in \mathbb{R}_{\geq 0}$ such that every point of X is at distance at most r from some point in Y and vice versa.

1.7.6 Lemma *Let $f : X \to Y$ be a (λ, ϵ)-quasi-isometric embedding between geodesic metric spaces X, Y, and let p be a geodesic path in X. Then, for some λ', ϵ', r that depend only on λ and ϵ, there is a (λ', ϵ')-quasigeodesic path q in Y that is at Hausdorff distance at most r from $f(p)$.*

Proof Choose $h \in \mathbb{R}_{>0}$ (such as $h = 1$), and let $r := \lambda h + \epsilon$. If $\ell(p) < h$, then let $q = f(p)$, which is a $(0, r)$-quasigeodesic. Otherwise, choose points $x_0, x_1, \ldots, x_n$ in order along p, where x_0 and x_n are its two endpoints, and $h/2 \leq d(x_{i-1}, x_i) \leq h$ for $1 \leq i \leq n$. Let $y_i = f(x_i)$ for $0 \leq i \leq n$, Choose any geodesic path in Y from y_{i-1} to y_i for $1 \leq i \leq n$, and let q be the union of these n geodesic paths taken in order.

We claim that q is a (λ', ϵ')-quasigeodesic, with $\lambda' := 4r\lambda/h$ and $\epsilon' = 4(\epsilon + 2r)\lambda r/h + 2r$. Note that, by choice of the x_i, we have $d(y_{i-1}, y_i) \leq r$ and $d(y_{i-1}, y_j) > (j - i)h/(2\lambda) - \epsilon$ for $1 \leq i \leq j \leq n$. Let t, u be points on q and suppose that $t \in [y_i, y_{i+1}]$ and $u \in [y_j, y_{j+1}]$ with $i \leq j$. Then $d_q(t, u) \leq r(j-i-1)+2r$ and $d(t, u) \geq (j - i - 1)h/(2\lambda) - \epsilon - 2r$. So, if $j - i - 1 \leq 4(\epsilon + 2r)\lambda/h$ then $d_q(t, u) \leq \epsilon'$, and otherwise $d_q(t, u) \leq \lambda' d(t, u) + 2r$, which proves the claim.

Any point z of p is at distance at most h from some x_i and hence, since $y_i \in q$, we have $d(f(z), q) \leq r$. Similarly, any point of q is at distance at most r from $f(p)$. So the Hausdorff distance between $f(p)$ and q is at most r. □

1.8 Ends of graphs and groups

For a connected graph Γ in which every vertex has finite degree, we define the number of ends of Γ to be the smallest integer n such that, whenever a set F of finitely many vertices and their incident edges is deleted from Γ, the graph $\Gamma \setminus F$ that remains has at most n infinite connected components; if there is no such n, then Γ has infinitely many ends. We may take F to be a finite ball within Γ and, if Γ is a Cayley graph of a group, then we may take F to be a finite ball about the vertex representing the identity element.

For a finitely generated group $G = \langle X \rangle$, it turns out that the number of ends of the Cayley graph $\Gamma(G, X)$ is the same for any finite generating set X, and we call this the *number of ends of G*, and write it $e(G)$.

When G is not finitely generated the above definition does not work. In that case we define $e(G)$ to be the dimension of the cohomology group $H^2(G, \mathbb{F}_2 G)$ of G acting on its group algebra over the field with 2 elements; when G is finitely generated this gives the same value as above.

This topic was first studied by Hopf [160], where the following results are proved for finitely generated groups G.

(1) $e(G) \in \{0, 1, 2, \infty\}$.
(2) $e(G) = 0$ if and only if G is finite.
(3) $e(G) = 2$ if and only if G is virtually infinite cyclic.
(4) $e(\mathbb{Z}^k) = 1$ and $e(F_k) = \infty$ for all $k > 1$.

1.8.1 Theorem (Stalling's Ends Theorem [241]) *If G is finitely generated, then $e(G) > 1$ if and only if either $G = H *_A K$ or $G = H*_{A,t}$ for subgroups H, A and (in the first case) K of G, with A finite and distinct from H (and K).*

In particular, if G is torsion-free, and so cannot have a finite subgroup, Stallings' theorem implies that $e(G) > 1$ if and only if G decomposes non-trivially as a free product. The proof by Muller and Schupp that groups with context-free word problem are virtually free, which we shall present in Theorem 11.1.1, uses this result.

1.9 Small cancellation

Small cancellation theory extends the ideas of Dehn, who exploited properties of the natural presentations of surface groups in order to solve the word and conjugacy problems in these groups. The theory identifies particular features of a presentation that indicate restricted interaction between relators, from which various properties of groups possessing such presentations can be deduced.

For a set R of words over an alphabet $(X^\pm)^*$, we define the *symmetric closure* $\hat{R}$ of R to be the set of all cyclic conjugates of the elements of R and their inverses. The symmetric closure of a group presentation $\langle X \mid R \rangle$ is defined to be $\langle X \mid \hat{R} \rangle$, which of course defines the same group. For example, the symmetric closure of the presentation $\langle a, b \mid aba^{-1}b^{-1} \rangle$ of $\mathbb{Z}^2$ is

$$\langle a, b \mid abAB, bABa, ABab, BabA, baBA, aBAb, BAba, AbaB \rangle,$$

where $A := a^{-1}$, $B := b^{-1}$. We normally assume that the words in R are all cyclically reduced before taking the symmetric closure.

We define a *piece* in a presentation $\langle X \mid R \rangle$ to be a word u that is a prefix of at least two distinct elements of $\hat{R}$. Now, for p, q positive integers and $\lambda \in (0, 1)$, the presentation is said to satisfy:

$C(p)$ (with $p \in \mathbb{N}$) if no element of $\hat{R}$ is a product of fewer than p pieces;

$C'(\lambda)$ (with $\lambda \in (0, 1)$) if whenever a piece u is a prefix of $r \in \hat{R}$, then $|u| < \lambda|r|$;

$T(q)$ (with $q \in \mathbb{N}$) if, whenever $3 \leq h < q$ and $r_1, r_2, \ldots, r_h \in \hat{R}$ with $r_i \neq r_{i+1}^{-1}$ for $1 \leq i < h$ and $r_1 \neq r_h^{-1}$, then at least one of the products $r_1 r_2, r_2 r_3, \ldots, r_h r_1$ is freely reduced without cancellation.

Note that $C'(\lambda)$ implies $C(p+1)$ whenever $\lambda \leq 1/p$, and that all presentations satisft $T(q)$ for $q \leq 3$. The conditions $C(p)$ and $T(q)$ might be best understood in the context of van Kampen diagrams, which will be defined in Section 3.2. The condition $C(p)$ requires that, in a reduced van Kampen diagram for the presentation, no internal region has fewer than p consolidated edges (where edges separated by a vertex of degree 2 are regarded as being part of the same consolidated edge), while $T(q)$ requires that in such a diagram no internal vertex has degree less than q.

1.9.1 Example In the above presentation of $\mathbb{Z}^2$, the pieces are a, b, a^{-1}, b^{-1}, and the presentation satisfies $C'(1/3), C(4)$ and $T(4)$. More generally, all right angled Artin groups (see 1.10.4) have presentations satisfying $C(4)$ and $T(4)$, and in particular free groups do so vacuously, since they have no relations.

1.9.2 Example The involutory relations in Coxeter groups (see 1.10.2) obstruct good small cancellation conditions, but they have index 2 subgroups that do better. For example,

$$\mathrm{Cox}_{4,4,4} = \langle x_1, x_2, x_3 \mid x_1^2 = x_2^2 = x_3^2 = 1, (x_1 x_2)^4 = (x_1 x_3)^4 = (x_2 x_3)^4 \rangle$$

has an index 2 subgroup generated by $a = x_1 x_2$, $b = x_2 x_3$,

$$\langle a, b \mid a^4 = b^4 = (ab)^4 = 1 \rangle$$

that satisfies $C(4)$ and $T(4)$.

1.9.3 Results We call a group G a $C(p)$-*group*, or a $C(p) + T(q)$-*group*, if it has a presentation that satisfies $C(p)$, or both $C(p)$ and $T(q)$, respectively.

The earliest results on groups satisfying small cancellation conditions are due to Greendlinger, who used purely combinatorial arguments to prove in [110] that $C'(1/6)$-groups have Dehn presentations (Section 3.5) (and so their word problem is soluble in linear time), and in [111] that they have soluble conjugacy problem. In [112] he proved further that $C'(1/4) + T(4)$-groups have Dehn presentations and soluble conjugacy problem.

Lyndon [182] (or see [183, Section V.4]), used curvature arguments on van Kampen diagrams to prove the stronger results that $C(6)$-groups, $C(4) + T(4)$-groups and $C(3) + T(6)$-groups all have soluble word problem. Schupp [229] (or see [183, Section V.7]) proved that such groups also have soluble conjugacy problem.

We shall be studying the more recently defined classes of (bi)automatic and (word-)hyperbolic groups later in this book, in Chapters 5 and 6. If we restrict attention to finite presentations, then a variety of small cancellation conditions have been shown to imply biautomaticity, which implies word problem soluble in quadratic time and soluble conjugacy problem, and hyperbolicity, which implies word and conjugacy problems soluble in linear time. We summarise these results in 5.5.2.

1.10 Some interesting families of groups

In the final section of this introductory chapter, we introduce some of the specific families of groups that arise in combinatorial and geometric group theory, and which occur as examples in the book. We describe some of their properties, which in some cases, such as *automaticity*, may not be familiar to the reader at this stage, but will be discussed in more detail later in the book.

1.10.1 Surface groups The fundamental group T_k of the orientable surface of genus k can be presented as

$$T_k = \langle a_1, b_1, a_2, b_2, \ldots, a_k, b_k \mid [a_1, b_1][a_2, b_2] \cdots [a_k, b_k] \rangle.$$

In particular the fundamental group T_1 of the torus is isomorphic to $\mathbb{Z}^2$.

The fundamental group of a non-orientable surface of genus n (a sphere with n attached crosscaps) has an orientable surface of genus n as a double cover. The non-orientable surfaces exhibit some differences according to the parity of the genus. The genus $2k$ surface is a k-fold Klein bottle, and its group K_k has presentation

$$\langle a_1, b_1, a_2, b_2, \ldots, a_k, b_k \mid [a_1, b_1][a_2, b_2] \cdots [a_{k-1}, b_{k-1}]a_k b_k a_k^{-1} b_k \rangle,$$

while the genus $2k + 1$ surface has group P_{k+1} with presentation

$$\langle a_1, b_1, a_2, b_2, \ldots, a_k, b_k, c \mid [a_1, b_1][a_2, b_2] \cdots [a_k, b_k]c^2 \rangle.$$

By putting $c = a_k$ and $d = a_k^{-1} b_k$, we see that K_k has the alternative presentation

$$\langle a_1, b_1, a_2, b_2, \ldots, a_{k-1}, b_{k-1}, c, d \mid [a_1, b_1][a_2, b_2] \cdots [a_{k-1}, b_{k-1}]c^2 d^2 \rangle.$$

From the small cancellation properties of these presentations and the results summarised in 5.5.2, we find that the surface groups are all automatic, and that T_k ($k \geq 2$), P_k ($k \geq 1$), and K_k ($k \geq 2$) are hyperbolic.

1.10.2 Coxeter groups Coxeter groups are groups with presentations

$$\langle x_1, x_2, \ldots, x_n \mid x_i^2 = 1 \, (1 \le i \le n), \, (x_i x_j)^{m_{ij}} = 1 \, (1 \le i < j \le n) \rangle.$$

They can be described by their associated Coxeter matrices (m_{ij}); these are defined to be symmetric matrices with entries from $\mathbb{N} \cup \{\infty\}$, with $m_{ii} = 1$ for all i and $m_{ij} \ge 2$ for all $i \ne j$. When $m_{ij} = \infty$, the relation $(x_i x_j)^{m_{ij}} = 1$ is omitted.

A Coxeter matrix can also be represented by a Coxeter diagram, with n nodes, indexed by $\{1, \ldots, n\}$, and with an edge labelled m_{ij} joining i to j for each $i \ne j$. The diagram is generally simplified by modifying edges labelled 2, 3, 4 as follows: the edges labelled 2 are deleted, the edges labelled 3 are shown unlabelled and the edges labelled 4 are shown as double edges. The group is called *irreducible* if its (simplified) Coxeter diagram is connected as a graph; otherwise it is a direct product of such groups. The reader is warned that there are alternative labelling conventions in which edges labelled ∞ are omitted.

Any Coxeter group has a faithful representation [34, Chapter V, 4.3] as a group of linear transformations of $\mathbb{R}^n = \langle e_i : i = 1, \ldots, n \rangle$ that preserves the symmetric bilinear form B defined by $B(e_i, e_j) = -\cos(\pi/m_{ij})$. The image of x_i is the involutory map $x \mapsto x - 2B(x_i, x)e_i$. The Coxeter group is finite, and then called *spherical*, if and only if B is positive definite [34, Chapter V, 4.8], which is equivalent to the eigenvalues of the associated Cartan matrix $(-2\cos(\pi/m_{ij}))$ all being positive. It is *Euclidean* (and virtually abelian) if it is not spherical and the Cartan matrix has no negative eigenvalues, and *hyperbolic* if the Cartan matrix has $n - 1$ positive and 1 negative eigenvalue. These are the only three possibilities for 3-generator Coxeter groups, but not in general.

In the 3-generator case, putting $a := m_{12}$, $b := m_{13}$, and $c := m_{23}$, the Cartan matrix is

$$2 \begin{pmatrix} 1 & -\cos\frac{\pi}{a} & -\cos\frac{\pi}{b} \\ -\cos\frac{\pi}{a} & 1 & -\cos\frac{\pi}{c} \\ -\cos\frac{\pi}{b} & -\cos\frac{\pi}{c} & 1 \end{pmatrix},$$

which has positive trace 6, and determinant

$$1 - \cos^2\frac{\pi}{a} - \cos^2\frac{\pi}{b} - \cos^2\frac{\pi}{c} - 2\cos\frac{\pi}{a}\cos\frac{\pi}{b}\cos\frac{\pi}{c}.$$

The type of the group now just depends on the angle sum

$$\frac{\pi}{a} + \frac{\pi}{b} + \frac{\pi}{c}.$$

The group is hyperbolic if and only if the angle sum is less than π, Euclidean if the angle sum is equal to π, and spherical if the angle sum is greater than π.

The reader is undoubtedly familiar with Escher's artwork displaying tessellations of the Poincaré disc (e.g. by angels and devils); these tessellations are preserved by 3-generated hyperbolic Coxeter groups. Similarly, tessellations of the Euclidean plane (e.g. by equilateral triangles or by squares) are preserved by 3-generated Euclidean Coxeter groups, and tessellations of the sphere (or, equivalently, Platonic solids) are preserved by 3-generated spherical Coxeter groups.

The subgroups of a Coxeter group generated by subsets of the standard generating set $\{x_1, \ldots, x_n\}$ are called its *standard subgroups*, and are also Coxeter groups, with presentations given by the appropriate subdiagrams. The term *parabolic subgroup* is also used, with exactly the same meaning.

The list of spherical Coxeter diagrams, of types A_n, $B_n = C_n$, D_n, E_6, E_7, E_8, F_4, G_2, H_3, H_4, $I_2(m)$, that correspond to the finite Coxeter groups is well-known. The Coxeter groups of type A_n are the symmetric groups Sym_{n+1}. The related affine diagrams, of types $\tilde{A}_n$, $\tilde{B}_n$, ..., correspond to the affine Coxeter groups, which are all infinite and virtually abelian.

The reader is warned that hyperbolicity of a Coxeter group does not correspond to *word-hyperbolicity*, which is the subject of Chapter 6. It is proved by Moussong [197] that a Coxeter group is word-hyperbolic if and only if it contains no subgroup isomorphic to $\mathbb{Z}^2$ or, equivalently, if and only if it contains no affine standard subgroup of rank greater than 2, and there are no pairs of infinite commuting standard subgroups. The word-hyperbolic Coxeter groups are all hyperbolic as Coxeter groups, and in the 3-generator cases the two properties coincide, but in higher dimensions there are hyperbolic Coxeter groups that are not word-hyperbolic.

In general, for letters x, y, define $(x, y)_m$ to be the word of length m that starts with x and then alternates between y and x. So, for example, $(x, y)_5 = xyxyx$. Then the Coxeter group defined by (m_{ij}) has the alternative presentation

$$\langle x_1, x_2, \ldots, x_n \mid x_i^2 = 1 \ \forall i, \ (x_i, x_j)_{m_{ij}} = (x_j, x_i)_{m_{ji}} \ \forall i, j, i < j \rangle.$$

The relations $x_i^2 = 1$ are called the *involutory relations*, and the relations $(x_i, x_j)_{m_{ij}} = (x_j, x_i)_{m_{ji}}$ the *braid relations*.

In a Coxeter group any word is clearly equivalent in the group to a *positive word*; that is, to a word that contains no inverses of generators. It was shown by Tits [246] that any positive word can be reduced to an equivalent geodesic word by application of braid relations and deletion of subwords of the form x_i^2. Hence the word problem is soluble in Coxeter groups. The conjugacy problem was proved to be soluble (in cubic time) by Krammer [173], using the action of the group on a suitable cubical complex. Brink and Howlett [49] proved that all Coxeter groups are shortlex automatic.

1.10.3 Artin groups The presentations formed by deleting the involutory relations from presentations for Coxeter groups define the Artin groups:

$$\langle x_1, x_2, \ldots, x_n \mid (x_i, x_j)_{m_{ij}} = (x_j, x_i)_{m_{ji}} \forall i, j, i < j \rangle.$$

These groups are all infinite (there is a homomorphism to $\mathbb{Z}$ mapping all generators to 1), and they map onto their corresponding Coxeter groups.

An Artin group is said to be of dihedral type if the associated Coxeter group is dihedral, of finite or spherical type if the associated Coxeter group is finite, and of large or extra-large type if all of the m_{ij} are at least 3 or at least 4, respectively.

It is proved by van der Lek [247] that a subgroup of any Artin group that is generated by a subset of the standard generating set is also an Artin group, with presentation defined by the appropriate subdiagram. Such a subgroup is called a *standard* or *parabolic subgroup*, and shown by Charney and Paris [67] to be convex within the original group; that is, all geodesic words in the group generators that lie in the subgroup are words over the subgroup generators.

It is proved by Brieskorn and Saito [47] that Artin groups of spherical type are torsion-free; but in general the question of whether or not Artin groups are torsion-free is open. It is not known which Artin groups are word-hyperbolic; but certainly an Artin group whose Coxeter diagram contains two vertices that are not connected cannot be hyperbolic, since it contains a standard $\mathbb{Z}^2$ subgroup. A number of open questions for Artin groups are listed by Godelle and Paris [104].

Artin groups of spherical and of extra-large type are known to be biautomatic, and Artin groups of large type are automatic; see 5.5.4 for references.

1.10.4 Right angled Artin groups The free group of rank n, F_n, and the free abelian group $\mathbb{Z}^n$ are Artin groups. In fact they are examples of right-angled Artin groups: those Artin groups for which all entries in the Coxeter matrix are in the set $\{2, \infty\}$, where 2 indicates a commuting relation, and ∞ the lack of a relation between two generators. Such groups are also called graph groups, since the commuting relations in the group can be indicated using a finite graph. They can be be described equivalently as graph products (see 1.5.7) of infinite cyclic groups.

1.10.5 Braid groups The spherical Artin groups of type A_n are equal to the braid groups B_{n+1}, found as the mapping class groups of $(n + 1)$-punctured disks, with presentations

$$B_{n+1} = \langle x_1, x_2, \ldots, x_n \mid x_i x_{i+1} x_i = x_{i+1} x_i x_{i+1} \, (1 \le i \le n),$$
$$x_i x_j = x_j x_i \, (1 \le j < i \le n, \, i - j \ge 2) \rangle.$$

1.10.6 Garside groups Spherical type Artin groups provide examples of Garside groups. Indeed a Garside group is defined by Dehornoy and Paris [78] to be the group of fractions of a monoid with certain properties of divisibility that had already been identified in spherical type Artin monoids. Many other examples are described in [78, Section 5]. They include torus knot groups (i.e. $\langle x, y \mid x^m = y^n \rangle$ for $m, n > 1$), the fundamental groups of complements of systems of complex lines through the origin in $\mathbb{C}^2$, and some 'braid groups' associated with complex reflection groups.

The definition of a Garside group depends on the concept of an *atomic monoid*. An element m of a monoid M is called *indivisible* if $m \neq 1$ and $m = ab$ implies $a = 1$ or $b = 1$. Then M is called *atomic* if it is generated by its indivisible elements and, for each $m \in M$, the supremum of the lengths of words $a_1 a_2 \cdots a_r$ equal to m in M and with each a_i atomic is finite. We can define a partial order with respect to left divisibility on any atomic monoid by $a \leq_L b$ if $ac = b$ for some $c \in M$. The finiteness of the supremum of lengths of words for elements in M implies that we cannot have $a = acd$ for $a, c, d \in M$ unless $c = d = 1$, and hence $a \leq_L b$ and $b \leq_L a$ if and only if $a = b$. We can do the same for right divisibility, and so define a second partial order $\leq_R$. A *Garside group G* is now defined to be a group having a submonoid G^+ that is atomic, and for which the following additional properties hold:

 (i) any two elements of G^+ have least common left and right multiples and greatest common left and right divisors;
(ii) there exists an element Δ of G^+ with the property that the sets of left and right divisors of Δ coincide; that set forms a finite generating set for G^+ as a monoid and G as a group. We call Δ the *Garside element*.

We note that a group may have more than one Garside structure, and indeed any Artin group of spherical type has at least two quite distinct such structures.

An elementary example of a Garside group is provided by the braid group on three strings generated by a and b, for which $\Delta = aba = bab$ and $X = \{a, b, ab, ba, aba\}$. Notice that the set X of divisors of Δ is generally much larger than the natural minimal generating set of the group. For the braid group B_n on n strings generated by $n - 1$ simple crossings, we have $|X| = n! - 1$. The $n - 1$ simple crossings are of course the atomic elements.

1.10.7 Knot groups A *knot* is the image of an embedding of the circle into $\mathbb{R}^3$; we generally envisage the knot as being contained within a portion of $\mathbb{R}^3$ of the form

$$\{(x, y, z); -\delta < z < \delta\}.$$

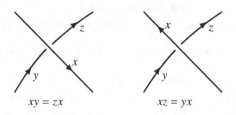

$$xy = zx \qquad xz = yx$$

Figure 1.4 Wirtinger relations

A *link* is the image of an embedding of finitely many disjoint circles into $\mathbb{R}^3$. The *fundamental group*, commonly called the *knot group*, of a knot, or link L, is defined to be the fundamental group of its complement in $\mathbb{R}^3$. Its *Wirtinger presentation* is defined as follows.

Let D be an oriented diagram for L, with n arcs. The arcs are labelled with the symbols from X, a set of cardinality n. Then for each of the n crossings, R contains the rule as shown in Figure 1.4.

Basically these are the relations satisfied by homotopy classes of loops in the complement of the knot, where the loop corresponding to a given oriented arc is the loop from a base point above the knot (e.g on the positive z-axis) that comes down to the slice of $\mathbb{R}^3$ containing the knot, crosses under the arc from left to right (these being defined by the orientation) and then returns to the base point. We call a loop that corresponds to an oriented arc in this way a *latitudinal loop*.

1.10.8 Exercise Show that the trefoil knot, depicted below, has fundamental group isomorphic to the braid group on three strands.

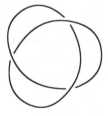

1.10.9 Flats in knot groups We can always find a $\mathbb{Z}^2$ subgroup within a knot group, generated by a longitudinal and a latitudinal loop. A longitudinal loop is defined by a path that goes from the base point above the knot to a point just above the knot, then follows the knot sitting just above it until it reaches the point at which it originally reached the knot, and then returns to the base point. For example, we can take the latitudinal loop b together with the longitudinal loop $a^{-1}b^{-1}c^{-1} = (cba)^{-1}$ that first reaches the knot just after the lowest of the three crossings. Now we see easily from our three relations that

$$cba.b = cb(ab) = cb(ca) = c(bc)a = c(ab)a = (ca)ba = (bc)ba = b.cba.$$

Hence the two generators commute, and we have a $\mathbb{Z}^2$ subgroup. The existence of this subgroup means that a knot group can never be word-hyperbolic, even when the knot complement has a hyperbolic structure; that structure has finite volume but is not compact. However it was proved by Epstein [84, Theorem 11.4.1] that the group of a hyperbolic knot is biautomatic, and more generally that the fundamental group of any geometrically finite hyperbolic manifold is biautomatic.

1.10.10 Baumslag–Solitar groups The Baumslag–Solitar groups are defined by the presentations

$$\mathrm{BS}(m,n) := \langle x,y \mid y^{-1}x^m y = x^n \rangle,$$

where $m,n \in \mathbb{Z} \setminus \{0\}$. They are examples of HNN-extensions, which were the topic of 1.5.15. This means, in particular, that their elements have normal forms, a fact that facilitates their study.

The group $\mathrm{BS}(1,n)$ is metabelian, and isomorphic to the subgroup

$$\left\{ \begin{pmatrix} n^k & an^l \\ 0 & 1 \end{pmatrix} : a,k,l \in \mathbb{Z} \right\}$$

of $\mathrm{GL}_2(\mathbb{Q})$.

The groups $\mathrm{BS}(m,n)$ were first studied by Baumslag and Solitar [21], who proved that, for certain m and n, they are *non-Hopfian*; that is, they are isomorphic to proper quotients of themselves. To be precise, they are non-Hopfian provided that $|m|, |n| > 1$ and m and n do not have the same sets of prime divisors. For example, it is not hard to show that $G := \mathrm{BS}(2,3) \cong G/N$, where N is the normal closure of $r := (x^{-1}y^{-1}xy)^2 x^{-1}$ in G. The normal form for HNN-extensions (Proposition 1.5.17) can be used to show that $r \neq_G 1$.

When $|m| \neq |n|$, the groups are asynchronously automatic but not automatic; see 5.5.7. The complexity of the word and co-word problems of these groups is addressed in Proposition 12.4.4, in 13.1.7 and in 14.2.10.

1.10.11 Higman–Thompson groups The *Higman–Thomson* groups $G_{n,r}$, for $n \geq 2$ and $r \geq 1$, are finitely presented infinite groups, whose commutator subgroups are simple and of index 2 if n is odd and equal to the whole group otherwise. They were defined as automorphism groups of certain algebras by

Higman [142], as generalisations of Richard Thompson's group V, which is $G_{2,1}$.

These groups have a number of different but equivalent definitions, and we summarise here the definition that is most convenient for their connections with formal language theory. Let $Q = \{q_1, q_2, \ldots, q_r\}$ and $\Sigma = \{\sigma_1, \sigma_2, \ldots, \sigma_n\}$ be disjoint sets. Define Ω to be the set of all infinite sequences of the form $\omega = q_i \sigma_{i_1} \sigma_{i_2} \sigma_{i_3} \cdots$, where $q_i \in Q$ and $\sigma_{i_j} \in \Sigma$ for all $j \geq 1$. Henceforth, by a *prefix* of ω, we mean a nonempty finite prefix. Then, roughly speaking, the group $G_{n,r}$ is the group that acts on Ω by replacing prefixes.

More precisely, a finite complete anti-chain in Ω is a finite subset C of $Q\Sigma^*$ such that every element of Ω has precisely one element of C as a prefix. Any bijection $\phi \colon B \to C$ between two finite complete anti-chains B and C induces a permutation of Ω, also called ϕ, by setting $\phi(\omega) = \phi(b)\omega'$ where $\omega = b\omega'$ with $b \in B$. The group $G_{n,r}$ is the group of all such permutations of Ω.

It is important to note that each element of $G_{n,r}$ has infinitely many representations by bijections between finite complete anti-chains, because we can replace B by $B' := (B \setminus \{b\}) \cup b\Sigma$, similarly C by $C' := (C \setminus \{\phi(b)\}) \cup \phi(b)\Sigma$ and ϕ by $\phi' \colon B' \to C'$, where ϕ' agrees with ϕ on $(B \setminus \{b\}) \subseteq B'$ and $\phi'(b\sigma) = \phi(b)\sigma$ for $\sigma \in \Sigma$. Using this process of expansion, the group operations can be carried out within the set of bijections between finite complete anti-chains, because any two finite complete anti-chains have a common expansion. This is easy to see, if one views the set $Q\Sigma^*$ as a disjoint union of r rooted trees with roots in Q and edges between v and $v\sigma$ for all $v \in Q\Sigma^*$ and $\sigma \in \Sigma$. Then Ω is the set of ends of this forest and finite complete anti-chains are in bijection with leaf sets of finite subforests that are finite with set of roots equal to Q. In order to find the common expansion of two finite complete anti-chains, take the leaves of the forest that is the union of the two given forests.

If we equip Ω with the lexicographic order induced by total orders on Q and Σ (see 1.1.4), then the group $G_{n,r}$ has an order preserving subgroup $F_{n,r}$ and a cyclic order preserving subgroup $T_{n,r}$. The famous Thompson group F equals $F_{2,1}$. The standard reference for F is still [62]. We will encounter some of these groups again in 9.1.4, Section 12.2, 14.2.5 and 14.2.9.

1.10.12 Exercise Given that $\langle x_i \, (i \in \mathbb{N}_0) \mid x_i x_j x_i^{-1} = x_{j+1} \text{ for } 0 \leq i < j \rangle$ is a presentation for Thompson's group F, show that F is an HNN-extension of a group isomorphic to F. Show also that $F = \langle x_0, x_1 \mid [x_1^{x_0}, x_1^{-1}x_0], [x_1^{x_0^2}, x_1^{-1}x_0] \rangle$.

2

Formal languages and automata theory

In this chapter, we present some basic material from formal language theory, and we concentrate on those topics that arise and have been studied in connection with group theory. These include, for example, real-time and indexed languages, and 2-variable automata, which do not generally merit extensive coverage in textbooks on the subject.

Although we do not assume any prior knowledge of the subject from the reader, some familiarity with the basics would be helpful. As a general reference, we recommend the first edition of the standard text by Hopcroft and Ullman [159]. The third edition of this book (co-authored also by Motwani) [158] serves as a more leisurely but less complete introduction to the subject. For many of the basic results of formal language theory, we provide only sketch proofs here (which we nevertheless hope are enough to allow the reader to fill in the details) and refer to [159] or [158] for full details and further discussion.

2.1 Languages, automata and grammars

Let A be a finite alphabet. Recall from 1.1.2 that A^* denotes the set of all strings or words over A, including the empty word ε. We call a subset of A^* a *language* over A. We shall sometimes study *families* of languages; the languages in a family are not normally all defined over the same alphabet.

A language over an alphabet A may be defined in various ways, in particular by an *automaton* that accepts it, or by a *grammar* that generates it. We use the term automaton in a general sense that encompasses the range from finite state automata to Turing machines. We use the term grammar where other authors use the terms *Type 0 grammar*, *phrase structure*, *semi-Thue system* or *unrestricted* grammar.

36

2.1.1 Automata We consider an *automaton* M over an alphabet A to be a device (with a finite description) that reads input strings over A from a tape, and accepts some of them; the language of M, which is denoted by $L(M)$, consists of those strings that are accepted. In this chapter we describe many of the different types of automata that can be defined, of varying complexity, ranging from the most basic finite state automata to the most general types of Turing machines, operating either deterministically or non-deterministically, and we examine properties of their associated languages. We are particularly interested in finite state automata, pushdown automata, nested stack automata, and in Turing machines where space is linearly bounded, or that operate in real-time. We describe all of these types of automata and some others below.

2.1.2 Grammars A *grammar* describes a mechanism for generating a language using:

- a finite set V of variables, with a specified start symbol S from V;

- a finite set T of terminals, disjoint from V; and

- a finite set P of production rules, each of the form $u \to v$, where u, v are strings over $V \cup T$ and $u \notin T^*$.

The language generated by the grammar is the set of strings over T that can be derived, using the production rules to specify allowable string substitutions, from the start symbol S.

More precisely, for strings α, β over $V \cup T$, we write $\alpha \Rightarrow \beta$ if $\alpha = w_1 u w_2$ and $\beta = w_1 v w_2$, where $u \to v$ is a production and w_1, w_2 are arbitrary strings over $V \cup T$. We then define $\Rightarrow^*$ to be the reflexive and transitive closure of $\Rightarrow$. In other words, $\alpha \Rightarrow^* \beta$ if either $\alpha = \beta$ or, for some $n \geq 1$, there exist $\alpha = \alpha_0, \alpha_1, \alpha_2, \ldots, \alpha_n = \beta$, with $\alpha_{i-1} \Rightarrow \alpha_i$ for $1 \leq i \leq n$. The language generated by the grammar is defined to be the set of strings $w \in T^*$ with $S \Rightarrow^* w$.

A grammar is actually a type of *rewriting system*, which is the subject of Chapter 4, but our study of rewriting systems in that chapter is not related to the use of grammars to generate languages. We study grammars in a little more detail in Section 2.3 and in the sections following that, which investigate specific types of automata and their associated grammars.

2.2 Types of automata

In this section we define briefly the different types of automata that we shall be considering. More details are given in later sections; here we aim to introduce the different devices and to give an intuitive understanding of their powers.

2.2.1 Introducing finite state automata The most basic automata in the well-known Chomsky hierarchy of models of computation are the finite state automata. The languages accepted by these are known as *regular* or *rational* languages. We denote the family of all regular languages by $\mathcal{R}eg$, and the family of regular languages over the alphabet A by $\mathcal{R}eg(A^*)$.

The set of regular languages over a finite alphabet A is easily seen to contain all finite subsets of A^*; infinite regular languages might well be considered to be the most elementary non-finite subsets of A^*.

A *finite state automaton* over A is an automaton with bounded memory and a read-only input-tape. The memory is represented by finitely many states that form a set Σ. Some states are called *accepting* states, and one is designated the *start* state. Initially the automaton is in the start state. Then the input string (in A^*) is read once only from left to right, and as each letter is read, the automaton changes state. The transition from one state to another is determined only by the current state and the letter that has just been read. The input string is accepted if, once it has been read, the automaton is in an accepting state. The language $L(M)$ of an automaton M is defined to be the set of strings that it accepts.

Throughout this book we abbreviate 'finite state automaton' (and also its plural) as fsa. It is common to use the term 'final state' instead of 'accepting state'.

2.2.2 Graphical representation An fsa can be visualised as a finite directed graph whose edges are labelled by elements of A. The states of the automaton are represented by vertices. Transition from state σ to state σ' on letter a is represented by a directed edge labelled by a from the vertex representing σ to the vertex representing σ'. An edge could be a loop (that is, join a vertex to itself) and two distinct edges could join the same ordered pair of vertices. A word $a_1 a_2 \ldots a_n$ over A is accepted by the automaton if (and only if) it labels a directed path from the start state to an accepting state; that is, if there is a sequence of states/vertices $\sigma_0, \sigma_1, \ldots \sigma_n$, where σ_0 is the start state, σ_n is an accepting state, and for each i there is a transition/edge from σ_{i-1} to σ_i labelled a_i.

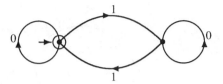

Figure 2.1 fsa recognising all binary strings containing an even number of 1's

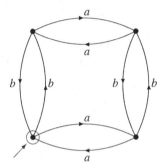

Figure 2.2 An fsa recognising the word problem of the Klein 4-group $\langle a, b \mid a^2 = b^2 = (ab)^2 = 1 \rangle$

2.2.3 Examples Figure 2.1 shows an fsa recognising all binary strings containing an even number of 1's. The left-hand state is the start state and sole accepting state. Following convention, we mark that vertex as the start state by pointing at it with a short arrow, and as an accepting state by ringing it.

Figure 2.2 shows an fsa that accepts the word problem (that is, the set of words representing the identity element; see Section 3.1) of the Klein 4-group $\langle a, b \mid a^2 = b^2 = (ab)^2 = 1 \rangle$. The fsa is constructed out of a Cayley graph (see Section 1.6) of the group, by identifying the vertex representing the identity element as the start state and sole accepting state. Given any finite group, an fsa recognising its word problem can be constructed in this way from its Cayley graph.

The 6-state fsa depicted in the upper half of Figure 2.3 accepts the language of all freely reduced words in $a^{\pm 1}, b^{\pm 1}$; that is all minimal length representatives of the elements of the free group on two generators a, b. To make the diagram more readable we have written $\bar{a}, \bar{b}$ rather than a^{-1}, b^{-1}. Notice that the state drawn highest is a non-accepting sink state; once the machine enters this state it stays there and any string that leads into that state is rejected. We call such a state a failure state, and represent it as fail. When an fsa has a failure state it is often convenient to delete it and transitions into it from the graphi-

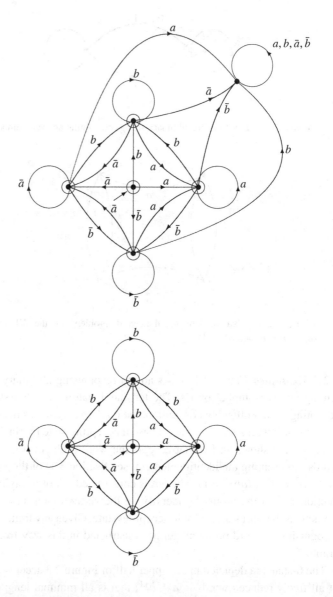

Figure 2.3 fsa recognising the reduced words in $F_2 = \langle a, b \mid \rangle$

cal description of the automaton. The diagram we then draw (as in the lower half of Figure 2.3) depicts an fsa for which transitions are undefined for some pairs (σ, a). We call such a device a *partial* fsa. But we often interpret the di-

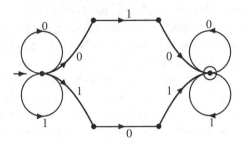

Figure 2.4 fsa recognising binary strings containing 010 or 101 as a substring.

agram of a partial fsa as the diagram of a *complete* fsa with one further state, fail, the target of all undefined transitions. Given that interpretation, the second diagram of Figure 2.3 represents the same fsa as the first.

2.2.4 Deterministic and non-deterministic fsa   We define an fsa to be *deterministic* if, for any state σ and any input letter a, there is at most one transition labelled by a out of σ; otherwise it is *non-deterministic*. A word is accepted by a non-deterministic automaton if it labels at least one path through that automaton from the start state to some accepting state. In addition, it may label paths from the start state to non-accepting states. We abbreviate 'deterministic fsa' by dfsa. The fsa depicted in Figures 2.1 and 2.2 and both halves of Figure 2.3 are all deterministic.

We shall see in Section 2.5 that, if a set of words arises as the language of a non-deterministic fsa, then a dfsa can also be constructed that accepts the same language. In general this dfsa has more states than the original non-deterministic fsa, and it may well be easier to construct a non-deterministic rather than deterministic automaton, in the first instance. The fsa represented in Figure 2.4 recognises the set of all binary strings that contain either 010 or 101 as a substring. This automaton is non-deterministic. There is a dfsa with the same number of states that accepts the same language.

2.2.5 Introducing pushdown automata   We can extend the power of a finite state automaton through the addition of an (unbounded) memory device, operating as a stack; the automaton then becomes a *pushdown automaton*. We abbreviate 'pushdown automaton' (and also its plural) as pda. We consider a *stack* as a discrete tape with one end (the bottom) fixed, used to store strings over a finite alphabet; a read-write head points to the top symbol on the stack.

When writing the contents of a stack as a word, it is conventional for the word to start at the top of the stack and finish at the bottom. So the first symbol of the word is the one to which the read-write head points.

During a move, the automaton reads and deletes a symbol from the top of the stack and sometimes (but not always) reads a letter from the input-tape. Then it changes state, choosing a new state on the basis of its current state and the symbols it has just read, and writes a (possibly empty) string of symbols to the top of the stack. As for fsa, some states are accepting. At the beginning of its operation, the machine is in a designated start state, with a designated initial string on its stack, and the input string on its input-tape. The input string is accepted if, at some stage after it has all been read, the machine enters an accepting state, and again the language $L(M)$ of a pda M is the set of words that it accepts. Note that the machine may continue to process the stack after it has read the last symbol on the input-tape, and might even run indefinitely without entering an accepting state (in which case it does not accept). If, before it has reached the end of the input, the machine is in a state where no move is specified for the current symbols under the heads, or if there is no symbol for it to read on the stack, then it halts without accepting.

Later on we describe a different, but equivalent, model of pda, in which a string is accepted if and only if it leads the automaton into a configuration with empty stack.

2.2.6 Determinism Both deterministic and non-deterministic pushdown automata can be defined. A non-deterministic pda is one for which there is more than one sequence of moves possible for some input strings; in that case a string is accepted if at least one of the sequences of possible moves leads to its acceptance. For pda (in constrast to fsa) there is a genuine difference between deterministic and non-deterministic models. In other words, there are languages that are accepted by non-deterministic pda that cannot be accepted by any deterministic machines. We shall show later, in Example 2.6.20, that the set of all palindromes over a given finite alphabet is an example of such a language.

2.2.7 Context-free language We call a language *context-free* if it is accepted by a pushdown automaton, and *deterministic context-free* if it is accepted by a deterministic pushdown automaton. The term context-free is derived from the type of grammar that corresponds to languages of this type (see Section 2.6), and we prove in that section that a language is accepted by a pda if and only if it can be generated by a context-free grammar. We denote the families of all context-free and deterministic context-free languages by $\mathcal{CF}$ and $\mathcal{DCF}$, respectively.

2.2.8 Example The word problem of a finitely generated free group is (deterministic) context-free; it can be recognised by a pda consisting essentially of a stack onto which each symbol in the input word is placed unless its inverse is observed at the top of the stack, in which case that is then deleted. If the stack is empty once the last input symbol has been read and processed, then the machine accepts the input; otherwise it is rejected.

The word problem of a finitely generated virtually free group is also deterministic context-free, and indeed this property characterises virtually free groups [198] (see Chapter 11). Let G be such a group, with a finite index free subgroup F, and let T be a transversal for F with $1 \in T$. Here is a slightly informal description of a deterministic pda with $|T| + 1$ states that accepts the word problem of G. As we read the input word w from left to right, we use the states of the pda to keep track of the coset Ft of the prefix $w(i)$ of w that has been read so far. At the same time, we use the stack, as above, to store a reduced representative of the element $w(i)t^{-1}$ of F. There is one state representing each non-identity element of T and two states representing the identity element. Of these, one is the start state and the only accepting state, and the other is non-accepting. When the current coset is F, the pda is in the accepting state if and only if the stack is empty. (In fact, to enable a pda to know when it is emptying its stack, we need to introduce an extra stack symbol for each generator, which is used only for the bottom symbol of the stack.) We give a more formal proof of this result in Chapter 11.

Note that both of the above pda read a letter from the input string on every move. We shall see examples later where this is not the case; that is, where in some configurations the machine takes input only from the stack.

2.2.9 One-counter automata A pushdown automaton whose stack alphabet consists of just one element (apart from a special symbol reserved for marking the bottom of stack) is called a *one-counter automaton*, and the language accepted by such is a *one-counter language*. We denote the families of all one-counter and deterministic one-counter languages by $\mathcal{OC}$ and $\mathcal{DOC}$, respectively. It is proved by Herbst and Thomas [135] that the word problem of a group G is one-counter if and only if G is virtually cyclic; see Theorem 11.2.1.

2.2.10 Introducing Turing machines We can extend the power of a pushdown automaton by replacing its stack by a more sophisticated memory device. This can be done in many ways. The power of even a single stack is increased if the tape-head can move down the stack in read-only mode without deletion; in that case we have a stack automaton. Nested stack automata, which accept indexed languages, are considered in 2.2.21. Once a second independent basic stack has been added to a pushdown automaton it has the full power of a Turing machine. *Church's Thesis* [167] (also known as the *Church–Turing Thesis*) posits that Turing machines are the most general model of computability, although it appears not to be possible to make this statement precise enough to allow proof or disproof.

There are many distinct but equivalent descriptions of Turing machines. We define as our basic model a device essentially formed by adding a bi-infinite tape to a finite state automaton. However, it is usually more convenient to assume that the input word is written on this infinite tape at the start of the computation, rather than being on a separate input-tape, so we take this as our basic model.

So we have a device that can exist in finitely many states from a set Σ, together with a single bi-infinite tape with associated read-write head. The input alphabet A is assumed to be a subset of the tape-alphabet B, and B contains a special symbol $b_0 \in B \setminus A$ known as the *blank symbol*. At the beginning of the computation, the tape is assumed to contain the input word, with the tape-head located at the first symbol of the input word. All other cells on the tape contain b_0 and, at any stage in the computation, all but finitely many cells contain b_0. In general, we think of the word currently on the tape as being the longest subword w on the tape for which the first and last symbols are not equal to b_0, or ε if all cells contain b_0.

Some states are designated accepting, and one state is designated as the start state, which is the machine's state at the beginning of the computation. In a single move of the machine, it first reads the symbol under its head on the work-tape. Based on what it has read and on its current state, it then changes state, writes a new symbol in that position on the work-tape, and either leaves the work-tape head in the same position or moves it one position to the right or left. An input string is accepted if at any stage during the computation it enters an accepting state; at this point it halts. If at some stage the machine enters a non-accepting state where no move is specified for the current symbol under the head, then it halts without accepting. Of course a computation may run indefinitely without accepting.

2.2.11 Determinism A Turing machine is deterministic if there are no choices in its operation; that is, the operation on any given input is always the same. Otherwise it is non-deterministic. As in the case of fsa, non-determinism does not increase the power of the device: if there is a non-deterministic Turing ma-

chine that accepts a language, then there is a deterministic one that accepts the
same language.

2.2.12 Variants It is often convenient to work with a model of a Turing machine that uses multiple work-tapes. This may result in a more natural construction, although it does not increase the power of the device. In that case,
each tape has an independent read-write head, and the symbols under all heads
are read before the machine decides into which state to move, which new symbols to write, and how to move each of its heads. Models are also defined
with one or more work-tapes that are semi-infinite rather than bi-infinite. All
of these models have the same power, but some may operate more efficiently
than others.

We shall briefly justify our assertions concerning the equality of the classes
of languages recognised by the different models of Turing machines at the beginning of Section 2.7, but the reader is advised to attempt this as an exercise,
and also to verify that Turing machines are capable of carrying out basic computational tasks, such as copying strings from one part of a tape to another
and performing elementary arithmetical operations. For further introductory
material, see for example [158, Chapter 8].

2.2.13 Instantaneous description The configuration of a Turing machine at
any point in time can be described by its *instantaneous description*. If the machine has n tapes, then this description consists of its current state σ, together
with the $2n$ strings $w_{i,1}, w_{i,2}$ for $1 \leq i \leq n$, where $w_{i,1}w_{i,2}$ is the string of symbols on the i-th tape, and the read-write head is set to point to the first symbol
of $w_{i,2}$. The set of instantaneous descriptions can be given the structure of a
directed graph, called the *configuration graph*; a directed edge joins two descriptions related by a single move of the machine.

2.2.14 Language of a Turing machine A set of strings is called *recursively
enumerable* if it is the language of a Turing machine, and *recursive* if both
it and its complement are recursively enumerable or, equivalently, if it is the
language of a Turing machine that must halt on all input strings. So, assuming
Church's Thesis (see 2.2.10), membership of strings in a language is decidable
if and only if the language is recursive.

2.2.15 Simulations An fsa can be modelled by a Turing machine with one
tape for which the read-write head always moves right after reading a symbol.
A pda is most easily modelled by a Turing machine with two tapes, one containing the input, and the other containing the stack. Alternatively a pda can

be modelled on a single-tape machine where the finite input string is stored on that tape to the left of the stack. The stack is built up from left to right, and so the bottom of stack marker marks the division between the input-tape and the stack. Each symbol of the input is deleted after it has been read.

2.2.16 Recursive functions Let A and B be any two sets whose elements can be represented by binary strings. A *partial function* $f: A \to B$ is a function with domain a subset of A and codomain B. It is called *total* if its domain is the whole of A.

The partial function $f: A \to B$ with domain $A' \subseteq A$ is called *recursive* if a Turing machine M exists which, when given the binary representation of any $a \in A$ as input, halts if and only if $a \in A'$ and, when it halts, has the binary representation of $f(a)$ on its tape. Total recursive functions $f: A \to B$ are often called *computable functions*. The meaning corresponds to it being possible to write a program to compute f in any standard general purpose programming language with no restrictions on time and space. See, for example, [159, Section 8.8] for more details.

2.2.17 Time and space complexity It is customary to define the complexity of familiar mathematical algorithms through their use of linear space, quadratic time, etc., and we shall be using this terminology in this book for algorithms such as the word problem of various types of group. Although we shall not generally feel the need to be excessively formal about this, the complexity can depend to some extent on the model of computation, and so it is worthwhile to discuss this briefly. The reader could consult, for example, [159, Chapter 12] for a more detailed treatment.

For a recursive language L and function $f: \mathbb{N}_0 \to \mathbb{N}_0$, we say that L can be computed in deterministic space f and write $L \in \mathrm{DSPACE}(f(n))$ if there is a deterministic Turing machine with a read-only input-tape and a single bi-infinite work-tape which, for any input word w of length n, decides whether $w \in L$ while using at most $f(n)$ cells of the work-tape. If there is a non-deterministic Turing machine that accepts words $w \in L$ of length n using space at most $f(n)$, then we write $L \in \mathrm{NSPACE}(f(n))$. The separate work-tape makes it possible for languages to have space complexity less than n. By increasing the tape-alphabet, the data stored on the tape can be compressed, and it can be proved that $\mathrm{DSPACE}(f(n)) = \mathrm{DSPACE}(cf(n))$ and $\mathrm{NSPACE}(f(n)) = \mathrm{NSPACE}(cf(n))$ for any constant $c > 0$. However, allowing more than one work-tape does not decrease the space requirements.

Similarly, we say that L can be computed in deterministic time f and write

$L \in \text{DTIME}(f(n))$ if there is a deterministic Turing machine with a read-only input-tape and a fixed number of bi-infinite work-tapes which, for any input word w of length n, decides whether $w \in L$ while making at most $f(n)$ moves. The class $\text{NTIME}(f(n))$ is defined analogously for non-deterministic Turing machines. In contrast with space complexity, it is possible that the time complexity of an algorithm can be decreased by increasing the number of work-tapes. Again it can be proved that $\text{DTIME}(f(n)) = \text{DTIME}(cf(n))$ and $\text{NTIME}(f(n)) = \text{NTIME}(cf(n))$ for any constant $c > 0$.

The following inclusions can be proved for any $f \colon \mathbb{N}_0 \to \mathbb{N}_0$:

$$\text{DTIME}(f(n)) \subseteq \text{NTIME}(f(n)) \subseteq$$
$$\text{DSPACE}(f(n)) \subseteq \text{NSPACE}(f(n)) \subseteq \text{DTIME}(k^{\log n + f(n)}),$$

where the constant k depends on the language L.

2.2.18 Context-sensitive languages We define a *linearly bounded Turing machine* to be one for which the length of the work-tape accessed during a computation with input string of length n is bounded by a linear function of n. The language accepted by a non-deterministic linearly bounded Turing machine is called a *context-sensitive language*, and that accepted by a deterministic linearly bounded Turing machine a *deterministic context-sensitive language*. It is not known whether these two language families are distinct. We denote the families of all context-sensitive and deterministic context-sensitive languages by CS and $\mathcal{DCS}$, respectively. In the terminology of the previous paragraph, we have $\mathcal{DCS} = \text{DSPACE}(n)$ and $CS = \text{NSPACE}(n)$.

2.2.19 Real-time languages We define a *real-time Turing machine* to be a multi-tape deterministic Turing machine that operates in real-time. More precisely, we define such an automaton to have an input-tape and a finite number of bi-infinite work-tapes. The head on the input-tape reads the input from left to right and, after each symbol has been read, only one operation (of reading, then writing, and shifting at most one position) is allowed on each of the work-tapes. Once the final input symbol has been read, and the subsequent set of single operations has been performed, the machine is obliged to halt. Clearly such a machine operates in linear time (that is, the time is bounded by a linear function of the length of the input), and thus is also linearly bounded, in the sense of the previous paragraph. The language accepted by a deterministic real-time Turing machine is called a *real-time language*. These were introduced and first studied by Yamada [258]. We denote the family of all real-time languages by $\mathcal{RT}$. We have $\mathcal{RT} \subseteq \text{DTIME}(n) \subseteq \text{DSPACE}(n) = CS$.

It is proved by Rabin [215] and Anderaa [1] that the families of languages accepted by real-time machines with k work-tapes grow strictly larger with increasing k. The language

$$\{0^{t_1}10^{t_2}1\cdots0^{t_k}11^j0^{t_{k+1-j}} : k > 0, t_i > 0, 1 \le j \le k\}$$

is presented by Rosenberg [222] as an example of a language in $\mathcal{DCF}$ that is not in $\mathcal{RT}$, but is recognisable in linear time. The non-deterministic version of $\mathcal{RT}$, the family of *quasi-real-time languages*, is studied by Book and Greibach [30]. More details on the properties of real-time languages can be found in Section 13.1.

2.2.20 Exercise Prove that each of the languages $\{0^a1^b0^{ab} : a, b \in \mathbb{N}_0\}$, $\{0^a1^{2^a} : a \in \mathbb{N}_0\}$ and $\{0^a1^{a!} : a \in \mathbb{N}_0\}$ lies in $\mathcal{RT}$.

2.2.21 Indexed languages The family $\mathcal{I}$ of *indexed languages*, which lies strictly between $\mathcal{CF}$ and $\mathcal{CS}$, is defined in [159, Section 14.3]. There are equivalent definitions in terms of grammars and automata. The automata accepting indexed languages are known as *nested stack automata*, which we now briefly define.

A *stack automaton* is a pda with the additional feature that the read-head can descend into the stack in read-only mode. So, for example, the language $\{a^nb^nc^n : n \in \mathbb{N}_0\}$ is not context-free but is accepted by a stack automaton.

A *nested stack automaton* M has the additional feature that, when the read-head of the main stack α is scanning a letter a in α, then M can open a new stack β on top of a. But now the read-head may not move above the letter a in α until the nested stack β has been deleted. It can similarly open new stacks based at letters in any existing stacks, subject to the same condition.

For example, we could have stacks $\alpha = a_1a_2a_3a_4$, $\beta_1 = b_1b_2$ based at a_2, $\beta_2 = b_3b_4b_5$ based at a_3 and $\gamma = c_1c_2c_3$ based at b_4, with the read-head at c_2. The read-write head can now only move along the word $c_1c_2c_3b_4b_5a_3a_4$ (and open new stacks of course). Also, before being able to access β_1, M first needs to delete γ and β_2, and before it can move to a_1 it has to delete β_1. (Recall from 2.2.5 that the bottom of a stack α is its right end.)

2.2.22 Poly-context-free languages A language L is said to be *k-context-free* for $k \in \mathbb{N}$ if it can be recognised by a set of k pda acting in parallel or, equivalently, if it is the intersection of k context-free languages L_i. If each of the pda is deterministic (or equivalently if each $L_i \in \mathcal{DCF}$), then L is *deterministic k-context-free*. It is *(deterministic) poly-context-free* if it is (deterministic) k-context-free for some $k \in \mathbb{N}$. The classes of poly-context-free and deterministic poly-context-free languages are denoted by $\mathcal{PCF}$ and $\mathcal{DPCF}$, re-

spectively. Note that the deterministic poly-context-free languages are called
parallel poly-pushdown languages by Baumslag, Shapiro and Short in [20].

It is proved by Gorun [109] (and in greater detail by Brough [54]) that the
language

$$\{a_1^{n_1} a_2^{n_2} \cdots a_k^{n_k} b_1^{n_1} b_2^{n_2} \cdots b_k^{n_k} : n_i \in \mathbb{N}_0\}$$

is deterministic k-context-free, but not l-context-free for any $l < k$.

2.2.23 Subclasses of $\mathcal{R}eg$ A number of subclasses of $\mathcal{R}eg$ have been defined.
The set of *star-free languages* over A is defined by Pin [214, Chapter 4, Definition 2.1] to be the closure of the finite subsets of A under concatenation and
the three Boolean operations of union, intersection, and complementation, but
not under Kleene closure (see 2.5.9 below).

A language L is k-*locally testable* if membership of a word in L is determined by the nature of its subwords of some bounded length k, and L is *locally
testable* if it is k-*locally testable* for some $k \in \mathbb{N}$. Any locally testable language
is star-free.

A language L over A is *locally excluding* if there is finite set W of words
over A such that a word $w \in L$ if and only if w has no subword that lies in W.
Note that this implies that L is k-locally testable, where k is the length of the
longest word in W.

2.2.24 Closure properties of languages The closure properties of languages
play an important role in their applications to group theory. Many of these
properties, such as closure under intersection and union are self-explanatory.
A language family $\mathcal{F}$ is closed under complementation if

$$L \in \mathcal{F}(A^*) \implies A^* \setminus L \in \mathcal{F}(A^*).$$

A family of languages $\mathcal{F}$ is said to be *closed under homomorphism* if,
whenever $\phi \colon A^* \to B^*$ is a monoid homomorphism and $L \in \mathcal{F}(A^*)$ then
$\phi(L) \in \mathcal{F}(B^*)$. It is *closed under inverse homomorphism* if $L \in \mathcal{F}(B^*) \implies$
$\phi^{-1}(L) \in \mathcal{F}(A^*)$, where $\phi^{-1}(L)$ is the complete inverse image of L.

2.2.25 Generalised sequential machines and mappings A more technical
condition is closure under (inverse) generalised sequential mappings. A *generalised sequential machine* is a finite state automaton M with output capacity.
Such machines do not have accepting states, and they do not accept or reject
their input words. Each has a set Σ of states, a start state σ_0, an input alphabet
A and a possibly different output alphabet B. A map $\tau \colon \Sigma \times A \to \Sigma \times B^*$ de-
termines transitions between states and output: if $\tau(\sigma, a) = (\sigma', \beta)$ then, when

a is read in state σ, the machine moves into state σ' and writes the string β to output.

So, for an input word $w = a_1 a_2 \cdots a_n \in A^*$, the output word is $w' = \beta_1 \beta_2 \cdots \beta_n$, where, for $i \geq 1$, the string β_i and the state σ_i after reading the prefix $a_1 \cdots a_i$ are defined inductively by $(\sigma_i, \beta_i) = \tau(\sigma_{i-1}, a_i)$. Then a map $\phi \colon A^* \to B^*$ is a *generalised sequential mapping* (GSM) if it is defined by $\phi(w) = w'$ for some generalised sequential machine M.

In fact what we have just defined is a *deterministic* generalised sequential machine, which is sufficient for our purposes, but non-deterministic versions can also be defined, in which case $\phi(w)$ could take multiple values.

Note that monoid homomorphisms are examples of generalised sequential mappings, so closure of a family of languages under (inverse) GSMs implies closure under (inverse) homomorphisms. The principal application to group theory is Proposition 3.4.7, which says that if a family $\mathcal{F}$ of languages is closed under both inverse GSMs and intersection with regular languages then the family of groups with word problem in $\mathcal{F}$ is closed under passing to finite index overgroups.

The closure properties of various families of languages are discussed in some detail and summarised in tabular form in [159, Section 11.6]. We do not aim for a comprehensive study of closure properties in this book, and we shall focus on those that are useful to us.

2.3 More on grammars

2.3.1 Parse trees We can depict a derivation $S \Rightarrow^* w$ of a word in a grammar by a *parse tree*, which is a tree in which the nodes are labelled by variables, terminals or ε, and the productions used in the derivation are indicated by downward arrows. The word w is obtained by reading the labels of the leaves of the tree from left to right.

For example, the language $\{a^m b^{m+n} a^n : m, n \in \mathbb{N}_0\}$ is generated by the grammar with productions

$$S \to vu, \quad v \to avb \mid \varepsilon, \quad u \to bua \mid \varepsilon,$$

(where the notation $v \to avb \mid \varepsilon$ means that there are productions $v \to avb$ and $v \to \varepsilon$). The parse tree for a derivation of *abbbaa* in this grammar is illustrated in Figure 2.5.

2.3.2 Equivalent, left-most and right-most derivations The productions applied in the derivation of a string of terminals in a given parse tree can generally

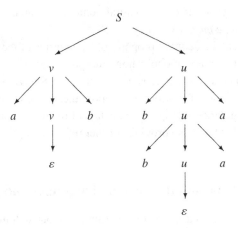

Figure 2.5 Parse tree for *abbbaa*.

be carried out in many different orders. Derivations arising from the same parse tree are called *equivalent*.

Among the equivalent derivations, we can define the *left-most* and *right-most* derivations to be those in which we always apply the next production to the left-most or right-most variable in the string, respectively.

So, in the example above, the left-most and right-most derivations are respectively

$$S \to vu \to avbu \to abu \to abbua \to abbbuaa \to abbbaa,$$

$$S \to vu \to vbua \to vbbuaa \to vbbaa \to avbbbaa \to abbbaa.$$

2.3.3 Ambiguity A grammar is said to be *ambiguous* if it contains inequivalent derivations of the same string of terminals. In some applications, such as compiler design, it is important to use unambiguous grammars, although this is not possible for all context-free languages. Since this topic does not appear to arise in applications to group theory, we shall not pursue it further here.

2.4 Syntactic monoids

Let $L \subseteq A^*$ be a language. We define a congruence $\approx_L$ on A^* by

$$w_1 \approx_L w_2 \iff (uw_1v \in L \iff uw_2v \in L \ \forall u, v \in A^*).$$

Equivalently, $\approx_L$ is the coarsest congruence on A^* such that L is the union of some of its classes. We call $\approx_L$ the *syntactic congruence* of L and $A^* / \approx_L$ the *syntactic monoid* Syn(L) of L. The natural homomorphism $\varphi_L\colon A^* \to \mathrm{Syn}(L)$ is called the *syntactic morphism* of L.

We can formulate the defining property of $\approx_L$ in terms of the syntactic morphism to see that, for any monoid homomorphism $\phi\colon A^* \to M$ such that $L = \phi^{-1}(S)$ for some $S \subseteq M$, there exists $\psi\colon \mathrm{im}(\phi) \to \mathrm{Syn}(L)$ with $\varphi_L = \psi \circ \phi$.

A good general source for information about the syntactic monoid, particularly when L is regular or belongs to a subclass of the regular languages, is the book by Pin [214], which also provides further references.

2.5 Finite state automata, regular languages and grammars

2.5.1 Finite state automata We recall from 2.2.1 our definition of a finite state automaton. We tend to use graph theoretical terminology to support our image of an fsa as a finite directed labelled graph. A word is accepted by an fsa if it labels a path in the associated directed graph from the start state to an accepting state.

A deterministic fsa (abbreviated dfsa) over A with state set Σ has a single start state $\sigma_0 \in \Sigma$ and, for each $a \in A$, out of any given state $\sigma \in \Sigma$, there is at most one transition/edge labelled a. In this situation, there is a well-defined partial function τ, which we call the *transition function*, from $\Sigma \times A$ to Σ, that maps the pair (σ, a) to the state that is the terminus of the directed edge labelled by a; we often abbreviate $\tau(\sigma, a)$ as σ^a, and we call this the *target* of σ under a. We extend τ in the obvious way to a partial function from $\Sigma \times A^*$ to Σ, and again abbreviate $\tau(\sigma, w)$ as σ^w, and call it the target of σ under w. Note that w is accepted by M (that is, $w \in L(M)$) if $\tau(\sigma_0, w)$ (i.e. σ_0^w) is an accepting state.

Note also that a dfsa is completely defined by the 5-tuple $(\Sigma, A, \tau, \sigma_0, F)$, where F is the set of accepting states, and is often referred to in this way. (The letter 'F' stands for 'final'.)

Many authors require $\tau\colon \Sigma \times A \to \Sigma$ to be a complete function as part of the definition of a dfsa; that is, that σ^a is defined for all $\sigma \in \Sigma$ and $a \in A$. We call an fsa with that property a *complete* dfsa. We can convert any dfsa to an equivalent complete one by adjoining an extra non-accepting state δ to Σ and defining $\sigma^a = \delta$ whenever σ^a is undefined, and $\delta^a = \delta$ for all $a \in A$. Note that the new state δ is a *dead state*, which means that there is no word $w \in A^*$ such that δ^w is accepting.

For a non-deterministic fsa we cannot define a transition function quite as above. However, we have a well-defined function from $\Sigma \times A$ to 2^Σ, the set of subsets of Σ. The image of (σ, a) under this function is the set of states that may be reached from σ by transition on a. We can extend our definition of a non-deterministic fsa to allow transitions on an additional symbol ε not in A. Such a transition is called an ε-*transition* or ε-*move*. Now, paths in M are labelled by words over $A \cup \{\varepsilon\}$, but we are more interested in the words obtained from these by deleting all ε's within them. An automaton with ε-moves accepts a string w over A if there is a directed path from σ to an accepting state whose label, once all ε's in it have been deleted, becomes w. For a non-deterministic fsa allowing ε-moves, it is natural to define a transition function as above, from $\Sigma \times (A \cup \{\varepsilon\})$ to 2^{Σ}.

The following result is well-known, and straightforward to prove. See, for example, [158, Theorem 2.12] for a more detailed discussion and proof.

2.5.2 Proposition *Given any non-deterministic* fsa *over a finite alphabet, there is a* dfsa *over the same alphabet accepting the same language.*

Proof Let M be the given non-deterministic fsa, which might include ε-moves, let Σ be its set of states, and let τ be its transition function. We define a dfsa M' with state set $\Sigma' := 2^{\Sigma}$ and transition function τ' as follows. The single start state of M' is the set consisting of the start state of M together with all states that can be reached from it by sequences of ε-moves. We define a state of M' to be accepting if, as a subset of Σ, it contains an accepting state of M. Then, for any $S \in \Sigma'$ and $a \in A$, we define $\tau'(S, a)$ to be the subset of states in Σ that are at the ends of directed paths in M that start in S and have labels collapsing to a once all ε's have been deleted. The function τ' is the transition function of the dfsa M' over A with state set Σ', and $L(M') = L(M)$. $\square$

As a consequence of this result, we may use the term *regular language* to denote the set of strings accepted by any finite state automaton, deterministic or otherwise. Recall also that we use $\mathcal{R}eg$ to denote the family of all regular languages, and that $\mathcal{R}eg(A^*)$ denotes the regular languages with alphabet A.

2.5.3 Minimising an fsa In fact there is a 'unique smallest' dfsa accepting any given regular language. In order to state this result precisely, we need to define a notion of equivalence of fsa. Suppose that M_1, M_2 are fsa over the same alphabet A, with state sets Σ_1, Σ_2 and transition functions τ_1, τ_2. We say that M_1, M_2 are *equivalent* if there is a bijection $f \colon \Sigma_1 \to \Sigma_2$, such that for any $\sigma \in \Sigma_1$, and any $a \in A \cup \{\varepsilon\}$, $\tau_2(f(\sigma), a) = f(\tau_1(\sigma, a))$.

See, for example, [158, Section 4.4] for a more detailed discussion of the following result. We present the result for complete dfsa, but the corresponding result holds also for partial dfsa, and we leave the necessary modifications to the reader. Note that the minimal complete dfsa accepting a language usually (but not always) has one more state than the minimal partial dfsa, namely a dead state.

2.5.4 Theorem (Myhill–Nerode)   *For any $L \in \mathcal{R}eg$ there is a minimal complete* dfsa *accepting L; that is, there is a unique (up to equivalence) complete* dfsa M_L *with* $L(M_L) = L$, *for which the number of states of* M_L *is no more than that of any other complete* dfsa *with language L.*

Proof   Let A be the alphabet of L. We define an equivalence relation $\sim_L$ on A^* by the rule $u \sim_L v$ if, for all $w \in A^*$, the words uw and vw are either both in L or both not in L.

Now let M be any complete dfsa with $L(M) = L$, and let σ_0 be its start state. For any state σ, let A_σ be the set of strings in A^* under which the target of σ_0 is σ. For two strings u, v in the same set A_σ, we have $\sigma = \sigma_0^u = \sigma_0^v$ and so $u \sim_L v$. Hence $\sim_L$ has finitely many equivalence classes. We write $[u]$ for the equivalence class of u.

When $u \sim_L v$ and $a \in A$, we see that $ua \sim_L va$. Hence we can define a complete dfsa M_L whose states are the equivalence classes of $\sim_L$, with $[\varepsilon]$ as the start state, where the target of the state $[u]$ under a is the state $[ua]$, and where $[u]$ is an accept state precisely when $u \in L$. Since $[u]$ is the target of $[\varepsilon]$ under u, it is clear that $L(M_L) = L$.

Let $\sigma_u = \sigma_0^u$ and $\sigma_v = \sigma_0^v$ be the targets of σ_0 in M under u, v. Then we see that $u \sim_L v$ precisely when, for all $w \in A^*$, the states σ_u^w and σ_v^w are either both accepting or both non-accepting. Hence M_L can also be defined by factoring out an equivalence relation on the states of M. That M_L is the unique minimal complete dfsa accepting L follows from the fact that M_L can be constructed in this way for any complete dfsa M with language L. $\square$

2.5.5 The transition monoid of an fsa   Suppose that M is a complete dfsa, with state set Σ, and let $\mathrm{Map}(\Sigma)$ be the monoid of all maps $\Sigma \to \Sigma$ under composition, where the order of composition is from left to right. For $w \in A^*$, define $\tau_w \in \mathrm{Map}(\Sigma)$ by $\tau_w \colon \sigma \mapsto \sigma^w$. The *transition monoid* $\mathrm{Trans}(M)$ of M is the submonoid $\{\tau_w : w \in A^*\}$ of $\mathrm{Map}(\Sigma)$.

Note that the map $w \mapsto \tau_w$ defines a monoid homomorphism $\phi \colon A^* \to \mathrm{Map}(\Sigma)$ with image $\mathrm{Trans}(M)$.

2.5.6 Proposition   *A language $L \subseteq A^*$ is regular if and only if there is a finite monoid $\mathcal{T}$ and a monoid homomorphism $\phi \colon A^* \to \mathcal{T}$ such that $L = \phi^{-1}(F_\mathcal{T})$ for some subset $F_\mathcal{T} \subseteq \mathcal{T}$.*

Proof   Assume first that L is defined from $\phi \colon A^* \to \mathcal{T}$ as described. Then L is the language of the dfsa M defined by $(\mathcal{T}, A, \tau, \phi(\varepsilon), F_\mathcal{T})$ with transition function τ defined by $\tau(t, a) = t \cdot \phi(a)$ for $t \in \mathcal{T}$ and $a \in A$ (the product being that of the monoid).

Conversely, suppose that $L = L(M)$ is regular, where M is a complete dfsa, defined by $(\Sigma, A, \tau, \sigma_0, F)$. Let $\mathcal{T} := \text{Trans}(M)$, and let $\phi_M : A^* \to \mathcal{T}$ be the monoid homomorphism defined by $w \mapsto \tau_w$. Then $L = \phi_M^{-1}(F_{\mathcal{T}})$ with $F_{\mathcal{T}} = \{\tau_w \in \mathcal{T} : \sigma_0^{\tau_w} \in F\}$. □

A monoid $\mathcal{T}$ with the property described in this proposition is said to *recognise* the language L. So the proof of the proposition tells us that $L(M)$ is recognised by $\text{Trans}(M)$.

2.5.7 Corollary *A language $L \subseteq A^*$ is regular if and only if its syntactic monoid $\text{Syn}(L)$ is finite. Furthermore, if L is regular, then $\text{Syn}(L)$ is (up to isomorphism) the unique smallest monoid that recognises L.*

Proof Suppose first that $\text{Syn}(L)$ is finite. We observed in Section 2.4 that L is a union of congruence classes of the syntactic congruence $\approx_L$; equivalently L is recognised by $\text{Syn}(L)$, and hence regular.

Conversely, suppose that L is regular, and recognised by the finite monoid $\mathcal{T}$; so let $L = \phi^{-1}(F_{\mathcal{T}})$ with $F_{\mathcal{T}} \subseteq \mathcal{T}$ and $\phi : A^* \to \mathcal{T}$, as in Proposition 2.5.6. It follows from the defining properties of $\approx_L$, as explained at the end of Section 2.4, that φ_L can be expressed as a composite $\psi \circ \phi$, for some (surjective) monoid homomorphism $\psi : \text{im}(\phi) \to \text{Syn}(L)$, and hence $\text{Syn}(L)$ is finite, being a quotient of the submonoid $\text{im}(\phi)$ of $\mathcal{T}$. □

2.5.8 Theorem *Let $L \subseteq A^*$ be a regular language, and let M be the minimal complete dfsa with $L(M) = L$, as in Theorem 2.5.4. Then $\text{Syn}(L)$ and $\text{Trans}(M)$ are isomorphic monoids.*

Proof The proof of Proposition 2.5.6 supplies a surjective monoid homomorphism $\phi_M : A^* \to \text{Trans}(M) =: \mathcal{T}$, such that $L = \phi_M^{-1}(F_{\mathcal{T}})$ for some $F_{\mathcal{T}} \subseteq \text{Trans}(M)$. We observed in the proof of Corollary 2.5.7 the existence of a surjective monoid homomorphism $\psi : \text{Trans}(M) \to \text{Syn}(L)$ with $\psi \circ \phi_M = \varphi_L$.

So we just need to prove that ψ is injective. Let $v_1, v_2 \in A^*$ with $\phi_M(v_1) \neq \phi_M(v_2)$. So there is a state σ of M with $\sigma^{v_1} \neq \sigma^{v_2}$. Recall from the proof of Theorem 2.5.4 that the states of M can be identified with the equivalence classes $[u]$ of words $u \in A^*$ under the relation

$$u_1 \sim_L u_2 \iff (u_1 v \in L \Leftrightarrow u_2 v \in L \,\forall v \in A^*),$$

with start state $[\varepsilon]$ and transitions defined by $[u]^v = [uv]$. So $\sigma = [u]$ for some $u \in A^*$ and $uv_1 \not\sim_L uv_2$. Then there exists $w \in A^*$ with exactly one of $uv_1 w$ and $uv_2 w$ in L. In other words $v_1 \not\approx_L v_2$, so

$$\psi(\phi_M(v_1)) = \varphi_L(v_1) \neq \varphi_L(v_2) = \psi(\phi_M(v_2)),$$

and ψ is injective, as claimed. □

In the 1960s, Schützenberger [230] established some more detailed connections between the language theoretic properties of regular languages and the algebraic structure of their syntactic monoid. For example, a regular language L is star-free (see 2.2.23) if and only if $\mathrm{Syn}(L)$ is aperiodic (that is, satisfies a rule $x^N = x^{N+1}$).

2.5.9 String operations We need a little more notation to state the next result. Let A be a finite alphabet. For subsets S, S' of A^*, we define SS', the concatenation of S and S', to be the set of all strings that are concatenations ww', for $w \in S$ and $w' \in S'$. Then we define $S^0 := \{\varepsilon\}$, $S^1 := S$, $S^2 := SS$, more generally $S^i := SS^{i-1}$, and finally we define S^* (the *Kleene closure of S*) to be the union $\bigcup_{i=0}^\infty S^i$.

2.5.10 Proposition *Let A be a finite alphabet. Then A^* is regular, any finite subset of A^* is regular and, for any regular subsets L and L' of A^*, each of $L \cup L'$, $L \cap L'$, $A^* \setminus L$, LL', and L^* are regular.*

In other words, Reg is closed under the operations of union, intersection, complementation, concatenation, and Kleene closure.

Proof A^* is recognised as the language of an fsa with a single accepting state, and an arrow labelled by each $a \in A$ looping from that state back to itself.

Any finite subset F of A^* is recognised by a non-deterministic fsa modelled by a finite directed tree, whose root is the start state, with distinct, disjoint, directed paths from the root labelled by the nonempty words in F. The accepting states are the leaves of the tree, together with the root if ε is in F.

Suppose now that M and M' are fsa with $L(M) = L$ and $L(M') = L'$. Provided that M is deterministic (which we may assume, by Proposition 2.5.2) and complete, the complement of L is recognised by an fsa constructed from M by changing the category of all accepting states to non-accepting and vice versa.

An fsa M'' accepting $L \cup L'$ is constructed by taking the union of fsa accepting L and L' together with one extra state, the start state of M'', from which there are ε-moves to the start states of M, M'. A state of M'' is defined to be accepting if it is accepting in either M or M'. Now we can construct an fsa to recognise $L \cap L'$ by using De Morgan's law.

Alternatively, automata to recognise both $L \cup L'$ and $L \cap L'$ can be derived from a direct product of M and M', whose states are ordered pairs (σ, σ') of states from L, L', with transitions on each coordinate derived from those in L, L'. The state (σ, σ') is accepting when either (in the first case) or both (in the second case) of σ, σ' are accepting in M, M'.

The concatenation LL' is recognised by a non-deterministic automaton $M^{LL'}$

formed by placing M, M' 'in series'. This fsa has as its state set the disjoint union of the states of M, M', as its transitions, all the transitions of either machine, together with ε-moves joining each accepting state of M to the start state of M'. The start state of $M^{LL'}$ is the start state of M, and the accepting states of $M^{LL'}$ are all of the accepting states of M'.

Finally, a non-deterministic fsa M^* with $L(M^*) = L^*$ can be constructed by modifying M as follows. M^* contains all of the states and all of the transitions of M, and all accepting states of M are accepting in M^*. In addition, M^* has an extra state σ_0, joined by an ε-move to the start state of M. Further ε-moves join each accepting state of M back to σ_0. The state σ_0 is the start state of M^* and is accepting; apart from σ_0, a state is accepting in M^* precisely when it is accepting in M. The state σ_0 is included in M^* to ensure that ε is accepted by M^*, even if it is not in L. □

2.5.11 Rational expressions Proposition 2.5.10 points us in the direction of an alternative characterisation of regular languages, via rational expressions. The term *regular expression* is often used with exactly the same meaning.

We define a *rational expression* for a subset of A^* to be a description of that subset that constructs it out of finite subsets of A^* using finitely many operations of set union, concatenation and Kleene closure. For example, where $A = \{a, b\}$, the language $\{a^i b^j : i, j \geq 0\}$ is defined by the rational expression, $\{a\}^* \{b\}^*$ (which is often abbreviated as $a^* b^*$), while the set of strings of alternating as and bs is defined by the rational expression $(ab)^* \cup (ab)^* a \cup (ba)^* \cup (ba)^* b$.

2.5.12 Theorem *Let A be a finite alphabet. Then a subset of A^* is the language of an fsa if and only if it can be described by a rational expression over A.*

Proof That a subset of A^* described by a rational expression over A is the language of an fsa is a direct consequence of Proposition 2.5.10

Our proof of the converse is based on that in [159, Theorem 2.4]. We suppose that L is a regular language, accepted by a dfsa with states σ_0 (the start state), $\sigma_1, \ldots \sigma_n$. Then, for $k \geq -1$, we define A_{ij}^k to be the set of strings over A that label paths from σ_i to σ_j with the property that no intermediate vertex on that path has label greater than k. Then

$$A_{ij}^k = A_{ij}^{k-1} \cup A_{ik}^{k-1} (A_{kk}^{k-1})^* A_{kj}^{k-1},$$
$$A_{ij}^{-1} = \{a \in A : \sigma_i^a = \sigma_j\}, \quad i \neq j, \quad \text{and}$$
$$A_{ii}^{-1} = \{a \in A : \sigma_i^a = \sigma_i\} \cup \{\varepsilon\}.$$

The sets A_{ij}^{-1} and A_{ii}^{-1} are finite, and hence can be defined by rational expressions. Now, using the recursive formulae, we can derive by induction on m a rational expression for each A_{ij}^m. Then, since L can be written as a union of sets A_{0j}^n for which σ_j is an accepting state, we can derive a rational expression for L.   □

This result gives us an alternative and equivalent definition of regular languages; that is, a subset of A^* is regular if and only if it can be represented by a rational expression over A.

Note that by definition a rational expression involves only the operations of set union, concatenation and Kleene closure. However Proposition 2.5.10 and Theorem 2.5.12 together show that any set that can be constructed out of finite subsets of A^* using finitely many of the five operations of set union, intersection, complementation in A^*, concatenation and Kleene closure can also be described by a rational expression. Sometimes it will be convenient to recall this.

Given this result it is now easy to add further to our collection of closure properties of regular languages.

2.5.13 Proposition *Suppose that L is a regular language over A, then so is L^R, the set of all strings over A whose reverse is in L.*

Proof A rational expression for L^R is formed by performing a simple reversing operation on a rational expression for L.

Alternatively we can simply 'reverse the arrows' on the transitions of an fsa accepting L (having first ensured that the fsa has a single accepting state), and then interchange the start and accept state.   □

2.5.14 Proposition *The family $\mathcal{R}eg$ of regular languages is closed under homomorphism and inverse homomorphism.*

Proof Let $\phi\colon A^* \to B^*$ be a homomorphism. For closure under homomorphism, just apply ϕ to the components of A in a rational expression defining a language $L \subseteq \mathcal{R}eg(A^*)$, to obtain a rational expression defining $\phi(L)$.

For closure under inverse homomorphism we start with a dfsa M accepting $L \subseteq \mathcal{R}eg(B^*)$ and define a dfsa M' accepting $\phi^{-1}(L)$ as follows. All states, including the start state and accepting states, are the same in M' as in M. For each state σ of M' and $a \in A$, we define $\tau'(\sigma, a) := \sigma^{\phi(a)}(= \tau(\sigma, \phi(a)))$.   □

2.5.15 Regular grammars Recall from 2.1.2 the definition of a grammar with set of variables V and set of terminals T. A grammar is called

- *left-regular* if all productions have the form $\alpha \to \beta w$ or $\alpha \to w$, with $\alpha, \beta \in V$ and $w \in T^*$;

- *right-regular* if all productions have the form $\alpha \rightarrow w\beta$ or $\alpha \rightarrow w$, with $\alpha, \beta \in V$, and $w \in T^*$; and
- *regular* if it is either left-regular or right-regular.

Regular grammars are also known as *Type 3 grammars*.

An alternative definition of a regular set is provided as a consequence of the following theorem. The proof comes from [159, Theorems 9.1, 9.2].

2.5.16 Theorem   *A subset of A^* is regular if and only if it can be derived using a regular grammar with terminal set A.*

Proof   If $L = L(M)$ with $M = (\Sigma, A, \tau, \sigma_0, F)$, then we can define a right-regular grammar that generates L as follows. The variables of the grammar consist of the set Σ together with the start variable S. The grammar contains a production rule $\sigma_i \rightarrow a\sigma_j$ corresponding to each transition $\sigma_i^a = \sigma_j$ of M. In addition there are productions $S \rightarrow \sigma_0$ and $\sigma_i \rightarrow \varepsilon$ for each $\sigma_i \in F$. Productions of this grammar that generate a given word in L are found by tracing an accepting path through the automaton labelled by that word. Also, the generation of any word $w = a_1 \cdots a_n \in A^*$ by the grammar must correspond to an accepting path of the word through M that ends with a production of the form $\sigma_n \rightarrow \varepsilon$, where $\sigma_0^w = \sigma_n$.

Conversely, suppose that we are given a right-regular grammar with terminal set A that defines a language L. By introducing additional variables and associated production rules as necessary, we can decompose production rules of the form $\alpha \rightarrow w\beta$ with $|w| > 1$ or $\alpha \rightarrow w$ with $|w| > 0$, and thereby assume that all rules have the form $\alpha \rightarrow \varepsilon$, or $\alpha \rightarrow x\beta$ with $x \in A \cup \{\varepsilon\}$. For example, we can replace a production $\alpha \rightarrow ab\beta$ with $a, b \in A$ by the two productions $\alpha \rightarrow a\alpha_0$ and $\alpha_0 \rightarrow b\beta$, and we can replace $\alpha \rightarrow a$ by $\alpha \rightarrow a\alpha_1$ and $\alpha_1 \rightarrow \varepsilon$, where α_0 and α_1 are new variables.

Now we can define a non-deterministic fsa that accepts L as follows. The states are in bijection with the (enlarged) set of variables of the grammar, where the start state corresponds to the start symbol of the grammar, and the accepting states α are those for which there is a production rule $\alpha \rightarrow \varepsilon$. For each rule $\alpha \rightarrow x\beta$, there is a transition from α to β labelled x.

We have now established that a language is regular precisely if it can be defined by a right-regular grammar.

The full result follows once we observe that whenever a language L is defined by a right-regular grammar, its reverse $L^R = \{w : w^R \in L\}$ is defined by a left-regular grammar, and conversely. Alternatively we can define a left-regular grammar defining the language of an fsa by tracking paths in the automaton back from the accepting states to the start state in the same way that we constructed a right-regular grammar by tracking paths from the start states to accepting states. □

The well-known *pumping lemma* for regular languages provides a very useful mechanism, which can often be applied to prove that a language is not regular.

2.5.17 Theorem (Pumping Lemma) *If $L \in \mathcal{R}eg$ then, for some fixed positive integer N, any word in L of length at least N can be written in the form uvw, where uv has length at most N, v is nonempty and, for any $n \geq 0$, $uv^n w \in L$.*

Proof Let M be an fsa with $L = L(M)$, and let its start state be σ_0, let N be the number of states in M, and let $z \in L$ with $|z| \geq N$. Then z labels a path in M from σ_0 to an accepting state σ_z, which passes through at least $N + 1$ states, and hence passes through at least one of them twice. We let uv be the shortest prefix of z that passes through a state twice, call that state σ, and suppose that u labels the prefix of uv from σ_0 to σ. Since uv is chosen to be shortest, it has length at most N. Then v labels a loop in M from σ to σ, and hence for any $n \geq 0$, $uv^n w$ labels a path in M from σ_0 to σ_z that follows the loop from σ to σ (at least) n times. Since the target of the path is the accepting state σ_z, the string $uv^n w$ is in L. □

2.6 Pushdown automata, context-free languages and grammars

2.6.1 Pushdown automata We have already introduced pushdown automata (pda) briefly in 2.2.5. In order to examine these automata properly we need to define them more formally than we have done so far.

We define a pda over an input alphabet A to consist of a set Σ of states, an input-tape, a semi-infinite stack, a finite stack alphabet B, and a finite set of possible moves, each of which consists of a read phase followed by an action. During each read phase, a symbol from B is read and removed from the stack and at most one symbol from A is read from the input-tape; during each action phase, a (possibly empty) string is written to the stack and the machine may change state.

One state in Σ is designated the start state, and some are designated accepting states. Initially the automaton is in its start state σ_0, the input is written on the input-tape and a read-only head is set to point at the first symbol on that tape. A designated string $\beta_0 \in B^*$ is written onto the stack, and a read-write head is set to point at the top stack symbol.

In each move, the read-write head reads and deletes the top stack symbol. The machine may or may not read a symbol from the input-tape; if it does so, then the read-only head moves one position to the right. The machine then moves into a new state, and writes a string to the top of the stack. The choice of action depends on the current state, and the symbols taken during this move from the stack and input-tape. So there are two types of moves, those that read a symbol from the input word, and those that do not. A move that does not take a symbol from the input word is called an ε-move.

By convention we assume that when a string is written to the stack the left-most symbol of the string is added last, and so becomes the top stack symbol. So initially the top stack symbol is the first symbol of the string β_0.

We specify the moves of a pda by means of a map τ from $\Sigma \times (A \cup \{\varepsilon\}) \times B$ to $2^{\Sigma \times B^*}$, which we call the *transition function*. If (σ', β) is in $\tau(\sigma, a, b)$, then there is a move in which b is taken from the stack, a is read from input, the string β is written to the stack (with the first letter of β becoming the new symbol at the top of the stack), and the pda moves from the state σ to σ'. We set a to be ε in order to describe an ε-move and β to be ε if b is taken from the stack but nothing new is written to it. We may write $(\sigma, b)^a$ rather than $\tau(\sigma, a, b)$.

In general we define our pda to accept words 'on final state'. In this case a word w over A is accepted by the pda if there is a sequence of permissible moves that take the automaton from its initial configuration to one where the entire input word has been read, and the automaton is in an accepting state. We do not require that the automaton can process all strings.

Note that in certain configurations of the pda there might be no possible moves, either because $\tau(\sigma, a, b)$ is empty, or because the stack is empty and so there is no symbol b to be read. In that case the computation halts, and the input word is accepted if and only if the whole word has been read and the pda is in an accept state.

An alternative (but equivalent) definition of pda defines acceptance 'on empty stack'; that is, a word is accepted, irrespective of the state of the pda, if at some stage after the entire input word has been read, the stack is empty. We prove in Theorems 2.6.6 and 2.6.7 that a pda accepting on final state can be transformed into one that accepts on empty stack and vice versa. Hence it is valid to define a context-free language to be a set of strings that is accepted by a pda in either model.

2.6.2 Variations in the definition of a pda There are many minor variations of our definition of a pda in the literature, and it is generally easy to show that they are equivalent to ours, in the sense that a language is accepted by one of our pda if and only if it is accepted by one of the variants. For example, the

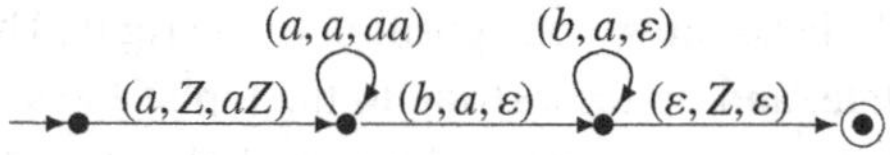

$$(a, a, aa) \quad (b, a, \varepsilon)$$

$$\longrightarrow \bullet \xrightarrow{(a, Z, aZ)} \overset{\circlearrowright}{\bullet} \xrightarrow{(b, a, \varepsilon)} \overset{\circlearrowright}{\bullet} \xrightarrow{(\varepsilon, Z, \varepsilon)} \circledcirc$$

Figure 2.6 A deterministic pda accepting $\{a^n b^n : n \geq 1\}$; the initial stack is Z

definition in [159] allows only a single symbol b_0 to be on the stack initially rather than a string β_0.

As we observed earlier, with our definition no transitions are possible when the stack is empty, and so the computation must halt in that situation. Some authors effectively enable such transitions by defining a special bottom-of-stack symbol, which is never deleted, and never occurs anywhere else on the stack. Since we may introduce such a symbol with this behaviour anyway whenever we want to, we prefer not to make it part of the definition. One situation when we need it is when we are studying one-counter languages (that is, pda with $|B| = 1$, cf. 2.2.9).

2.6.3 Representing pushdown automata graphically We can represent a pda by a directed graph much as we do an fsa. Vertices correspond to states, the start state is indicated by an arrow, and accepting states are ringed. When $(\sigma', \beta) \in \tau(\sigma, a, b)$, the directed edge from the vertex representing σ to the vertex representing σ' is labelled by the triple (a, b, β). See, for example, Figure 2.6.

2.6.4 Deterministic pushdown automata Recall that we define a Turing machine to be deterministic if, for any given input, its operation is completely determined. For a pda, whether it accepts by final state or by empty stack, this translates to the following condition. We can take this as an equivalent definition of determinism for a pda.

Suppose that M is a pda with input alphabet A, stack alphabet B, state set Σ, and transition function τ. We say that M is *deterministic* if, for any pair $(\sigma, b) \in \Sigma \times B$, one of the following is true:

(i) $\tau(\sigma, a, b) = \emptyset$ for all $a \in A$, and $|\tau(\sigma, \varepsilon, b)| = 1$;
(ii) $|\tau(\sigma, a, b)| \leq 1$ for all $a \in A$, and $\tau(\sigma, \varepsilon, b) = \emptyset$.

We define a context-free language to be deterministic if it is accepted by a deterministic pda that accepts by final state. We recall that the family of context-free and deterministic context-free languages are denoted by $\mathcal{CF}$ and $\mathcal{DCF}$, respectively.

2.6.5 Remark The family of context-free languages that are accepted by deterministic pda that accept by empty stack is strictly smaller than $\mathcal{DCF}$. A language L is said to have the *prefix property* if no proper prefix of any element of L lies in L. It is shown in [158, Theorem 6.19] that this family of languages consists of the deterministic context-free languages with the prefix property, and the reader could attempt to prove this as an exercise.

The following two results show that our two models of pda are essentially equivalent.

2.6.6 Theorem *Suppose that M_1 is a* pda *that accepts by final state. Then there exists a* pda *M_2 that accepts by empty stack with $L(M_1) = L(M_2)$.*

Proof Our proof is modelled on that of [159, Theorem 5.1]. Suppose that M_1 has input alphabet A, stack alphabet B, state set Σ, and transition function τ_1. Then we define M_2 with input alphabet A, stack alphabet $B \cup \{b_0\}$, state set $\Sigma \cup \{\sigma_\varepsilon\}$, and transition function τ_2 as follows.

The role of b_0 is to mark but also to cover up the bottom of the stack, and the state σ_ε is a state that we use while we empty the stack of M_2. Initially the stack of M_2 contains the concatenation of the initial stack string of M_1 and b_0.

For all $\sigma \in \Sigma, a \in A, b \in B$, we define

$$\tau_2(\sigma, a, b) = \tau_1(\sigma, a, b).$$

In addition, for each $b \in B$, we define

$$\tau_2(\sigma, \varepsilon, b) = \tau_1(\sigma, \varepsilon, b),$$

for each non-accepting state σ of M_1, and

$$\tau_2(\sigma, \varepsilon, b) = \tau_1(\sigma, \varepsilon, b) \cup \{(\sigma_\varepsilon, \varepsilon)\},$$

for each accepting state σ of M_1.

Finally, we define

$$\tau_2(\sigma_\varepsilon, \varepsilon, b) = \{(\sigma_\varepsilon, \varepsilon)\} \text{ for all } b \in B \cup \{b_0\},$$

and the image of τ_2 is empty otherwise.

Note that with this construction, M_2 is almost always non-deterministic, even when M_1 is deterministic. The idea is that, whenever M_2 is in an accepting state of M_1, it has the option of guessing that it has read all of its input, and emptying the stack. If it guesses correctly, then this results in the word being accepted by M_2. □

2.6.7 Theorem   *Suppose that M_1 is a* pda *that accepts by empty stack. Then there exists a* pda *M_2 that accepts by final state, with $L(M_1) = L(M_2)$.*

Proof   Suppose that M_1 has input alphabet A, stack alphabet B, state set Σ, and transition function τ_1. Then M_2 has stack alphabet $B \cup \{b_0\}$, state set $\Sigma \cup \{\sigma_F\}$, and transition function τ_2.

Again, the role of b_0 is to mark and cover up the bottom of the stack, and the state σ_F is the sole accepting (i.e. final) state. Initially the stack of M_2 contains the concatenation of the initial stack string of M_1 and b_0.

For $a \in A \cup \{\varepsilon\}, b \in B, \sigma \in \Sigma$, we define $\tau_2(\sigma, a, b) = \tau_1(\sigma, a, b)$. In addition we define $\tau_2(\sigma, \varepsilon, b_0) = \{(\sigma_F, \varepsilon)\}$, and the image of τ_2 is empty otherwise.

So, whenever the stack of M_1 is empty, M_2 has the option of guessing that the input word has been read, moving into the accepting state σ_F, and halting.   □

2.6.8 Inspecting multiple stack symbols   In some proofs and examples, it would be convenient if a pda were allowed to inspect more than one symbol from the top of the stack on each move. Provided that there is a bound k on the number that can be read, this behaviour can be simulated by a standard pda.

For fixed $k \geq 1$, let us call a machine M a k-pda if, on each move, it can remove and read up to k symbols from the top of the stack in turn, and then move into a new state and write a string to the stack, both of which may depend on the current input symbol and the stack symbols read. As usual, M is *deterministic*, if there is only one possible move for a given input symbol and stack symbols read. It is customary to reserve one or more stack symbols for use only at the bottom of the stack, so that M can avoid attempting to remove more symbols than are currently on the stack.

2.6.9 Proposition   *If a language L is accepted by a (deterministic) k-pda then it is also accepted by a standard (deterministic)* pda, *and hence lies in* CF *(or DCF).*

Proof   Let M be a (deterministic) k-pda with state set Σ and stack alphabet B, that accepts L by final state. The state set of the (deterministic) pda M' that simulates M is $\Sigma \times \Lambda$, where Λ is the set of strings in B^* of length at most k, and the initial state of M' is (σ_0, ε).

Then M' simulates each single move of M by a sequence of moves as follows. It begins the simulation of a move of M whenever the second component of its state is the empty word. It then makes up to k ε-moves, each of which removes one symbol from the top of the stack, and stores it as part of the second state component. (It makes fewer than k-moves only when the stack height is less than k, which is typically detected by means of special bottom of stack symbols.) The length of the second state component or the detection of b_0 allows M' to know when these ε-moves have been completed. It then makes the same move as the move of M that is being simulated, and replaces its second state component by the empty word.   □

Although the use of states can simplify the description of a pda, they are not strictly necessary, as we see from the following result. The proof is adapted from that of [159, Theorem 5.4].

2.6.10 Theorem   *Any context-free language is recognised by a pda with one state, accepting by empty stack.*

Proof   Let $L \in C\mathcal{F}$ be the given language, defined over an alphabet A. We suppose that we have a pda M accepting L by empty stack, with state set Σ and stack alphabet B. We suppose that σ_0 is the start state of Σ, and β_0 the initial contents of its stack. Our aim is to construct a pda M' with just one state, accepting by empty stack and with $L(M') = L$.

Making appropriate adjustments to the stack alphabet and transition function as necessary, we can make the following assumptions about M. The stack initially contains a single symbol b_0 (that is, $\beta_0 = b_0$). At the end of a computation that accepts a string w, an ε-move takes the pda into a configuration with empty stack and in a final state σ_F. The pda is never in state σ_F when the stack is nonempty.

We now define a new pda M' whose stack alphabet is

$$B' = \{(\sigma, b, \sigma') : \sigma, \sigma' \in \Sigma, b \in B\}.$$

The pda M' operates non-deterministically. Initially the stack consists of the single symbol $(\sigma_0, b_0, \sigma_F)$.

For each computation in M there are corresponding computations in M' for which the stack of M' has the same height as that of M throughout the computation. When M is in state σ with b at the top of its stack, the pda M' has a symbol (σ, b, σ') at the top of its stack. Furthermore, the symbol below (if any) has the form (σ', b', σ''). This ensures that at any stage, the first component of the top symbol on the stack records the state of M in a parallel computation.

Let τ be the transition function of M. We now define the transition function τ' of M'. Note that τ' needs only two, not three arguments, since there is no choice of state. Its output is the set of strings that may be written to the stack.

Suppose that $a \in A \cup \{\varepsilon\}$, $b \in B$. It is convenient to consider the transitions that delete from the stack but do not write to it (that is, write ε) separately from the rest. To each such transition of M is associated a single transition of

M' that also deletes from the stack without writing; that is, whenever (σ', ε) is in $\tau(\sigma, a, b)$, then ε is in the set $\tau(a, (\sigma, b, \sigma'))$. Note that after such a stack-deleting move of M', the first component of the top stack symbol is σ', the target of σ in the parallel computation of M.

For each transition of M that writes a nonempty string to the stack, we define many possible corresponding transitions of M' that write to the stack. For $n \geq 1$ and each transition of M writing a string $b_1 \ldots b_n$ to the stack (that is, for each $(\sigma', b_1 \cdots b_n) \in \tau(\sigma, a, b)$), and for each $\sigma'' \in \Sigma$, the set $\tau'(a, (\sigma, b, \sigma''))$ contains the strings

$$(\sigma', b_1, \sigma_1)(\sigma_1, b_2, \sigma_2) \cdots (\sigma_{n-1}, b_n, \sigma'')$$

for all choices of $\sigma_1, \ldots, \sigma_{n-1} \in \Sigma$. (So, for example, if $n = 1$, then there is a single such string (σ', b_1, σ'').)

We claim that, accepting by empty stack, M' recognises L. In order to justify this claim, we need to verify first that every string in L is accepted by M', and secondly that every string accepted by M' is in L. These verifications are for the most part lengthy but routine, and we leave most of the details to the reader.

Any string in L is formed by deleting ε's from a string $a_1 \ldots a_n$, which defines a successful computation of M as follows. Initially M is in state σ_0 with b_0 on top of the stack. In n successive moves it takes $b_0, b_1, \ldots, b_{n-1}$ from the top of the stack, $a_1, \ldots, a_n$ from input, moves through states $\sigma_1, \ldots, \sigma_n = \sigma_F$ and writes $\beta_1, \ldots, \beta_n$ to the stack. Here b_i is the top symbol of the stack after β_i has been written, and σ_i, β_i are defined to satisfy

$$(\sigma_i, \beta_i) \in \tau(\sigma_{i-1}, a_i, b_{i-1}).$$

After the n-th transition M is in state σ_F and the stack is empty.

To verify the existence of corresponding moves in M' that accept the same string, we prove the following assertion, for $0 \leq i \leq n$, by induction on i. Suppose that immediately after move number i in the above computation the stack of M contains precisely $c_1 \cdots c_k$. (So $c_1 = b_i$ is at the top of the stack for $i < n$, and $k = 0$ when $i = n$. Also, $k = 1$ and $c_1 = b_0$ when $i = 0$.) Then, for any sequence $\sigma'_1, \ldots, \sigma'_{k-1}$ of states in Σ, there is a valid computation of the input string $a_1 \cdots a_i$ in M' in which after move number i the stack contains precisely $(\sigma_i, c_1, \sigma'_1)(\sigma'_1, c_2, \sigma'_2) \cdots (\sigma'_{k-1}, c_k, \sigma_F)$. This assertion with $i = n$ says exactly that there is a valid computation in M' that accepts the input word.

Since initially the stack of M' contains precisely $(\sigma_0, b_0, \sigma_F)$, the assertion is true with $i = 0$, so fix i and assume by induction that it is true for smaller i.

We present the details of the argument only for the case that the ith move of M writes nothing to the stack. We apply the inductive hypothesis for $i - 1$ to M'. If $i = n$ then, after move $n - 1$, M has just b_n on its stack, and there is a corresponding computation in M' with just $(\sigma_{n-1}, b_n, \sigma_F)$ on its stack, so we can read a_n (in fact $a_n = \varepsilon$ by our assumption on M), and empty the stack of M' to complete the computation. If $i < n$ and $c_1 \cdots c_k$ is on the stack of M after move $i - 1$ then, for any $\sigma'_2, \ldots, \sigma'_{k-1} \in \Sigma$, there is a computation in M' in which after move $i - 1$ the stack contains precisely $(\sigma_{i-1}, c_1, \sigma_i)(\sigma_i, c_2, \sigma'_2) \cdots (\sigma'_{k-1}, c_k, \sigma_F)$, and again we may read a_i in M' and remove the top stack symbol, completing the inductive step.

We leave the verification of the completion of the inductive step when M writes a nonempty string to the stack on move i to the reader.

It is not difficult to verify the converse, that to each successful computation of M' there corresponds a parallel successful computation of M, and so we also leave that to the reader. □

2.6.11 Example We consider the language of binary strings with equal numbers n_0 and n_1 of 0's and 1's.

This language can be recognised by a non-deterministic **pda** M accepting by empty stack, with states σ_+, σ_- and σ_F, and two stack symbols b_0 and X. The X's on the stack record the difference $n_1 - n_0$ or $n_0 - n_1$, whichever is positive, while the states σ_+ and σ_- keep track of the sign of $n_1 - n_0$. The machine starts in state σ_+ with only b_0 on the stack. The transitions are described as follows:

$$
\begin{aligned}
\tau(\sigma_+, 0, b_0) &= \{(\sigma_-, Xb_0)\}, & \tau(\sigma_-, 0, b_0) &= \{(\sigma_-, Xb_0)\}, \\
\tau(\sigma_+, 1, b_0) &= \{(\sigma_+, Xb_0)\}, & \tau(\sigma_-, 1, b_0) &= \{(\sigma_+, Xb_0)\}, \\
\tau(\sigma_-, 0, X) &= \{(\sigma_-, XX)\}, & \tau(\sigma_+, 1, X) &= \{(\sigma_+, XX)\}, \\
\tau(\sigma_+, 0, X) &= \{(\sigma_+, \varepsilon)\}, & \tau(\sigma_-, 1, X) &= \{(\sigma_-, \varepsilon)\}, \\
\tau(\sigma_+, \varepsilon, b_0) &= \{(\sigma_F, \varepsilon)\}, & \tau(\sigma_-, \varepsilon, b_0) &= \{(\sigma_F, \varepsilon)\}.
\end{aligned}
$$

The single state **pda** M' has input alphabet $\{0, 1\}$ and a stack alphabet of size 18, consisting of all triples (σ, b, σ') with $b \in \{b_0, X\}$, $\sigma, \sigma' \in \{\sigma_+, \sigma_-, \sigma_F\}$. Initially the stack contains the single stack symbol $(\sigma_+, b_0, \sigma_F)$. The transitions of M' are now described by the transition function τ', listed in the same order as the corresponding transitions of M above.

$$\forall \sigma'' \in \Sigma, \ \tau'(0, (\sigma_+, b_0, \sigma'')) = \{(\sigma_-, X, \sigma_1)(\sigma_1, b_0, \sigma'') : \sigma_1 \in \Sigma\},$$

$$\forall \sigma'' \in \Sigma, \ \tau'(0, (\sigma_-, b_0, \sigma'')) = \{(\sigma_-, X, \sigma_1)(\sigma_1, b_0, \sigma'') : \sigma_1 \in \Sigma\},$$

$$\forall \sigma'' \in \Sigma, \ \tau'(1, (\sigma_+, b_0, \sigma'')) = \{(\sigma_+, X, \sigma_1)(\sigma_1, b_0, \sigma'') : \sigma_1 \in \Sigma\},$$

$$\forall \sigma'' \in \Sigma, \ \tau'(1, (\sigma_-, b_0, \sigma'')) = \{(\sigma_+, X, \sigma_1)(\sigma_1, b_0, \sigma'') : \sigma_1 \in \Sigma\},$$

$$\forall \sigma'' \in \Sigma, \ \tau'(0, (\sigma_-, X, \sigma'')) = \{(\sigma_-, X, \sigma_1)(\sigma_1, X, \sigma'') : \sigma_1 \in \Sigma\},$$

$$\forall \sigma'' \in \Sigma, \ \tau'(1, (\sigma_+, X, \sigma'')) = \{(\sigma_+, X, \sigma_1)(\sigma_1, X, \sigma'') : \sigma_1 \in \Sigma\},$$

$$\tau'(0, (\sigma_+, X, \sigma_+)) = \{\varepsilon\},$$

$$\tau'(1, (\sigma_-, X, \sigma_-)) = \{\varepsilon\},$$

$$\tau'(\varepsilon, (\sigma_+, b_0, \sigma_F)) = \{\varepsilon\},$$

$$\tau'(\varepsilon, (\sigma_-, b_0, \sigma_F)) = \{\varepsilon\}.$$

We can construct from M' a single state machine M'' with a rather simpler description as follows.

First we observe that we only need stack symbols whose first component is $\sigma_\pm$, since there are no transitions defined when other stack symbols are read. So, in the above description, σ_1 need only take the values $\sigma_\pm$. Further, we note that the only transitions that write onto the stack a symbol of the form $(\sigma_1, b_0, \sigma_\pm)$ or (σ_1, X, σ_F) for some $\sigma_1 \in \Sigma$ are those that read a stack symbol of the same form. So in fact these stack symbols are never written to the stack, and we can delete them from the stack alphabet together with all transitions that read them.

Then we note that, once symbols (σ_+, X, σ_-) or (σ_-, X, σ_+) are on the stack they are never deleted, and so any transitions involving those stack symbols cannot be part of a route to an empty stack; hence those symbols and all transitions involving them may as well be removed. And finally we see that the machine reacts in the same way to the two stack symbols $(\sigma_\pm, b_0, \sigma_F)$, so we can replace that pair of symbols by a single symbol.

So we can define pda M'', with $L(M'') = L(M') = L(M)$, with a stack alphabet B'' of size 3. Relabelling the two stack symbols $(\sigma_\pm, b_0, \sigma_F)$ as b_0, and the stack symbols $(\sigma_+, X, \sigma_+), (\sigma_-, X, \sigma_-)$, as s_+ and s_-, we write

$$B'' = \{b_0, s_+, s_-\}.$$

We put b_0 onto the stack initially, and define τ'' as follows:

$$\tau''(0, b_0) = \{s_- b_0\}, \quad \tau''(1, b_0) = \{s_+ b_0\},$$

$$\tau''(0, s_-) = \{s_- s_-\}, \quad \tau''(1, s_+) = \{s_+ s_+\},$$

$$\tau''(0, s_+) = \{\varepsilon\}, \quad \tau''(1, s_-) = \{\varepsilon\}, \quad \tau''(\varepsilon, b_0) = \{\varepsilon\}.$$

2.6.12 Context-free grammars A grammar with variable set V, terminal set T, and set P of production rules (see 2.1.2) is called a *context-free grammar* if all left-hand sides of rules in P are variables from V. It is clear that every regular grammar (see 2.5.15) satisfies this condition. We have already seen that a set of strings is accepted by an fsa if and only if it can be defined by a regular grammar. In fact we have a similar result for context-free languages. We shall prove soon that a set of strings can be recognised by a pda if and only if it can be constructed using a context-free grammar. Context-free grammars are also known as *Type 2 grammars*.

2.6.13 Greibach and Chomsky normal forms A context-free grammar in which all productions have the form either $x \rightarrow yz$ or $x \rightarrow a$, for $x, y, z \in V$, $a \in T$, is said to be in *Chomsky normal form*. Additionally, such a grammar may contain the production $S \rightarrow \varepsilon$, provided that S does not appear in the right-hand side of any production.

A context-free grammar in which all productions have the form $x \rightarrow a\xi$, where $x \in V$, $a \in T$ and $\xi \in V^*$, is said to be in *Greibach normal form*. Here too, we also allow the production $S \rightarrow \varepsilon$, provided that S does not appear in the right-hand side of any production.

The next result states that any context-free grammar can be replaced by a grammar generating the same language that is in either Chomsky or Greibach normal form. This replacement process can be made into an algorithm (although we have not explained exactly how to do that in the proof), but the resulting grammar can be exponentially larger than the original.

These normal forms are probably of theoretical rather than of practical importance, because it is useful in certain proofs relating to context-free languages, such as that of the Pumping Lemma (Theorem 2.6.17), to assume that the grammar is in normal form. We note also that, in the proof of the Muller–Schupp Theorem (Theorem 11.1.1) that groups with context-free word problem are virtually free, we assume that the word problem is generated by a context-free grammar in Chomsky normal form.

We call a variable in a context-free grammar *useful* if it occurs in the derivation of a word $w \in T^*$ from S. Otherwise it is *useless*.

2.6.14 Theorem *Let L be a language over a finite alphabet T. Then the following are equivalent.*

(i) *L is generated by a context-free grammar.*
(ii) *L is generated by a grammar in Chomsky normal form with no useless variables.*

(iii) *L is generated by a grammar in Greibach normal form with no useless variables.*

Proof Any grammar in Chomsky or Greibach normal form is context-free by definition. We present an outline proof of (i) $\implies$ (ii) and refer the reader to [159, Theorem 4.5] or [158, Theorem 7.16] for further details. (i) $\implies$ (iii) is proved in [159, Theorem 4.6].

Starting with an arbitrary context-free grammar, we describe a sequence of transformations that result in a grammar in Chomsky normal form with no useless variables that generates the same language. Recall from 2.1.2 that, for a variable x and a word $w \in (V \cup T)^*$, we write $x \Rightarrow^* w$ to mean that there is a (possibly empty) sequence of productions that transforms x to w.

To avoid the problem of the start symbol S occurring on the right-hand side of a production when there is a production $S \to \varepsilon$, we start by introducing a new variable S', replacing S by S' in all productions, and then introducing the new production $S \to S'$.

The next step is to eliminate all productions of the form $x \to \varepsilon$ except for $S \to \varepsilon$. Call a variable z *nullable* if $z \Rightarrow^* \varepsilon$. Let $x \to w$ be a production, and write $w = w_1 z_1 w_2 z_2 \cdots w_k z_k w_{k+1}$, where $z_1, z_2, \ldots, z_k$ are the nullable variables that occur in z. Then, for each nonempty subset I of $\{1, 2, \ldots, k\}$, we introduce a new production $x \to w_I$, where w_I is defined by deleting the variables z_i with $i \in I$ from w. (So this process can potentially result in an exponential increase in the number of productions.) We can now remove all productions of the form $x \to \varepsilon$ without changing the language, except that we may have removed ε from L. So, if $\varepsilon \in L$, then we adjoin the production $S \to \varepsilon$.

Our next aim is to eliminate all productions of the form $x \to y$, with $x, y \in V$. These are called *unit productions*. For $x, y \in V$, we call (x, y) a *unit pair* if $x \Rightarrow^* y$. In particular, (x, x) is a unit pair for all $x \in V$. For each unit pair (x, y) and each non-unit production $y \to w$, we introduce the new production $x \to w$ (if it is not there already). We can now remove all unit productions without changing the language.

We now eliminate any useless variables as follows. Call the variable x *reachable* if $S \Rightarrow^* w$, where x occurs somewhere in w. First eliminate all unreachable variables and all productions that contain them. Then eliminate all variables x for which there is no word $w \in T^*$ with $x \Rightarrow^* w$, and all productions containing them.

The next step is to eliminate terminals from the right-hand sides of productions $x \to w$ with $|w| > 1$. For each such terminal a, introduce a new variable x_a, a production $x_a \to a$, and replace a by x_a in the right-hand side of the production $x \to w$.

Finally, for productions $x \to y_1 y_2 \cdots y_k$ with $y_i \in V$ and $k > 2$, introduce new variables $z_1, z_2, \ldots, z_{k-2}$ and replace the production by the $k - 1$ productions

$$x \to y_1 z_1, \ z_1 \to y_2 z_2, \ \ldots, \ z_{k-3} \to y_{k-1} z_{k-2}, \ z_{k-2} \to y_{k-1} y_k.$$

The grammar is now in Chomsky normal form. □

2.6.15 Theorem *A subset $L \subseteq A^*$ is the language of a* pda *if and only if it can be generated by a context-free grammar with terminal set A.*

Proof Our proof is essentially a summary of the more detailed treatment in [158, Section 3.6]. Suppose first that L is generated by a context-free grammar with terminal set A and variable set B.

We construct a pda M over A with one state, and stack alphabet $A \cup B \cup \{b_0\}$ that accepts L by empty stack as follows. Note that the transition function needs only two arguments, and its value is a set of strings. This construction is a little easier if we start with the grammar in Greibach normal form, but we do not assume that here, because we did not prove (i)$\Rightarrow$(iii) of Theorem 2.6.14.

The symbol b_0 is used to cover the bottom of the stack, to ensure that M only accepts properly terminated strings. Initially the string $S b_0$ is on the stack, where S is the start symbol of the grammar, and we define $\tau(\varepsilon, b_0) = \{\varepsilon\}$. If there is a production $S \to \varepsilon$ then we define an ε-move of M that simply removes S from the top of the stack; that is we define $\tau(\varepsilon, S) = \{\varepsilon\}$.

For each $x \in B$, we define $\tau(\varepsilon, x)$ to be the set of strings w for which $x \to w$ is a production. And, for each $a \in A$, we define $\tau(a, a) = \{\varepsilon\}$. Otherwise, the image of τ is empty.

This completes the construction of the pda M, and we leave it to the reader to verify that $L(M) = L$. In fact, for each $w \in L$, we get an accepting path in M corresponding to the left-most derivation of w, as described in 2.3.2.

Conversely, suppose that M is given as the language of a pda. By Theorem 2.6.10, we may assume that M has a single state, and is constructed as in the proof of that theorem. Recall that M starts with a single symbol b_0 on the stack, and accepts by empty stack. We let B be the stack alphabet for M.

The corresponding grammar that generates L has variable set $B \cup \{S\}$ and terminal set A. It contains the rule $S \to b_0$ together with all rules $b \to a\beta$ for which $\beta \in \tau(a, b)$ with $a \in A \cup \{\varepsilon\}$, $b \in B$. □

2.6.16 Example We consider again the language of binary strings with equal numbers of 0's and 1's. From the description of the one state pda M'' of Example 2.6.11 we derive immediately the grammar with the following rules (listed here in the same order as the transitions of M'').

$$S \;\rightarrow\; b_0,$$

$$b_0 \;\rightarrow\; 0s_-b_0, \qquad b_0 \;\rightarrow\; 1s_+b_0,$$

$$s_- \;\rightarrow\; 0s_-s_-, \qquad s_+ \;\rightarrow\; 1s_+s_+,$$

$$s_+ \;\rightarrow\; 0, \qquad s_- \;\rightarrow\; 1, \qquad b_0 \;\rightarrow\; \varepsilon.$$

For example, we can generate the string 110001 as follows:

$$S \;\rightarrow\; b_0 \qquad\qquad \rightarrow\; 1s_+b_0 \qquad \rightarrow\; 11s_+s_+b_0 \;\rightarrow\; 1100b_0 \;\rightarrow$$
$$11000s_-b_0 \;\rightarrow\; 110001b_0 \;\rightarrow\; 110001.$$

2.6.17   Theorem (The Pumping Lemma for context-free languages)   *Let $L \in C\mathcal{F}(A^*)$. Then there exists a constant p such that, if $z \in L$ and $|z| \geq p$, then we can write $z = uvwxy$ with $u, v, w, x, y \in A^*$, such that*

(i) *$|vwx| \leq p$;*
(ii) *$vx \neq \varepsilon$;*
(iii) *for all $i \geq 0$, $uv^iwx^iy \in L$.*

Proof   Let $\mathcal{G}$ be a grammar in Chomsky normal form (2.6.13) that generates L, where $\mathcal{G}$ has terminals A and variables B. Let $m = |B|$ and $p = 2^m$, and let $z \in L$ with $|z| \geq p$. Note that $p > 0$ so, if there is a production $S \rightarrow \varepsilon$, then it is not used in the derivation of z in $\mathcal{G}$.

Since the productions of $\mathcal{G}$ that do not generate terminals all have the form $b \rightarrow b_1 b_2$ with $b, b_1, b_2 \in B$, we see that the parse tree (see 2.3.1) of any derivation of z in $\mathcal{G}$ must contain a path from S to a terminal a that contains at least m productions of this form, and therefore contains at least $m + 1$ variables (including S) among the labels of its nodes.

So at least one such variable $b \in B$ must be repeated on the path. Hence we can write $z = uvwxy$ where, in the derivation, the first b on the path generates vwx and the second b generates w; see Figure 2.7. The condition $vx \neq \varepsilon$ follows from the fact that the two occurrences of b are distinct and (since $\mathcal{G}$ is in Chomsky normal form), all variables generate nonempty strings.

By choosing the occurrences of b to be the final two repeated variables in the path to a, we can ensure that they occur among the final $m + 1$ variables in the path, and hence that $|vwx| \leq 2^m = p$.

Finally, we observe that the parts of the derivation of z between the two b's can be repeated an arbitrary number i of times, including $i = 0$, and hence we have $uv^iwx^iy \in L$ for all $i \geq 0$.   □

2.6.18   Example   The language $L = \{a^m b^n a^m b^n : m, n \in \mathbb{N}_0\}$ is not context-free.

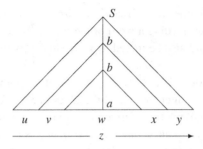

Figure 2.7 The derivation of $z = uvwxy$ in $\mathcal{G}$

Proof Suppose that L is context-free, let p be the constant defined in the Pumping Lemma for L, and apply this lemma to $z = a^m b^n a^m b^n$ with $m, n > p$, to give $z = uvwxy$. If either v or x had non-trivial intersection with two of the four subwords a^m, b^n, a^m, b^n of z, then $uv^2 wx^2 y$ would not lie in the regular set $a^* b^* a^* b^*$ so would not be in the language.

So each of v and x lies entirely within one of these subwords, and $m, n > p$ forces them to lie either in the same or in adjacent subwords. In either case, since at least one of v and x is nonempty, uwy is not in the language, contradicting the Pumping Lemma. □

2.6.19 Exercise Use the Pumping Lemma to show that the language

$$\{a^{2n} b^2 a^{2n} b^2 c^{2n} : n \in \mathbb{N}\}$$

is not context-free.

2.6.20 Example Let L be the language of palindromes (i.e. words w with $w = w^R$) over the alphabet $\{a, b\}$. Then L is accepted by a non-deterministic pda that starts by putting the letters read onto the stack, guesses where the middle of the word is, and then checks that $w = w^R$ by removing letters from the stack that match the letters read. So $L \in \mathcal{CF}$, but $L \notin \mathcal{DCF}$.

Proof If $L \in \mathcal{DCF}$ then, as we shall prove in Proposition 2.6.27, so is its intersection K with the regular set

$$(aa)^+ b^2 (aa)^+ \cup (aa)^+ b^2 (aa)^+ b^2 (aa)^+,$$

and

$$K = \{a^{2n} b^2 a^{2n} : n \in \mathbb{N}\} \cup \{a^{2n} b^2 a^{2m} b^2 a^{2n} : m, n \in \mathbb{N}\}.$$

Let M be a deterministic pda with $L(M) = K$. Then, for each even $n = 2k > 0$, after reading the word $a^n b^2 a^n$, M must enter an accepting state σ_k before attempting to read another letter. Let $F' := \{\sigma_k : k > 0\}$. Then M cannot enter a state in F' after reading a word of the form $a^{2n}b^2a^{2m}b^2a^{2n}$, because if it did that then it would also be forced to accept longer words, that contain an additional b^2 and are not in $L(M)$.

We define a new pda M' on the alphabet $\{a, b, c\}$ as follows. M' behaves in the same way as M until it enters one of the states in F'. After entering any of these states, M' will not read any further letters labelled a until it has read the word b^2 (that is, M' will halt without accepting unless the next two symbols that it reads are both b), and after reading b^2, every transition of M labelled a is replaced by the same transition labelled c. Then

$$L(M') = \{a^{2n}b^2a^{2n} : n \in \mathbb{N}\} \cup \{a^{2n}b^2a^{2n}b^2c^{2n} : m, n \in \mathbb{N}\},$$

and by intersecting this with the regular language $(aa)^+b^2(aa)^+b^2(cc)^+$, we see that this contradicts the statement in the exercise above.   □

The following generalisation of the Pumping Lemma is proved in [105]. Notice that we have added more detail to the statement of the theorem, which is present only in the proof of Theorem 2.6.17. This is to enable us to apply it in the proof of Parikh's theorem (2.6.23).

2.6.21 Theorem   *Let $L \in \mathcal{CF}(A^*)$ be generated by the context-free grammar $\mathcal{G}$ in Chomsky normal form. Then there exists a constant p such that, for any integer $k \geq 1$, given $z \in L$ with $|z| \geq p^k$ and any derivation of z in $\mathcal{G}$, there is a variable b and an equivalent derivation of the form*

$$S \Rightarrow^* ubv \Rightarrow^* uv_1bx_1y \Rightarrow^* uv_1v_2bx_2x_1y \Rightarrow^* \cdots$$
$$\Rightarrow^* uv_1 \cdots v_kbx_k \cdots x_1y$$
$$\Rightarrow^* uv_1 \cdots v_kwx_k \cdots x_1y = z,$$

with $u, v_i, w, x_i, y \in A^*$, such that

(i) $|v_1 \cdots v_kwx_k \cdots x_1| \leq p^k$;
(ii) $v_ix_i \neq \varepsilon$ for all i with $1 \leq i \leq k$;
(iii) $uv_1^{i_1} \cdots v_k^{i_k}wx_k^{i_k} \cdots x_1^{i_1}y \in L$ for all $i_1, \ldots, i_k \geq 0$.

Proof   The proof is a straightforward generalisation of that of Theorem 2.6.17. When $|z| > p^k = 2^{mk}$, the parse tree of the given derivation of z must contain a path from S to a terminal that contains at least mk productions of the form $b \to b_1b_2$, which therefore contains at least $mk + 1$ variables among the labels of its nodes. So some variable $b \in B$ must occur at least $k + 1$ times among the final $mk + 1$ variables on the path. From this, we obtain the equivalent derivation (it is equivalent, because it has the same parse tree) described in the conclusion.   □

2.6.22 Parikh's theorem In addition to the Pumping Lemma, a result due to Parikh [211] has turned out to be useful for proving that certain types of languages are not context-free. In particular, it is used in Chapter 14 to show that the complement of the word problem (the *co-word problem*) in certain types of groups is not context-free.

The proof that we give here is due to Goldstine [105], and uses the above generalisation of the Pumping Lemma. We first need some definitions.

For $n \geq 0$, a subset N of $\mathbb{N}_0^n$ is called *linear* if it is a coset of a finitely generated submonoid of $\mathbb{N}_0^n$. In other words, there exists $m \geq 0$, $c \in \mathbb{N}_0^n$, and $\alpha_i \in \mathbb{N}_0^n$ for $1 \leq i \leq m$, with

$$N = \{c + n_1\alpha_1 + \cdots + n_m\alpha_m : n_i \in \mathbb{N}_0\}.$$

A subset of $\mathbb{N}_0^n$ is *semilinear* if it is a union of finitely many linear subsets.

For a fixed finite alphabet $A = \{a_1, \ldots, a_n\}$, define $\psi \colon A^* \to \mathbb{N}_0^n$ by

$$\psi(w) = (|w|_{a_1}, \ldots, |w|_{a_n}),$$

where $|w|_{a_i}$ denotes the total number of occurrences of a_i in w. Then ψ is a monoid homomorphism from the multiplicative monoid A^* to the additive monoid $\mathbb{N}_0^n$. We say that the two languages L and L' are *letter-equivalent* if $\psi(L) = \psi(L')$, and we say that L has a *semilinear image* if $\psi(L)$ is a semilinear subset of $\mathbb{N}_0^n$.

2.6.23   Theorem   *Each $L \in \mathcal{CF}(A^*)$ is letter-equivalent to a regular language, and has a semilinear image.*

Proof   Let $\mathcal{G}$ be a context-free grammar in Chomsky normal form generating L, and let p be the constant from Theorem 2.6.21. For each set U of variables of $\mathcal{G}$ containing the start symbol S, let L_U be the subset of L consisting of those $z \in L$ that have at least one derivation in $\mathcal{G}$ in which U is the set of variables that occur in that derivation. Since L is the finite union of the sets L_U, it suffices to fix U and to prove that L_U is letter-equivalent to a regular language, and has a semilinear image.

Let $|U| = k$ and define

$$F := \{z \in L_U : |z| < p^k\},$$
$$G := \{vx \in A^* : 1 \leq |vx| \leq p^k \text{ and } b \Rightarrow^* vbx \text{ for some } b \in U\}.$$

We claim that $\psi(L_U) = \psi(FG^*)$.

To show that $\psi(L_U) \subseteq \psi(FG^*)$, let $z \in L_U$. If $|z| < p^k$ then $z \in F \subseteq FG^*$, so assume that $|z| \geq p^k$. Then there is a derivation $S \Rightarrow^* z$ that uses exactly the variables in U and which, by Theorem 2.6.21, is equivalent to a derivation

$$S \Rightarrow_{d_0}^* uby \Rightarrow_{d_1}^* uv_1 b x_1 y \Rightarrow_{d_2}^* \cdots$$
$$\Rightarrow_{d_k}^* uv_1 \cdots v_k b x_k \cdots x_1 y \Rightarrow_{d_{k+1}}^* uv_1 \cdots v_k w x_k \cdots x_1 y = z,$$

with the properties specified in Theorem 2.6.21. Note that we have labelled the subderivations as $d_0, d_1, \ldots, d_k, d_{k+1}$.

For each of the $k-1$ variables $c \in U \setminus \{b\}$, choose some d_i such that c occurs in the subderivation labelled d_i. Then there exists a j with $1 \leq j \leq k$ such that d_j is not chosen for any c. Hence, if we omit the subderivation labelled d_j, then we are left with a derivation of a word z' that still involves all variables in U, so $z' \in L_U$ and, furthermore, $\psi(z) = \psi(z' v_j x_j)$. Since $|z'| < |z|$, we may assume inductively that $\psi(z') \in \psi(FG^*)$, and then $\psi(z) \in \psi(FG^*)$ because $v_j x_j \in G$.

Conversely, to show that $\psi(FG^*) \subseteq \psi(L_U)$, let $z \in FG^*$. If $z \in F$ then $z \in L_U$. Otherwise, we can write $z = z_0 t$ with $z_0 \in FG^*$ and $t \in G$. So $t = vx$, and $b \Rightarrow^* vbx$ for some $b \in U$. Since $|z_0| < |z|$, we may assume inductively that $\psi(z_0) = \psi(z')$ for some $z' \in L_U$. So there is a derivation $S \Rightarrow^* z'$ in which every variable in U occurs. In particular, b occurs so, for some $u, w, y \in A^*$, we have

$$S \Rightarrow^* uby \Rightarrow^* uwy = z' \quad \text{and hence}$$
$$S \Rightarrow^* uby \Rightarrow^* uvbxy \Rightarrow^* uvwxy.$$

Putting $z'' = uvwxy$, we have $z'' \in L_U$ and

$$\psi(z) = \psi(z_0 t) = \psi(uwy) + \psi(vx) = \psi(z'') \in \psi(L_U).$$

Now $\psi(L_U) = \psi(FG^*)$ and FG^* is regular. Let $G = \{t_1, \ldots, t_m\}$ and $F = \{z_1, \ldots, z_l\}$. Then $\psi(FG^*) = \psi(Ft_1^* \cdots t_m^*)$, which is the union over j of the linear sets $\{\psi(z_j) + n_1 \psi(t_1) + \cdots + n_m \psi(t_m) : n_i \in \mathbb{N}_0\}$, and hence is semilinear. $\square$

2.6.24 Strengthening Parikh's theorem There are some stronger versions of Parikh's theorem that are useful in applications. We state them here without proof. A linear set $\{c + n_1 \alpha_1 + \cdots + n_m \alpha_m : n_i \in \mathbb{N}_0\}$ is called *stratified* if each α_i has at most two nonzero components and, if α_i and α_j have nonzero components in positions i_1, i_2 and j_1, j_2, respectively, then it is not the case that $i_1 < j_1 < i_2 < j_2$.

For example $\{(i, j, j+k, i+k) : i, j, k \in \mathbb{N}_0\}$ is stratified, but $\{(i, i, i) : i \in \mathbb{N}_0\}$ and $\{(i, j, i, j) : i, j \in \mathbb{N}_0\}$ are not. A semilinear set is stratified if it is a finite union of stratified linear sets.

Let $A = \{a_1, \ldots, a_n\}$, and let L be a subset of the regular language $a_1^* a_2^* \cdots a_n^*$. So any $w \in L$ has the form $a_1^{k_1} \cdots a_n^{k_n}$, and $\psi(w) = (k_1, \ldots, k_n)$. It is proved in [101, Theorem 2.3] that L is context-free if and only if $\psi(L)$ is a stratified semilinear set.

2.6.25 Exercise Show, without using the above result, that

$$\{a^i b^j c^{j+k} d^{i+k} : i, j, k \in \mathbb{N}_0\} \in C\mathcal{F},$$

but

$$\{a^i b^i c^i : i \in \mathbb{N}_0\}, \{a^i b^j c^i d^j : i, j \in \mathbb{N}_0\} \notin C\mathcal{F}.$$

(*Hint*: Use the Pumping Lemma for the negative results.)

2.6.26 Closure properties of context-free languages The closure properties of context-free languages are not as straightforward as for regular languages. By the exercise above, $L = \{a^i b^i c^i : i \in \mathbb{N}_0\} \notin C\mathcal{F}$, but $L = L_1 \cap L_2$, with $L_1 = \{a^i b^i c^j : i, j \in \mathbb{N}_0\} \in C\mathcal{F}$ and $L_2 = \{a^i b^j c^j : i, j \in \mathbb{N}_0\} \in C\mathcal{F}$, so $C\mathcal{F}$ is not closed under intersection.

Also $L^c = \{a, b, c\}^* \setminus L = L_1^c \cup L_2^c$ is context-free, by the next result, so $C\mathcal{F}$ is not closed under complementation. On the positive side, we have the following facts.

2.6.27 Proposition *$C\mathcal{F}$ is closed under union, intersection with regular languages, concatenation and Kleene closure.*

Proof   This is similar to the proof of Proposition 2.5.10 for regular languages. Suppose that M_1, M_2 are pda accepting context-free languages L_1, L_2 by final state.

To recognise the union $L_1 \cup L_2$, we define a new machine M with stack alphabet equal to the union of those of M_1 and M_2, which contains copies of the states and transitions of M_1 and M_2. It has its own separate start state σ_0 and ε-moves from σ_0 to the start states of M_1 and M_2 that put the initial stack strings of M_1 and M_2 on the stack. So any computation starts by jumping non-deterministically to the initial configuration of either M_1 or M_2, and simulating M_1 or M_2 thereafter.

To recognise the intersection $L_1 \cap L_2$ of a context-free and a regular language, assume that M_2 is a complete deterministic fsa. Now we construct M whose state set is the product of those of M_1 and M_2, with a state accepting if and only if both of its components are. A transition $(\sigma_1, a, b) \rightarrow (\sigma_1', \beta)$ of M_1 is replaced by the transition $((\sigma_1, \sigma_2), a, b) \rightarrow ((\sigma_1', \sigma_2^a), \beta)$, where we define $\sigma_2^\varepsilon = \sigma_2$.

To recognise the concatenation $L_1 L_2$, we construct a machine M with stack alphabet equal to the union of those of M_1 and M_2, and which contains copies of the states and transitions of M_1 and M_2. The start state is that of M_1, and the accepting states are those of M_2. For each accepting state of M_1, there are additional ε-moves that clear the stack and move to the start state and the initial stack configuration of M_2.

The construction of a machine to recognise the Kleene closure of L_1 is similar, but here we introduce ε-moves from the accepting states of M_1 to the start state and stack configuration of M_1. We must also ensure that ε is accepted. □

2.6.28 Improvements of deterministic pda   It seems intuitively clear that $\mathcal{DCF}$ is closed under complementation, because we ought simply to be able to interchange the accepting and non-accepting states of a deterministic pda M accepting the language in question to produce a pda M' accepting its complement.

Unfortunately this construction does not work, for a number of reasons, and the proof of the result, which comes from [159, Theorem 10.1], is surprisingly tricky. Firstly, M may execute a number of ε-moves after reading the final letter of the input word, and M accepts if any of these moves lead to an accepting state, so M' needs to handle this situation correctly. Secondly, it is possible that M halts before it has finished reading its input word, in which case we want M' to accept the word. Thirdly, it is possible that M goes into an infinite sequence of ε-moves, which could happen before or after it has finished reading its input. To circumvent these problems, we show first that we can adjust M to avoid the second and third of these eventualities. We call a deterministic pda *complete* if, for all $\sigma \in \Sigma$, $a \in A$ and $b \in B$, one of $\tau(\sigma, a, b)$ and $\tau(\sigma, \varepsilon, b)$ is nonempty.

2.6.29 Lemma   *Let M be a deterministic* pda. *Then there exists a complete deterministic* pda M_0 *with $L(M) = L(M_0)$ such that M_0 never halts before it has finished reading its input word, and never executes an infinite sequence of ε-moves.*

Proof   The most difficult problem is eliminating infinite sequences of ε-moves, and we attend to that first. Call a (finite) sequence of ε-moves of M *repeating* if the state of M and the top stack symbol are the same at the beginning and at the end of the sequence, and if, during the course of the sequence, the height of the stack is never less than what it was at the beginning of the sequence. So a repeating sequence, once entered, will repeat indefinitely and, conversely, it is not hard to see that any infinite sequence of ε-moves of M must contain a repeating sequence.

We can examine each ε-move of M in turn and check whether it is the first move of a repeating sequence. If so, then we replace it by an ε-move to a new state σ, from which there are no transitions, defined to be an accepting state of M_0 precisely when the eliminated repeating sequence contains an accepting state. Then the behaviour of M_0 is the same as that of M, except that it moves to this new state and halts whenever M is about to enter a repeating sequence.

To prevent $M = (\Sigma, A, \tau, \sigma_0, F)$ from halting before it has finished reading its input word, we introduce a new non-accepting state ω and some new transitions. For any $a \in A$ and $(\sigma, b) \in \Sigma \times B$ such that $\tau(\sigma, a, b)$ and $\tau(\sigma, \varepsilon, b)$ are both empty, we introduce a transition $(\sigma, b)^a = (\omega, b)$. We also define transitions $(\omega, b)^a = (\omega, b)$ for all $b \in B$ and $a \in A$. So whenever M halts before reading all of its input word, the modified pda M_0 enters the state ω and reads the rest of the word before halting without accepting. Note that M_0 is complete. □

2.6.30 Proposition $\mathcal{DCF}$ *is closed under complementation.*

Proof Let $L \in \mathcal{DCF}$. Then, by the lemma, there is a complete deterministic pda $M = (\Sigma, A, \tau, \sigma_0, F)$ with $L(M) = L$ that never executes an infinite sequence of ε-moves, and always reads the whole of its input word before halting. But M may still execute a finite number of ε-moves after reading its input word, and we want the (deterministic) pda M' that accepts $A^* \setminus L(M)$ to accept the word if and only if M does not enter an accepting state after it has read the last letter of the input word.

To achieve this, we let $M' = (\Sigma', A, \tau', \sigma_0', F')$, where $\Sigma' = \Sigma \times \{0, 1, \omega\}$, $F' = \Sigma \times \{\omega\}$, and $\sigma_0' = (\sigma_0, 1)$ or $(\sigma_0, 0)$ when $\sigma_0 \in F$ or $\sigma_0 \notin F$, respectively. The general idea is for the second component of a state of M' to be 1 when M has entered an accepting state since reading its last input letter and 0 when it has not. Whenever the second component is 0 and there is no ε-transition defined in M, we introduce an ε-transition in M' to an accepting state, with second component ω.

More precisely, we define τ' as follows, and we leave the reader to verify that $L(M') = A^* \setminus L$. Let $\sigma \in \Sigma$, $a \in A$ and $b \in B$.

If $\tau(\sigma, a, b)$ is nonempty, and equal to $\{(\sigma', v)\}$, then we define

$$\tau'((\sigma, 1), a, b) := \tau'((\sigma, \omega), a, b) := \begin{cases} \{((\sigma', 1), v)\}, & \text{if } \sigma' \in F \\ \{((\sigma', 0), v)\}, & \text{if } \sigma' \notin F \end{cases}$$

and $\tau'((\sigma, 0), \varepsilon, b) := \{((\sigma', \omega), b)\}$.

Otherwise $\tau(\sigma, \varepsilon, b) := \{(\sigma', v)\}$ for some σ', v, and we define

$$\tau'((\sigma, 1), \varepsilon, b) := \{((\sigma', 1), v)\},$$

and

$$\tau'((\sigma, 0), \varepsilon, b) := \begin{cases} \{((\sigma', 1), v)\}, & \text{if } \sigma' \in F \\ \{((\sigma', 0), v)\}, & \text{if } \sigma' \notin F \end{cases}$$

All other images under τ' are empty. □

We observe also that $\mathcal{DCF}$ is not closed under union. In the example in 2.6.26 above, L_1^c and L_2^c are both in $\mathcal{DCF}$, but their union is not (because its complement $L_1 \cap L_2$ is not context-free).

2.6.31 Exercise Show that $\mathcal{DCF}$ is not closed under concatenation or Kleene closure, but it is closed under intersection with regular languages.

2.6.32 Proposition *CF is closed under homomorphism.*

Proof Let $\phi \colon A^* \to B^*$ be a homomorphism. Let $\mathcal{G}$ be a context-free grammar generating $L \subseteq A^*$. By replacing every terminal a in every production of $\mathcal{G}$ by $\phi(a)$, we obtain a context-free grammar generating $\phi(L)$. $\square$

More generally, CF is closed under GSMs [159, Theorem 11.1]; GSMs are defined in 2.2.25.

2.6.33 Exercise Show that $\mathcal{DCF}$ is not closed under homomorphism.

2.6.34 Proposition *CF and $\mathcal{DCF}$ are closed under inverse GSMs (and hence under inverse homomorphisms).*

Proof Let $\phi \colon A^* \to B^*$ be a GSM defined by the machine M' with state set Σ', and let M be a pda with state set Σ with $L(M) = L \subseteq B^*$. Let k be the maximum length of an output string in a transition of M'. Then we can construct a k-pda M'' (see 2.6.8) with state set $\Sigma \times \Sigma'$ and stack alphabet the same as M that accepts $\phi^{-1}(L)$, and M'' is deterministic if M is. The machine M'' reads an input word, using the second state component to read it through M'. On each move, the output word of M' is passed to M. Since this word has length up to k and must be processed in a single move of M'', we must allow M'' to inspect its top k stack symbols. $\square$

We shall see later in the book, in Proposition 14.2.7, that if $L \in CF$ then so is the language of all cyclic permutations of the words in L. This result will be applied in some proofs involving the co-word problem of groups.

2.6.35 Closure properties of one-counter languages Recall from 2.2.9 that the classes OC and $\mathcal{DOC}$ of one-counter and deterministic one-counter languages consist of languages accepted by a (deterministic) pda with a single stack symbol, together with a bottom of stack marker.

The same example as for CF (2.6.26) shows that OC is not closed under intersection. The class OC is closed under union, intersection with regular languages, concatenation, Kleene closure and under inverse GSMs. Once again, OC is not closed under complementation, but $\mathcal{DOC}$ is. The class $\mathcal{DOC}$ is closed under intersection with regular languages and under inverse GSMs. The proofs of these assertions are essentially the same as those for $\mathcal{CF}$.

2.7 Turing machines, recursively enumerable languages and grammars

2.7.1 The equivalence of the models of Turing machines We start with a brief explanation of why the various models of Turing machines accept the same classes of languages. See [159, Theorems 7.1, 7.2] for details. It is easy to see that a bi-infinite tape can be replaced by one that is infinite in one direction only. If the tape-symbol in position n of the bi-infinite tape is a_n for $n \in \mathbb{Z}$, then we store the contents of this tape on the one-way infinite tape by putting a_n in position $2n$ for $n \geq 0$ and a_{-n} in position $2n - 1$ for $n > 0$. By increasing the states, it is routine to adjust the transitions to complete the simulation.

Replacing k bi-infinite tapes by a single bi-infinite tape is a little more complicated. Let T be the k-tape machine. Then we construct a single-tape machine T' with $L(T') = L(T)$ that simulates T as follows. Suppose that, at some stage in the computation, cell n in tape number i (with $1 \leq i \leq k$) of T contains a_{in}. We can store this information on the single tape of T' by putting a_{in} in position $kn + i$. The problem is that T has k separate read-write heads, while T' has only one. So the current positions of the tape-heads of T must be marked on the tape of T'. We can do that by increasing the number of tape-symbols, and using the new symbols to mark the k tape-head positions.

To simulate a move of T, the machine T' must start by moving up and down its tape, searching for all k symbols currently being scanned by T. (It can improve the efficiency of this process slightly by remembering in which direction they all lie from its current position.) Having located them all, T' can decide which transition of T it is simulating, and carry it out, which involves a further journey up and down its tape so that the current scanned symbols can be changed as necessary and any moves of the tape-heads of T simulated.

Note that the space requirements of a computation increase only by a linear factor as a result of this simulation so, for example, the linear-boundedness of a computation is not affected. But the time complexity is potentially increased because, after n moves of T, the k tape-heads could be at distance up to $2n$ apart, so simulating the nth move of T in general requires up to Cn moves of T' for some constant C.

We state the equivalence of deterministic and non-deterministic Turing machines as a theorem. The proof is based on that of [159, Theorem 7.3].

2.7.2   Theorem   *Let T be a non-deterministic Turing machine. Then there is a deterministic Turing machine T' with $L(T) = L(T')$.*

Proof   We use our default basic model, where T has a single bi-infinite tape, and assume that its input word is initially stored on that tape, delineated as usual by blank symbols. Recall from 2.2.13 that an *instantaneous description* (ID) of T is a triple (α, w_1, w_2), where α is its state, and $w_1 w_2$ is the contents of the tape, where the tape-head is located at the first symbol of w_2. (But note that either or both of w_1 and w_2 could be empty.)

The simulating machine T' has two bi-infinite tapes. Its first tape is used to store a finite sequence $D_1, D_2, \ldots, D_n$ of IDs of T. An ID is stored as the word $\alpha w_1 \# w_2$, where $\#$ is a new tape symbol used as a separator, and the same symbol $\#$ can be used to separate D_i from D_{i+1} for all i. But, as we shall see shortly, at any stage in the computation, one of the D_i is currently being processed, so it is convenient to mark its beginning by another special symbol, so that it can be easily located.

If the initial state of T is an accepting state, then it accepts all words, so suppose not. Let us assume that the input word is initially on the second tape of T'. Then T' starts by calculating the ID of the initial configuration of T, and writing it to its first tape as D_1. Thereafter, T' carries out deterministically a sequence $\mu_1, \mu_2, \ldots$ of sequences of moves, where each μ_i simulates all possible moves of T from the configuration of T with ID D_i.

Suppose that there are k possible moves that T could make from the configuration D_i. Then for each of these moves T' does the following. It first replaces the contents of its second tape with the configuration described by D_i. It then makes the move of T in question on its second tape. If this leads to an accepting state of T, then T' moves into an accepting state and halts. Otherwise, T' calculates the ID of the new configuration of T and writes this ID to its first tape as D_{n+1} (where n is the current number of stored IDs).

We leave it to the reader to verify that every computation of T that leads to an accepting state is eventually simulated in its entirety by T', and hence T' also accepts the input word. As with the case of fsa, the computation with T' could take exponentially longer than that with T. The space requirement could also increase exponentially.   □

2.7.3   Turing machines and grammars   We have seen earlier that the family of languages accepted by finite state automata is equal to the family of languages that can be defined by regular grammars (2.5.12), and that the languages that are accepted by (non-deterministic) pushdown automata are precisely the languages that can be defined by context-free grammars (2.6.15).

In this section we prove corresponding results for the languages that are accepted by Turing machines (the recursively enumerable languages), and those that are accepted by linearly-bounded Turing machines (the context-sensitive languages). In our proofs, we concentrate on explaining the ideas involved, and omit the verification of technical details. See [159, Theorems 9-3–9.6] for further details.

2.7.4 Theorem *A language L is recursively enumerable if and only if it can be generated by a grammar.*

Proof The idea is to construct a Turing machine that mimics the generation of words by a grammar and vice versa. Since the Turing machine starts with the input word and performs a computation that may accept it, whereas the grammar starts with a fixed starting symbol and then generates a word in the language, the Turing machine performs its computation in the reverse direction from the application of productions in the corresponding grammar.

Suppose first that L is generated by a grammar G. By introducing a new variable if necessary, we may assume that the start symbol S of G does not occur in the right-hand side of any production. The corresponding (non-deterministic) Turing machine M with $L(M) = L$ starts with its input word on its input-tape delineated by blanks at both ends. It then repeatedly scans the word on the tape, never moving beyond the blanks that mark the two ends of the word. If it finds the right-hand side of a production of G as a substring of the word, then it has the option of replacing it by its left-hand side (which may involve changing the length of the word and moving parts of the word). If it sees the start symbol S with blanks on either side, then it halts and accepts the input word; otherwise it does not halt. Since computations of the Turing machine correspond exactly to the generation of words in L in reverse, we see that $L = L(M)$.

The other direction is more complicated. Let L be the language of a Turing machine M. It is convenient to assume some restrictions on the way M operates. As usual, Σ denotes the set of states of M and b_0 is the blank tape-symbol. As explained in 2.7.1, we may assume that the work-tape of M is infinite to the right only, and that the input word is initially at the left-hand end of the tape. We assume also that all transitions of M are of one of the following two types:

(i) writing transitions $(\sigma, a) \rightarrow (\tau, b)$ that go from state σ to state τ and replace the scanned tape-symbol a by b without subsequently moving the tape-head; or

(ii) tape-head moving transitions $(\sigma, a) \rightarrow (\tau, \mathsf{R})$ or $(\sigma, a) \rightarrow (\tau, \mathsf{L})$ that go from state σ to state τ and move the tape-head left or right without changing the scanned tape-symbol a.

We can reduce to this situation by introducing new states and breaking up into two separate transitions any transition that both writes to the tape and moves the tape-head.

The other assumptions we need are more technical and are designed to facilitate the mimicking of M by a grammar. We assume that the first move of M is to replace the first letter of the input word w (or the blank symbol under the tape-head if w is empty) by a new symbol Λ and that M never subsequently overwrites Λ or returns to its start state σ_0. Since M can remember the letter of w that it overwrote, this restriction can be achieved.

We assume also that M never writes the blank symbol b_0 onto the tape and that, whenever it reads b_0, it replaces it by a different symbol. A new tape-symbol can be introduced for that purpose. This assumption ensures that whenever M reads b_0, the tape-head is at the right-hand end of the word being processed.

Finally, we adjust M so that, when it is about to go into an accept state and halt, it first moves the tape-head left to Λ, goes into a new state σ_e, moves right replacing all non-blank tape-symbols scanned by a new symbol # and finally, when it reads b_0, it enters a unique accept state q_e, replaces that b_0 by #, and then halts.

We can now describe the grammar $\mathcal{G}$ that mimics M in reverse. The set T of terminals is equal to the input alphabet A of M. We define

$$V := \{S, S_0\} \cup (B \setminus A) \cup (\Sigma \times B).$$

We denote a variable in $\Sigma \times B$ by $[\sigma, a]$. This is intended to correspond to M being in state σ with the symbol a in the position scanned by the tape-head. The productions of $\mathcal{G}$ are as follows.

(1) $S \rightarrow \Lambda S_0$; $S_0 \rightarrow \#S_0 \mid [q_e, \#]$;
(2) $[\tau, b] \rightarrow [\sigma, a]$ for all writing transitions $(\sigma, a) \rightarrow (\tau, b)$ of M with $\sigma, \tau \in \Sigma$, $a, b \in B$;
(3) $a[\tau, b_0] \rightarrow [\sigma, a]$ for all right-moves $(\sigma, a) \rightarrow (\tau, \mathsf{R})$ of M with $\sigma, \tau \in \Sigma$, $a \in B$;
(4) $a[\tau, b] \rightarrow [\sigma, a]b$ for all right-moves $(\sigma, a) \rightarrow (\tau, \mathsf{R})$ of M with $\sigma, \tau \in \Sigma$, $a \in B$, and all $b \in B \setminus \{b_0\}$;
(5) $[\tau, b]a \rightarrow b[\sigma, a]$ for all left-moves $(\sigma, a) \rightarrow (\tau, \mathsf{L})$ of M with $\sigma, \tau \in \Sigma$, $a \in B$, and all $b \in B \setminus \{b_0\}$;
(6) $[\sigma_0, a] \rightarrow a$ for all $a \in A$; $[\sigma_0, b_0] \rightarrow \varepsilon$.

For a word $w \in L(M)$, the corresponding derivation of w in $\mathcal{G}$ starts by using the productions (1) to insert the variable # as many times as # is written in the final stage of the computation in M. It then uses the productions (2)–(5) to carry out the main part of the computation of M in reverse order. Recall that the first move of M is the writing transition $(\sigma_0, a) \to (\tau, \Lambda)$ for some $a \in A \cup \{b_0\}$ and $\tau \in \Sigma$, and the corresponding production in G is $[\tau, \Lambda] \to [\sigma_0, a]$, which is followed by an application of production (6) to complete the generation of w by G. Conversely, any generation of a word $w \in A^*$ by G corresponds to a computation in M that accepts w. □

2.8 Linearly bounded automata, context-sensitive languages and grammars

2.8.1 Context-sensitive and growing context-sensitive grammars We recall from 2.2.18 that a language is called (deterministic) context-sensitive if it is accepted by a (deterministic) linearly bounded Turing machine, and that the families of context-sensitive and deterministic context-sensitive languages are denoted by CS and DCS, respectively. (It is not known whether these families are distinct.)

A grammar is called *context-sensitive* if $|u| \le |v|$ for all productions $u \to v$. It is *growing context-sensitive* if, in addition, for all productions $u \to v$, S does not occur in v and either $u = S$ or $|u| < |v|$.

It is also customary, in both context-sensitive and growing context-sensitive grammars, to allow a single production $S \to \varepsilon$ that violates the $|u| \le |v|$ condition, provided that S does not occur in the right-hand side of any production. This is to enable the empty word ε to be contained in the language generated by a context-sensitive grammar. Context-sensitive grammars are also known as *Type 1 grammars*. The following proof comes from [159, Theorem 9.5].

2.8.2 Theorem *A language L is context-sensitive if and only if it can be generated by a context-sensitive grammar.*

Proof It is not hard to see that, for a language generated by a context-sensitive grammar, the corresponding Turing machine described in the proof of Theorem 2.8.2 that accepts the language uses space $|w| + 1$ for an input word w. (The +1 comes from the need to look at the blank space to the right of the input in order to check where the input ends.)

Conversely, suppose that L is the language of a linearly bounded Turing machine M. Then there is a constant k such that the space used by M in accepting a word w is at most $k|w| + 1$. (The +1 is again there for the blank space to the right of w.) By increasing the number of tape-symbols, we can replace M by a machine accepting the same language in which disjoint sequences of k symbols on the original tape are represented by single symbols on the new tape;

hence we may assume that the space M needs to accept w is at most $|w| + 1$. Furthermore, M only needs to look at the blank space to the right of w once because, having done that, it can use a special tape-symbol to mark the final letter of w and remember that letter. (This is similar to the use of Λ in the proof of Theorem 2.8.2 to mark the first letter of the word.)

To ensure that the corresponding grammar $\mathcal{G}$ constructed in the proof of Theorem 2.8.2 is context-sensitive, we need to avoid using the length reducing productions (3) and also the production $[\sigma_0, b_0] \to \varepsilon$ in (6). The second of these is only necessary when $\varepsilon \in L$ and we deal with that by adjoining the allowed production $S \to \varepsilon$ to $\mathcal{G}$.

The productions (3) arise only from transitions of M that look at the blank space beyond the end of w. To avoid having to use these, we adjust M to make it guess non-deterministically which is the last letter of w and to use the special tape-symbol to mark its position, as described above. We then disallow any moves of M that attempt to go beyond that marked position. At the end of the computation, M checks that the first blank space encountered when replacing symbols by # really does occur immediately to the right of the marked position and, if not, then it halts in a non-accepting state.   □

2.8.3 Closure properties of context-sensitive languages Using the fact that the context-sensitive languages are those accepted by a linearly bounded Turing machine, it is straightforward to show that $\mathcal{CS}$ and $\mathcal{DCS}$ are closed under union and intersection, and that $\mathcal{CS}$ is closed under concatenation and Kleene closure. Furthermore, $\mathcal{DCS}$ is closed under complementation and, perhaps surprisingly, $\mathcal{CS}$ is also closed under complementation. This is proved by Immerman [163].

If $\phi: A^* \to B^*$ is a monoid homomorphism (or, more generally, a GSM), and $L \subseteq B^*$ is accepted by the (deterministic) linearly bounded Turing machine M, then we can construct a machine M' with $L(M') = \phi^{-1}(L)$ by making M' start by replacing its input word w by $\phi(w)$ and then simulating M. So $\mathcal{CS}$ and $\mathcal{DCS}$ are closed under inverse homomorphisms and, more generally, under inverse GSMs.

The family $\mathcal{CS}$ is not closed under homomorphism, however. The problem arises when $\phi(a) = \varepsilon$ for some $a \in A$, since such a ϕ would not necessarily map a context-sensitive grammar to another one. For a proof, solve [159, Exercise 9.14].

2.9 Turing machines and decidability

2.9.1 Separating recursive from recursively enumerable languages The operation of our default model for a Turing machine M that contains a single bi-infinite tape can be described by a finite set of rules of the form

$$(\sigma, b) \mapsto (\sigma', b', \eta)$$

where $\sigma, \sigma' \in \Sigma$, $b, b' \in B$, and η describes the movement of the head on the bi-infinite tape, and takes one of three values R, L, C.

For the purpose of demonstrating the existence of recursively enumerable languages that are not recursive, it is sufficient to restrict our attention to the languages of Turing machines with $A = \{0, 1\}$ and $B = A \cup \{b_0\}$. We can specify a natural order on the rules of such machines, and write the rules in that order, to get a unique and complete description for M as a finite sequence of symbols from a fixed finite alphabet. That sequence can be encoded as a binary string $c(M)$, in such a way that M is unambiguously recoverable; we call this the *binary code* of the machine, and call the set of all such binary codes 'Codes'. We can easily ensure that Codes is a regular set.

Since there are only countably many binary strings, we see that there are only countably many Turing machines, and hence countably many recursively enumerable languages, which has the following consequence.

2.9.2 Proposition *There exist languages that are not recursively enumerable.*

In fact, following an argument of [159, Section 8.3], we can construct a language that is not recursively enumerable using the binary codes. We define

$$L_C = \{c(M) : M \text{ a Turing machine}, c(M) \notin L(M)\}.$$

Now suppose that M_C is a Turing machine with $L(M_C) = L_C$, and that $w_C = c(M_C)$. Then either of the two statements $w_C \in L_C$ or $w_C \notin L_C$ leads to a contradiction, and hence there can be no such Turing machine.

But the complement of L_C, which we call $\bar{L}_C$, is recursively enumerable. We can recognise $\bar{L}_C$ on a Turing machine $\bar{M}_C$ that, as a first step, reads and stores the input, and either recognises that the input is not a code and so halts accepting it, or identifies the input as the code of a Turing machine M. In the second situation, $\bar{M}_C$ then reads the input a second time operating as M, and accepts the input if M does.

With $\bar{L}_C$ as an example, we have now proved the following result.

2.9.3 Proposition *There exist languages that are recursively enumerable but not recursive.*

2.9.4 Closure properties It is straightforward to show that the classes of recursive and recursively enumerable languages are closed under inverse homomorphisms and GSMs, union and intersection. It follows directly from their definition that recursive sets are closed under complementation, and the example $\bar{L}_C$ above shows that the recursively enumerable languages are not.

By applying the homomorphism to the corresponding grammar, we see that the recursively enumerable languages are closed under homomorphism, but the recursive languages are not [159, Exercise 9.14].

2.9.5 The semigroup of a Turing machine We can use the Turing machine $\bar{M}_C$ above to find an example of a finitely presented semigroup with insoluble word problem, and hence prove the Markov–Post theorem. This example forms an important component of the proof of the famous Novikov–Boone theorem on the insolubility of the word problem for finitely presented groups, which we present in Chapter 10.

We first describe how to build a finitely presented semigroup $S(M)$ from a given Turing machine M, where the relations of $S(M)$ are derived from the configuration graph of M. Our account comes from Rotman's book [223, Chapter 12]. We use our default model of a Turing machine, with a single bi-infinite tape, and we recall that the input alphabet A is a subset of the tape-alphabet B, and that B contains the blank symbol b_0. We may assume by Theorem 2.7.2 that M is deterministic.

The contents of the tape at any time during the computation can be described by a word w, whose first and last letters are b_0, and which is of minimal length at least 2 subject to the conditions that (i) no symbol other than b_0 can be found on the tape either to the left or to the right of w, and (ii) at no point during the computation so far has the read-write head visited any cell either to the left or to the right of w. It follows, from the above and our definition of a Turing machine that, at the beginning of the computation, w is the word formed by attaching a symbol b_0 at each end of the input word, the read-write head points at the first letter of the input word (so the second letter of w) and the machine is in its initial state σ_0.

The configuration of the machine M at any point during the computation can be described by a string $u\sigma v$, where $uv = w$ with $u \in B^*$, $v \in B^+$, and the read-write head positioned at the first letter of v, and where $\sigma \in \Sigma$ is the current state of the machine.

The semigroup $S(M)$ is generated by the finite set

$$X = X(M) = B \cup \Sigma \cup \{h, \hat{\sigma}\},$$

where the symbol h plays the role of a marker at each end of the string uv on the bi-infinite tape of M. Following [223], we call a string over $S(M)$ of the form $h\alpha h$, where $\alpha = u\sigma v$ with $u \in B^*$, $\sigma \in \Sigma$ and $v \in B^+$, an *h-special string*. So, in particular, if $u\sigma v$ represents a configuration of M during a computation, then $hu\sigma vh$ is h-special. Note that the set of all h-special strings is a regular language. As we shall see, the relations of $S(M)$ ensure that two such strings are equal in the semigroup if they correspond to two configurations within a single computation of M.

The operation of M is defined by a finite set of rules each of the form

$$(\sigma, b) \mapsto (\sigma', b', \eta),$$

where $\sigma, \sigma' \in \Sigma$, $b, b' \in B$, and η takes one of the three values R, L, C (where C indicates no movement of the head). As usual, let σ_0 be the start state of M. We assume also that M has a unique accepting state σ_H, and that σ_H is a halting state, so there are no transitions starting from σ_H.

We now define the set $R(M)$ of defining relations for $S(M)$ as follows.

For each rule $(\sigma, b) \mapsto (\sigma', b', \mathsf{C})$: $\quad \sigma b = \sigma' b'$,

for each $b'' \in B$, and
each rule $(\sigma, b) \mapsto (\sigma', b', \mathsf{R})$: $\quad \begin{cases} \sigma b b'' &= b'\sigma' b'', \\ \sigma b h &= b'\sigma' b_0 h, \end{cases}$

for each $b'' \in B$, and
each rule $(\sigma, b) \mapsto (\sigma', b', \mathsf{L})$: $\quad \begin{cases} b''\sigma b &= \sigma' b'' b', \\ h\sigma b &= h\sigma' b_0 b', \end{cases}$

for each $b \in B$: $\quad \begin{cases} \sigma_H b &= \sigma_H, \\ b\sigma_H h &= \sigma_H h, \\ h\sigma_H h &= \hat{\sigma}. \end{cases}$

The following lemma is [223, Lemma 12.4].

2.9.6 Lemma *For $w \in A^*$, $w \in L(M) \iff h\sigma_0 w h =_{S(M)} \hat{\sigma}$.*

Proof Suppose first that $w \in L(M)$. The relations of $S(M)$ ensure that, for each step in a computation in M with input w leading to the accept state σ_H, there is a corresponding relation in $R(M)$, and these relations together transform the word $h\sigma_0 w h$ to $hu\sigma_H v h$ for some $u, v \in B^*$. Then the final three relations of $R(M)$ can be used to transform this word to $\hat{\sigma}$, so $h\sigma_0 w h =_{S(M)} \hat{\sigma}$.

Conversely, suppose that $h\sigma_0 w h =_{S(M)} \hat{\sigma}$. Then, since the only relation in $S(M)$ that involves $\hat{\sigma}$ is the final one, there exist words

$$w_1 = h\sigma_0 w h, w_2, \ldots, w_t = h\sigma_H h$$

such that, for each i with $1 < i \leq t$, w_i is derived from w_{i-1} by a single substitution using one of the relations in $S(M)$ other than $h\sigma_H h = \hat{\sigma}$. Since each of these substitutions replaces an h-special word by another one, each w_i is h-special. We may clearly assume that the w_i are all distinct.

A substitution $w_i \to w_{i+1}$ that uses a relation in $R(M)$ other than the final three corresponds to a move in M but, since these relations can be applied in either direction, that move could conceivably be in the reverse direction. However, since there are no transitions of M from the state σ_H, the final move of this type must be in the forward direction. Since M is deterministic, there is only one possible transition from each configuration, and so the final move in the reverse direction (if any) could only be followed by the same move in the forward direction, contradicting our assumption that the w_i are distinct.

So all of the moves of M corresponding to $w_i \to w_{i+1}$ go in the forward direction. Hence the final three relations in $R(M)$ must be used only at the end of the sequence of substitutions, and the moves of M corresponding to the remaining substitutions combine to give a successful computation of w by M. □

We use this construction, as in [223, Theorem 12.5], to prove the next result.

2.9.7 Theorem (Markov–Post, 1947) *There is a finitely presented semigroup*

$$S(M) = \mathrm{Sgp}\langle \hat{\sigma}, h, B, \Sigma \mid R(M) \rangle$$

with insoluble word problem. In particular, there is no decision process to determine, for an arbitrary h-special word $h\alpha h$, whether or not $h\alpha h =_{S(M)} \hat{\sigma}$.

Proof We let M be a Turing machine whose language $L = L(M)$ is recursively enumerable but not recursive. With $X = B \cup \Sigma \cup \{h, \hat{\sigma}\}$, as above, we define

$$[\hat{\sigma}]_{S(M)} := \{v \in X^* : v =_{S(M)} \hat{\sigma}\}.$$

We also define

$$H(A^*) := \{h\sigma_0 wh : w \in A^*\},$$
$$H(L) := \{h\sigma_0 wh : w \in L\};$$

these are clearly in bijective correspondence with A^* and L. Then $H(A^*) \in \mathcal{R}eg(X^*)$ and $H(L)$ is non-recursive. By Lemma 2.9.6, $H(L) = [\hat{\sigma}]_{S(M)} \cap H(A^*)$, and a non-recursive language cannot be the intersection of a regular and a recursive language, by 2.9.4. So $[\hat{\sigma}]_{S(M)}$ is also non-recursive, and hence the semigroup $S(M)$ has insoluble word problem.

Since we can write $H(L)$ as the intersection $Y \cap H(A^*)$, where Y is the set of h-special words equal to $\hat{\sigma}$ in $S(M)$, we see that Y is also non-recursive; that is, the solubility of the given equation is undecidable. □

For a proof of the Novikov–Boone theorem, the following corollary of the Markov–Post theorem is given in [223, Corollary 12.6].

2.9.8 Corollary   *There is a finitely presented semigroup*

$$S(M) = \mathrm{Sgp}\langle C \cup \Sigma \cup \{\hat{\sigma}\} \mid \phi_i \sigma_i \psi_i = \theta_i \sigma_i' \kappa_i, i \in I \rangle$$

with $\phi_i, \psi_i, \theta_i, \kappa_i \in C^*$, $\sigma_i, \sigma_i' \in \Sigma \cup \{\hat{\sigma}\}$, for which there is no decision process to decide, given arbitrary strings $u, v \in C^*$ and $\sigma \in \Sigma$, whether $u\sigma v =_{S(M)} \hat{\sigma}$.

Proof   The presentation of $S(M)$ takes the same form as that in Theorem 2.9.7 with $C = B \cup \{h\}$. If the set of $(u, v) \in C^* \times C^*$ satisfying $u\sigma v =_{S(M)} \hat{\sigma}$ were recursive, then the set $\{u\sigma v : u, v \in C^*, u\sigma v =_{S(M)} \hat{\sigma}\}$ would be recursive, and hence the intersection of this set with the set of all h-special words, which is regular, would be recursive. But this intersection is equal to the set of h-special words $h\alpha h$ with $\alpha =_{S(M)} \hat{\sigma}$, which is not recursive by Theorem 2.9.7   □

2.10 Automata with more than one input word

For some decision problems in group theory, the input consists of more than one word. One that springs to mind immediately is the word equality problem: do two given words u, v over the group generators represent the same element of the group? Since this is equivalent to uv^{-1} representing the identity element, this problem is usually regarded as being equivalent to the word problem in the group. But the word problem in monoids and semigroups is by definition the same as the word equality problem. The conjugacy problem in groups is another natural example and, as we shall see when we come to study automatic groups in Chapter 5, fsa with two input words form an essential component of that topic.

For a finite alphabet A and $n \in \mathbb{N}$, a subset of $(A^*)^n$ is called an *n-variable language* over A, and an automaton M of any type that takes input from $(A^*)^n$ and accepts an n-variable language is known as an *n-variable automaton*.

There are a number of different ways that such an automaton could read the two (or more) input words, and the choice that is made can, at least in the case of fsa and pda, have a significant influence on its capabilities. Perhaps the most obvious choice is to read the two words one after the other, with a recognisable separating symbol between them. Another possibility, which is motivated by the equivalence of $u =_G v$ and $uv^{-1} =_G 1$ in a group G, is to read the first word u followed by the reverse of the second word v, and this option has indeed been studied in connection with word problems of semigroups.

A different approach, which arises in the theory of automatic groups, is to read the two words together in parallel; so they would be read from two separate input-tapes. The automaton could read the two input words either synchronously or asynchronously. In the first case it must read one symbol from each tape on each move, and in the second case it may read more than one symbol from one tape before reading from the other. In the asynchronous case, the usual convention is for the state of the automaton to determine from which tape to read the next symbol. Again, one could conceivably read one of the two words in reverse order: for example, it is shown by Holt, Rees and Röver [153] that, in some situations, this is the most effective approach to the conjugacy problem (see Section 14.4). Note that reading one word after the other can be regarded as a special case of asynchronous reading from two tapes.

For Turing machines and linearly bounded Turing machines, the machine can read any of the letters of either word more than once, and so the initial configuration of the input words on the tape or tapes is less important, and could at most affect the time complexity of the computation by a polynomial factor. In particular, the solubility of the word problem in semigroups or the conjugacy problem in groups is not affected by these issues.

2.10.1 2-variable fsa For the remainder of this section, we restrict attention to 2-variable fsa, and to the properties that we shall need when we study automatic groups in Chapter 5.

So, suppose that A is a finite alphabet, and we want to read words $v, w \in A^*$ simultaneously and synchronously. We can almost do this by using an fsa with alphabet $A \times A$, but this does not work if $|v| \neq |w|$. To deal with words of unequal length, we introduce an extra alphabet symbol $\$$, known as the *padding symbol*, which is assumed not to be in A, and use it to pad the shorter of the two words at the end, so that the resulting padded word has the same length as the longer word.

To be precise, let $A^p := A \cup \{\$\}$, and let $v, w \in A^*$, where $v = a_1 a_2 \cdots a_l$ and $w = b_1 b_2 \cdots b_m$, with each $a_i, b_i \in A$. Then we define $(v, w)^p \in (A^p \times A^p)^*$ to be the word $(\alpha_1, \beta_1)(\alpha_2, \beta_2) \cdots (\alpha_n, \beta_n)$, where $n := \max(l, m)$, and

(i) $\alpha_i = a_i$ for $1 \leq i \leq l$ and $\alpha_i = \$$ for $l < i \leq n$;
(ii) $\beta_i = b_i$ for $1 \leq i \leq m$ and $\beta_i = \$$ for $m < i \leq n$.

For example, if $v = ab$ and $w = cdef$, then $(v, w)^p = (a, c)(b, d)(\$, e)(\$, f)$. We call $(v, w)^p$ a *padded pair* of words.

In applications to group theory, the alphabet A is typically an inverse-closed generating set for a group G. Whenever we need to interpret a word w involving the padding symbol as an element of G, we regard the padding symbol as mapping onto the identity element of G.

We define a *synchronous 2-variable* fsa (also called a *synchronous 2-tape* fsa) M over the alphabet A to be an fsa with alphabet $A^p \times A^p$ with the property that all words in $L(M)$ are of the form $(v, w)^p$ for words $v, w \in A^*$. Notice that $(\$, \$)$ never occurs in an accepted word of M, so we could, if we preferred, take the alphabet to be $A^p \times A^p \setminus \{(\$, \$)\}$. We can define a synchronous *n*-variable fsa over A for any $n > 1$ in a similar fashion.

It is straightforward, and left as an exercise for the reader, to prove that the set of all padded pairs of words over A is regular. (This is the language of an fsa with three states, which record whether or not it has read the padding symbol in either of the two input words.) It follows that we can adapt an arbitrary fsa with alphabet $A^p \times A^p$ to one that accepts padded pairs only by intersecting its language with that of all padded pairs.

2.10.2 Asynchronous 2-variable fsa As mentioned earlier, we may also define a type of fsa that can read two input words at different rates, and we call fsa of this type *asynchronous 2-variable automata* (or *asynchronous 2-tape automata*). The state of the fsa is used to decide from which of the two input words the next symbol will be read.

For example, the language $\{(a^n, a^{2n}) : n \in \mathbb{N}_0\}$ is accepted by the following fsa with $A = \langle a \rangle$. The states are $\{1, 2, 3\}$ with 1 the start state and the only accepting state. In State 1, a symbol is read from the left-hand word and we move to State 2. In States 2 and 3, a symbol is read from the right-hand word, and we move to States 3 and 1, respectively. As usual, the fsa halts if it fails to read a symbol as a result of having reached the end of one of the input words, and it accepts the input if it is in an accepting state after reading both words.

2.10.3 Exercise Let $A = \{a, b\}$ and L be the language of pairs (u, v) with $u, v \in A^*$, where $(u, v) \in L$ if and only if u and v contain either the same number of occurrences of a or the same number of b. Show that L is the language of a non-deterministic asynchronous 2-variable fsa, but not of a deterministic one. So Proposition 2.5.2 fails for asynchronous 2-variable fsa.

For the remainder of the section, we restrict attention to synchronous 2-variable fsa and their accepted languages. The logical operations on regular sets that were introduced earlier in Proposition 2.5.10 extend to corresponding operations on the languages accepted by 2-variable fsa. In some cases, we may need to intersect the result of the logical construction on the given fsa with the language of all padded pairs. For example, for the complement of the language of a 2-variable fsa M, the desired language is

$$\{(v, w)^p : v, w \in A^*, (v, w)^p \notin L(M)\},$$

which we get by intersecting $\neg L(M)$ with the set of all padded pairs.

Given an fsa M over A, it is straightforward to construct a 2-variable fsa with language $\{(v, w)^p : v, w \in A^*, v \in L(M)\}$. (The states are those of M together with three extra states to indicate that we have read the padding symbol in the left-hand word with M in an accepting or non-accepting state, or that we have read the padding symbol in the right-hand word.) We can of course do the same thing but with $w \in L(M)$. Hence, by using the intersection operation, given an fsa M over A and a 2-variable fsa T over A, we can construct an fsa with language

$$\{(v, w)^p : (v, w)^p \in L(T), v, w \in L(M)\}.$$

This construction will be used in the proof of Theorem 5.2.3.

Recall the definition of shortlex orderings from 1.1.4.

2.10.4 Proposition   *Let $\leq$ be any shortlex ordering on A^*. Then*

$$\{(v, w)^p : v, w \in A^*, v \leq w\}$$

is the language of a 2-variable fsa.

Proof   The required fsa has four states $\sigma_0, \sigma_1, \sigma_2, \sigma_3$, with start state σ_0, accepting states σ_1, σ_3, and the following transitions:

$\sigma_0^{(a,a)} = \sigma_0$ for all $a \in A$;

$\sigma_0^{(a,b)} = \sigma_1$ for all $a, b \in A$ with $a < b$;

$\sigma_0^{(a,b)} = \sigma_2$ for all $a, b \in A$ with $a > b$;

$\sigma_1^{(a,b)} = \sigma_1$ and $\sigma_2^{(a,b)} = \sigma_2$ for all $a, b \in A$;

$\sigma_0^{(\$,a)} = \sigma_1^{(\$,a)} = \sigma_2^{(\$,a)} = \sigma_3^{(\$,a)} = \sigma_3$ for all $a \in A$.

$\square$

2.10.5 Existential quantification   Given a 2-variable dfsa M over A, it is not difficult to construct an fsa $M_\exists$ over A with

$$L(M_\exists) = \{v \in A^* : \exists w \in A^* (v, w)^p \in L(M)\}.$$

To do this, we let the state set, the start state, and the final states be the same for $M_\exists$ as for M. We replace a transition $\sigma_1^{(\alpha,\beta)} = \sigma_2$ of M by a transition $\sigma_1^{\alpha'} = \sigma_2$ of $M_\exists$, where $\alpha' = \alpha$ if $\alpha \in A$ and $\alpha' = \varepsilon$ if $\alpha = \$$. Then, for any accepting path of arrows of $M_\exists$ with label $\alpha'_1 \cdots \alpha'_r$, there is a corresponding accepting path of M with label $(\alpha_1, \beta_1) \cdots (\alpha_r, \beta_r)$ for some $\beta_i \in A^p$. By definition of 2-variable fsa, M accepts only words of the form $(v, w)^p$ for $v, w \in A^*$, and so any occurrences of ε in the path $\alpha'_1 \cdots \alpha'_r$ must occur at the end of the word.

The corresponding accepted word of $M_\exists$ is then just $\alpha'_1 \cdots \alpha'_r$ with any trailing ε symbols removed, which is precisely the word v for which

$$(\alpha_1, \beta_1) \cdots (\alpha_r, \beta_r) = (v, w)^\mathrm{p}.$$

Hence the language of $M_\exists$ is as required.

The fsa $M_\exists$ is non-deterministic in general, and so we may want to apply the subset construction of Proposition 2.5.2 to get an equivalent dfsa. This process often results in a substantial increase in the number of states. In practice, even the minimised version of an equivalent dfsa for $M_\exists$ can have very many more states than the original non-deterministic version: see exercise below.

An fsa $M_\forall$ with language $\{ v \in A^* : \forall w \in A^* (v, w)^\mathrm{p} \in L(M) \}$ can be constructed by using the fact that $L(M_\forall) = \neg L(M'_\exists)$, where M' is a 2-variable fsa with language $\neg L(M)$ intersected with the padded pairs.

In some applications to be described later that involve reducing words to words in normal form, for a given v, we need to be able to find a $w \in A^*$ with $(v, w)^\mathrm{p} \in L(M)$ if such a w exists. We observe first that we can choose a w that is not much longer than v.

2.10.6 Proposition *Let M be a 2-variable fsa over A with C states, and let $v \in A^*$. If there exists $w \in A^*$ with $(v, w)^\mathrm{p} \in L(M)$, then there exists such a w with $|w| < |v| + C$.*

Proof If $|w| \geq |v| + C$ then, when M reads $(v, w)^\mathrm{p}$, a state must be repeated while reading $ in the left-hand word. We can then remove the corresponding section of the right-hand word w, thereby obtaining a shorter word w' with $(v, w')^\mathrm{p} \in L(M)$. □

Pseudocode for the algorithm described in the next theorem can be found on Page 449 of [144].

2.10.7 Theorem *Let M be a 2-variable dfsa over A with C states, and let $v \in A^*$. If there exists $w \in A^*$ with $(v, w)^\mathrm{p} \in L(M)$, then we can find such a w with $|w| < |v| + C$ in time $O(|v|)$.*

Proof Let $v = a_1 a_2 \cdots a_n$ and $v(i) = a_1 a_2 \cdots a_i$ for $0 \leq i \leq n$, and let σ_0 be the start state of M. We read v, one letter at a time and, for each i with $0 \leq i \leq n$, we calculate and store a subset S_i of the set of states of M. A state σ of M will lie in S_i if and only if there is a word $w \in A^*$ with $|w| \leq i$ and $\sigma_0^{(v(i), w)^\mathrm{p}} = \sigma$. So we start by setting $S_0 = \{\sigma_0\}$. For each $i > 0$ and $\sigma \in S_i$, we also store a state $\rho \in S_{i-1}$ and $a \in A^\mathrm{p}$ with $\rho^{(a_i, a)} = \sigma$. This will enable us to reconstruct a word w with $(v, w)^\mathrm{p} \in L(M)$ (if it exists) at the end of the process.

To construct S_i from S_{i-1} we just need to consider all transitions $\rho^{(a_i, a)} = \sigma$

with $\rho \in S_{i-1}$ and $a \in A^p$. The same state σ may arise more than once in this way, but we need only store it once in S_i. Since the number of such transitions is bounded, the process of constructing S_i from S_{i-1} takes place in bounded time for each i. If any S_i is empty, then there is no word w with $(v, w)^p \in L(M)$, and we can abort immediately. If, after reading v, S_n contains an accepting state of M, then we know that the required word w exists and has length at most n. If not, then we continue the process and construct sets S_{n+j} for $j = 1, 2, \ldots$ in the same way, but using transitions $\rho^{(\$,a)} = \sigma$ with $a \in A$. If any S_{n+j} contains an accepting state of M, then the required word w exists and has length $n + j$. By Proposition 2.10.6, if S_{n+C-1} contains no accepting state, where C is the number of states of M, then there is no w, and we can stop.

If some S_{n+j} with $0 \le j < C$ contains an accepting state σ of M, then we can use the stored information to reconstruct the word $w = b_1 b_2 \cdots b_{n+j}$ in reverse order. Assuming that we have found the suffix $b_{i+1} \cdots b_{n+j}$ of w and the state σ_i of S_i with $\sigma_0^{(v(i),w(i))^p} = \sigma_i$, we can use the stored information for S_i to find a state $\sigma_{i-1} \in S_{i-1}$ and $b_i \in A^p$ with $\sigma_{i-1}^{(a_i,b_i)} = \sigma_i$. Notice that if $j = 0$, then we could have $|w| < |v|$, in which case $b_i = \$$ for $i > |w|$, and these b_i should then be removed from w.

Since reconstructing w also takes constant time for each letter, the whole construction takes time proportional to $n + C$, and is therefore $O(|v|)$. □

2.10.8 Composite fsa By similar methods to those used above, given 2-tape fsa M_1 and M_2 over A, we can construct a 3-tape fsa with language

$$\{ (u, v, w)^p : (u, v)^p \in L(M_1), (v, w)^p \in L(M_2) \}$$

and from this we can in turn construct a 2-variable fsa with language

$$\{ (u, w)^p : \exists v \in A^* \text{ with } (u, v)^p \in L(M_1), (v, w)^p \in L(M_2) \}.$$

This fsa is known as the *composite* of M_1 and M_2, and this construction plays an important role in the practical algorithms to construct and prove the correctness of automatic structures, which will be described in Section 5.4.

3

Introduction to the word problem

3.1 Definition of the word problem

Let G be a finitely generated group, and let A be a set that generates G as a monoid. We define the *word problem for G over A* to be the set of words $\mathrm{WP}(G, A)$ that represent the identity; that is,

$$\mathrm{WP}(G, A) := \{w \in A^* : w =_G 1\}.$$

We are mainly interested in the case when A is finite, but much of what follows makes sense at least when A is countably infinite. In this book, we shall nearly always have $A := X^{\pm}$ (recall that $X^{\pm} := X \cup X^{-1}$), where X is a finite set that generates G as a group, and in that case we shall often abuse notation and also call $\mathrm{WP}(G, A)$ the word problem for G over X.

We say that the word problem is *soluble* if there exists a terminating algorithm that can decide whether an input string w over A is in the set $\mathrm{WP}(G, A)$. This is equivalent to $\mathrm{WP}(G, A)$ being recursive as a language over A, that is, both $\mathrm{WP}(G, A)$ and its complement are accepted by Turing machines.

The word problem is one of three famous decision problems posed by Max Dehn [75, 76], the others being the conjugacy problem and the isomorphism problem. Dehn solved the word problem for surface groups, using an algorithm now known as *Dehn's algorithm*.

3.1.1 Factorisation of a word In general we shall consider groups that are defined by presentations. A word w over X represents the identity element of $G = \langle X \mid R \rangle$ (and hence lies in $\mathrm{WP}(G, X^{\pm})$) if and only if, for some $k \geq 0$, and $r_i \in R$, $u_i \in F(X)$ for $1 \leq i \leq k$, we have

$$w =_F u_1 r_1^{\pm 1} u_1^{-1} \cdots u_k r_k^{\pm 1} u_k^{-1}.$$

97

We call the expression on the right-hand side a *factorisation* of w of length k over R.

3.1.2 Recursively presented groups Now suppose that X is a finite set. We call a presentation $\langle X \mid R \rangle$ a *recursive presentation* if the set R is recursive, and a *recursively enumerable presentation* if R is recursively enumerable. It can be shown [223, Exercise 12.12.] that any group G that can be defined by a recursively enumerable presentation $\langle X \mid R \rangle$ can also be defined by a recursive presentation on the same finite generating set X. (In the exercise, a new generator y is introduced, but we can replace y by any fixed word w over X with $w =_G 1$.) Hence a group is called *recursively presentable* if it can be defined by a presentation that is either recursive or recursively enumerable. When R is recursively enumerable, then so is the set of all factorisations of words in A^* over R (with $A := X^{\pm}$), and so we have the following result.

3.1.3 Proposition *If the group $G = \langle X \rangle$ is recursively presentable, then* WP(G, A) *is recursively enumerable.*

In other words, if $G = \langle X \rangle$ is recursively presentable and $w \in$ WP(G, A), then this property can be verified. Similarly, it can be verified that two elements are conjugate in G and, for two isomorphic groups defined by finite presentations, an explicit isomorphism can be found. Nonetheless even finitely presented groups can be defined for which the word problem, conjugacy problem and isomorphism problem are insoluble, as we shall see in Chapter 10.

It turns out also that for many properties that arise in group theory, while the properties themselves are undecidable, their verification is possible when they do hold. Examples of such properties, for a group defined by a finite presentation, include being trivial, finite, abelian, polycyclic, nilpotent, free, automatic, or word-hyperbolic.

3.2 Van Kampen diagrams

Let $G = \langle X \mid R \rangle$ be a group presentation, and let

$$w =_F u_1 r_1^{\pm 1} u_1^{-1} \cdots u_k r_k^{\pm 1} u_k^{-1}$$

be a factorisation of length k of the word w over R, as defined in 3.1.1.

Provided that w is a reduced word, the factorisation labels a closed path in $\Gamma := \Gamma(G, X)$, the Cayley graph of G (see 1.6.1), that starts and ends at the identity vertex, and which can be decomposed as a union of k loops. A path in the Cayley graph labelled by u_i leads from the identity vertex to a point on the *i*-th of those loops, and the closed path around the loop that starts at that point is labelled by the relator r_i.

This results in a planar diagram Δ, which is known as a *Dehn diagram* or a *van Kampen diagram* for w, over the given presentation. More formally, Δ is a planar combinatorial 2-complex in which the vertices, edges, and interior regions are the 0-, 1- and 2-cells, respectively. With the correct starting vertices, the boundaries of the 2-cells are labelled (reading clockwise around the region) by elements of $R \cup R^{-1}$, and the boundary of Δ is labelled by a word that freely reduces to w. As in the Cayley graph, we generally identify an edge labelled x with an edge labelled x^{-1} between the same two vertices but in the other direction.

3.2.1 Reducing diagrams Suppose that we have successive vertices v_1, v_2, v_3 on the boundary of Δ with edges $v_1 v_2$ and $v_2 v_3$ labelled by x and x^{-1} for some $x \in X^{\pm}$. Then we can identify v_1 and v_3, and move them together in Δ. If there are no other edges through v_2, then we remove v_2 and the two edges. Otherwise, we identify the two edges, and the single resulting edge becomes an internal edge. See Figure 3.1.

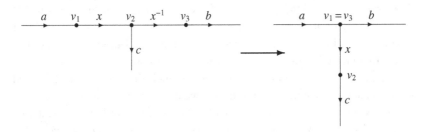

Figure 3.1 Amalgamating two vertices and two edges.

By performing such operations repeatedly, we can transform Δ to a diagram, which is still a planar 2-complex, of which the boundary label is w.

It is also possible to eliminate two internal 2-cells if their labels are $r^{\pm 1}$ for some $r \in R$, and they have a corresponding edge in common. The reader could consult [183, Chapter V, Section 1] or [243] for example, for a more detailed and accurate description of the reduction process. The diagram is said to be *reduced* if no further reductions of either of these types are possible.

3.2.2 Example Let $G = \langle x, y \mid y^{-1}xy = x^3, y^2 = x^2 \rangle$. Then $G \cong Q_8$, the quaternion group. This is easy to see if we include the additional relation $x^4 = 1$ but it turns out that $x^4 =_G 1$ is a consequence of the two existing relations. Let

$r_1 = y^{-1}xyx^{-3}$ and $r_2 = y^2x^{-2}$. Then it can be verified that

$$x^4 =_F (x^3 w_1 x^{-3})(w_2)(y^{-1}w_3 y)(w_4),$$

where $w_1 = w_2 = r_1^{-1}$, $w_3 = r_2^{-1}$, $w_4 = r_2$ and F is the free group on $\{x, y\}$.

In Figure 3.2, we show both the initial van Kampen diagram and its reduction.

The base point of each diagram is indicated by a larger black circle. Since we are identifying an edge labelled x with one in the opposite direction labelled x^{-1}, we have labelled all of the edges x or y and avoided using x^{-1} and y^{-1}.

We show now that the lengths of the conjugating elements u_i in factorisations of a given word can be bounded.

3.2.3 Proposition   *Let $G = \langle X \mid R \rangle$ with X, R finite, and suppose that there is a van Kampen diagram Δ for G with boundary label $w \in \mathrm{WP}(G, X^\pm)$ and k internal faces.*

Then there is a factorisation $u_1 r_1^{\pm 1} u_1^{-1} \cdots u_k r_k^{\pm 1} u_k^{-1}$ of w of length k, for which $|u_i| < (|w| + il)/2$ for $1 \le i \le k$, where l is the length of the longest relator in R.

Proof   We use induction on k. If $k = 0$, then Δ is a tree so $w =_F 1$ (with F the free group on X), and the result holds. So suppose that $k > 0$. It is clear that at least one edge on the boundary of Δ must border an internal face of Δ. So let u be a maximal prefix of w for which no edge of the path labelled u borders an internal face of Δ; then $w = u s_1 v$, with s_1 nonempty, where $s_1 s_2$ is a cyclic permutation of a relator in $r \in R \cup R^{-1}$, and r labels the internal face bordered by the boundary edge in question. Each edge on the path labelled by the prefix u must occur twice on the boundary of Δ (since it cannot border an internal face), and so $|u| < |w|/2$. We have $w =_F (u s_1 s_2 u^{-1})(u s_2^{-1} v)$, and we get a van Kampen diagram for $u s_2^{-1} v$ with $k - 1$ internal regions by removing from Δ the internal face labelled by r, together with those of its edges that label s_1.

By the induction hypothesis, $u s_2^{-1} v$ has a factorisation $u_2 r_2^{\pm 1} u_2^{-1} \cdots u_k r_k^{\pm 1} u_k^{-1}$ with $|u_i| < (|w| + il)/2$ for $2 \le i \le k$. Now $u s_1 s_2 u^{-1}$ is a conjugate of r by a word u_1 of length at most $(|u| + |r|)/2 \le (|w| + l)/2$. We append these two factorisations (of $u s_1 s_2 u^{-1}$ and $u s_2^{-1} v$) to get an appropriate factorisation of w.   □

3.2.4 Corollary   *Let $G = \langle X \mid R \rangle$ with X, R finite, and let*

$$w =_F u_1 r_1^{\pm 1} u_1^{-1} \cdots u_k r_k^{\pm 1} u_k^{-1}, \quad r_i \in R, u_i \in F,$$

be a factorisation of a reduced word $w \in F$ with $w =_G 1$.

Then there is a factorisation $u_1' r_1'^{\pm 1} u_1'^{-1} \cdots u_k' r_k'^{\pm 1} u_k'^{-1}$ of w having the same length k, in which $|u_i'| < (|w| + il)/2$ for $1 \le i \le k$, where l is the length of the longest relator in R.

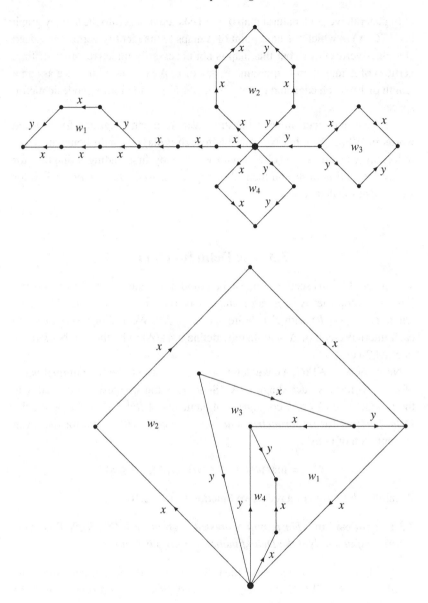

Figure 3.2 Reduction of van Kampen diagram for x^4 in $\langle x, y \mid y^{-1}xy = x^3, y^2 = x^2 \rangle$

Proof We saw in 3.2.1 that w is the boundary word of a van Kampen diagram with k internal faces, so the result follows from the proposition. □

In general, we get a natural map of the 1-skeleton of Δ into the Cayley graph $\Gamma = \Gamma(G, X)$ in which the base point of Δ maps to the identity vertex (or indeed any other vertex) of Γ, but this map is not necessarily injective, since distinct vertices of Δ may map to the same vertex of Γ. As we did with Γ, we assign a length of 1 to each edge and thereby make its 1-skeleton into a geodesic metric space.

We have explained how to construct van Kampen diagrams for reduced words in $\mathrm{WP}(G, X^{\pm})$, but we can obtain such a diagram with boundary label w for any word $w \in \mathrm{WP}(G, X^{\pm})$ with $w =_G 1$ by first finding a diagram for the free reduction of w, and then adjoining spikes to the boundary vertices for subwords of w that cancel.

3.3 The Dehn function

The material in this section can also be found, for example, in Gersten's paper [89]. As before, let Δ be a van Kampen diagram for a word $w \in \mathrm{WP}(G, A)$, where $G = \langle X \mid R \rangle$ with X, R finite and $A := X^{\pm}$. We assign an *area* of 1 to each internal 2-cell of Δ, and thereby define $\mathsf{area}(\Delta)$ to be the number k of its internal 2-cells.

Now, for $w \in \mathrm{WP}(G, A)$, we define $\phi(w) = \phi_{X,R}(w)$ to be the minimal value of $\mathsf{area}(\Delta)$ for any such diagram Δ. (So $\phi(w)$ is the smallest k such that w is the product in $F(X)$ of k conjugates of elements of $R^{\pm 1}$.) Then we define the *Dehn function* (or *isoperimetric function*) $\phi = \phi_{X,R} \colon \mathbb{N}_0 \to \mathbb{N}_0$ for the given presentation of G by

$$\phi(n) = \max\{\phi(w) : w \in \mathrm{WP}(G, A), |w| \leq n\}.$$

Recall the definition of a recursive function from 2.2.16.

3.3.1 Proposition *For a finitely presented group $G = \langle X \mid R \rangle$, $\mathrm{WP}(G, A)$ is recursive if and only if the Dehn function $\phi = \phi_{X,R}$ is recursive.*

Proof If $\mathrm{WP}(G, A)$ is recursive then, for any $n \geq 0$, we can compute a list of all words $w \in \mathrm{WP}(G, A)$ with $|w| \leq n$ and then, for each such w, find the shortest k such that w has a factorisation of length k, using the fact that we only need consider factorisations in which the conjugating elements have length bounded by the computable function of k, n and $l = \max\{|r| : r \in R\}$ specified in Corollary 3.2.4. So we can compute $\phi(n)$.

Conversely, if ϕ is recursive then, given $w \in F$ we can consider all factorisations of length at most $\phi(|w|)$, again using Corollary 3.2.4 to bound the length of the conjugating elements, and check whether any of them freely reduce to w. □

However, in [186, Theorem 4.2], Madlener and Otto proved that the Dehn function can, in a precise sense, be much larger than the (time) complexity of the word problem. In other words, the problem of expression of a word w as a product of conjugates of defining relators of the group can be arbitrarily more difficult than the problem of just deciding whether $w \in \mathrm{WP}(G, A)$. Section 4 of the survey article [226] by Sapir includes many further results on the complexity of the Dehn function.

Indeed it was proved by Novikov and Boone (independently) [206, 31, 32, 33] in the 1950s that the word problem is insoluble in general; in particular there exist finitely presented groups with insoluble word problem. We give an account of this proof in Chapter 10.

3.3.2 Invariance under change of generators, and passage to finitely generated subgroups and overgroups Our definition of the word problem is expressed in terms of a particular finite generating set X. However, whenever Y is a second finite generating set, a finite set of substitutions of generators by words of bounded length transforms words over X to words over Y, and conversely, and so an algorithm to recognise $\mathrm{WP}(G, X^{\pm})$ can easily be translated to one that recognises $\mathrm{WP}(G, Y^{\pm})$. Hence both the solubility of the word problem and the complexity of its solution are essentially independent of the choice of generating set for G. Similarly we can see that whenever G has soluble word problem, so does any finitely generated subgroup H of G, or any group L in which G has finite index. (We shall be considering such questions for general families of languages in the next section.)

We can also show that the Dehn functions with respect to quasi-isometric finitely generated groups (as defined in 1.7.2) are closely related. For functions $f, g: \mathbb{N}_0 \to \mathbb{N}_0$, we define $f \preceq g$ if there is a constant $C > 0$ with $f(n) \leq Cg(Cn + C) + Cn + C$ for all $n \in \mathbb{N}_0$, and $f \simeq g$ if $f \preceq g$ and $g \preceq f$. The result below is proved by Alonso [4] for Dehn functions of a more general class of metric spaces.

3.3.3 Proposition *Let $\langle X \mid R \rangle$ and $\langle Y \mid S \rangle$ be finite presentations of groups G and H and suppose that the Cayley graphs $\Gamma(G, X)$ and $\Gamma(H, Y)$ are quasi-isometric. Then $\phi_{X,R} \simeq \phi_{Y,S}$.*

Proof By symmetry, it is enough to show that $\phi_{X,S} \preceq \phi_{Y,R}$. Let $f: \Gamma(G, X) \to \Gamma(H, Y)$ be a quasi-isometry with parameters λ, ϵ, μ (see Section 1.7). It is convenient to choose $\mu \geq 1/2$.

Let $n \in \mathbb{N}_0$ and $w = a_1 a_2 \cdots a_m$ be a word over X with $w =_G 1$ and $|w| = m \le n$. For $0 \le i \le m$, let p_i be the vertex in $\Gamma(G, X)$ defined by $a_1 \cdots a_i$ (where $p_0 = p_m$ is the base point of $\Gamma(G, X)$). Then $f(p_i)$ might not be a vertex of $\Gamma(H, Y)$, but there exists a vertex q_i of $\Gamma(H, Y)$ with $d_{\Gamma(H,Y)}(q_i, f(p_i)) \le 1/2$ (and hence $d_{\Gamma(H,Y)}(q_i, f(p_i)) \le \mu$) and

$$d_{\Gamma(H,Y)}(q_i, q_{i+1}) \le d_{\Gamma(H,Y)}(f(p_i), f(p_{i+1})) + 1 \le A := \lambda + \epsilon + 1$$

for $0 \le i < m$.

Let $v_i \in (Y \cup Y^{-1})^*$ be the label of a path in $\Gamma(H, Y)$ of length at most A joining q_i to q_{i+1}, and let $v = v_1 v_2 \cdots v_m$. Then $v =_H 1$ with $|v| \le An$, so there is a Dehn diagram Δ_v for v with respect to the presentation $\langle Y \mid S \rangle$ having area at most $\phi_{Y,S}(An)$.

There is a uniquely defined (not necessarily injective) homomorphism of labelled graphs τ from the 1-skeleton of Δ_v to $\Gamma(H, Y)$ that maps the base point of Δ_v to q_0, and the vertices ρ_i at the ends of the paths of the boundary of Δ_v labelled $v_1 v_2 \cdots v_i$ to q_i.

For each vertex ρ of Δ_v, we can find $\hat{p} \in \Gamma(G, X)$ with $d_{\Gamma(H,Y)}(f(\hat{p}), \tau(\rho)) \le \mu$, and then find a vertex p of $\Gamma(G, X)$ with $d_{\Gamma(G,X)}(p, \hat{p}) \le 1/2$. In particular, we can choose $p = p_i$ when $\rho = \rho_i$, with $0 \le i \le m$. We define the map σ from the vertex set of Δ_v to the vertex set of $\Gamma(G, X)$ by putting $\sigma(\rho) = p$; so $\sigma(\rho_i) = p_i$ for each i. Now, for any two vertices ρ, ρ' of Δ_v that are joined by an edge of Δ_v, we have

$$d_{\Gamma(G,X)}(\sigma(\rho), \sigma(\rho')) \le B := \lambda(1 + 2\mu + \epsilon) + 1.$$

We now construct a van Kampen diagram $\Delta_{w'}$ for a word w' over X, such that $\mathsf{area}(\Delta_{w'}) = \mathsf{area}(\Delta_v)$, and the interior regions of $\Delta_{w'}$ are labelled by relators of G, each of length at most BC, where $C := \max\{|s| : s \in S\}$, as follows.

For each vertex ρ of Δ_v, we define a corresponding vertex of $\Delta_{w'}$ in the same position on the plane. Each edge of Δ_v from vertex ρ to vertex ρ' is replaced in $\Delta_{w'}$, in the same position on the plane, by a path of length at most B, labelled by a word over X that is equal in G to the label of some path of length at most B in $\Gamma(G, X)$ from $\sigma(\rho)$ to $\sigma(\rho')$. (We choose any such path; we know from above that there is one.) The internal regions of $\Delta_{w'}$ are now in the same position in the plane as those of Δ_v. This completes the construction of $\Delta_{w'}$. It is convenient to refer below to the vertices of $\Delta_{w'}$ by the same names as the vertices of Δ_v to which they correspond.

Now the vertices ρ_i of $\Delta_{w'}$ lie on the boundary of $\Delta_{w'}$, and the label of the path in $\Delta_{w'}$ joining ρ_{i-1} to ρ_i has length at most AB, and is equal in G to a_i for $1 \le i \le m$ (since $\sigma(\rho_{i-1}) = p_{i-1}, \sigma(\rho_i) = p_i$). We construct a Dehn diagram Δ_w with boundary label w by adjoining new edges labelled a_i joining ρ_{i-1} to ρ_i in the exterior region of $\Delta_{w'}$. This creates m new interior regions, each labelled by a relator of length at most $1 + AB$. Now

$$\text{area}(\Delta_w) = m + \text{area}(\Delta_{w'}) = m + \text{area}(\Delta_v) \leq n + \phi_{Y,S}(An),$$

and the interior regions of Δ_w are labelled by relators of G of length at most $D := \max(1 + AB, BC)$; each such relator has a factorisation of length at most $\phi_{X,R}(D)$ over R. So $\text{area}(w) \leq \phi_{X,R}(D)(n + \phi_{Y,S}(An))$ and, since Dehn functions are increasing functions, we have $\phi_{X,R}(n) \leq E(n + \phi_{Y,S}(En))$ with $E := \max(\phi_{X,R}(D), A)$, which proves the result. □

3.3.4 Corollary *If $H \leq G$ with $|G : H|$ finite, and G and H have the finite presentations $\langle X \mid R \rangle$ and $\langle Y \mid S \rangle$, then the Dehn functions associated with the presentations satisfy $\phi_{X,R} \simeq \phi_{Y,S}$. This applies in particular to different finite presentations of the same group.*

Proof This follows from Proposition 1.7.4. □

So we can unambiguously say, for example, that the Dehn function of a group is polynomial of degree $k \geq 1$. But, since free groups can have either a zero or a linear Dehn function, depending on the presentation, this equivalence does not extend to sublinear functions.

In fact it was announced by Gromov and subsequently proved by Ol'shanskii [207], Bowditch [35] and Papasoglu [209] that, if the Dehn function is subquadratic, then it is linear. We prove this later in Theorem 6.5.3. Groups with linear Dehn function are called *(word-)hyperbolic*, and are the subject of Chapter 6. We prove in Theorem 3.5.1 that groups with Dehn presentations have this property. It has also been proved by Brady and Bridson [37] that the set of $k \in [2, \infty)$ for which there exist groups with Dehn function equivalent to n^k is dense in $[2, \infty)$.

3.4 The word problem as a formal language

We already observed that the word problem is soluble if and only if the set $\text{WP}(G, X^{\pm})$ is recursive. Hence it is natural to study the word problem as a formal language. The following basic result is due to Anisimov [7].

3.4.1 Theorem *For a finitely generated group $G = \langle X \rangle$, $\text{WP}(G, X^{\pm})$ is regular if and only if G is finite.*

Proof When G is finite, its Cayley graph can be interpreted as a dfsa to accept its word problem (see 2.2.3). To prove the opposite implication we need only observe that, if M is a dfsa accepting the word problem of G with start vertex σ_0, then two words w, v with $\sigma_0^w = \sigma_0^v$ must represent the same group element, and hence the number of states of M is bounded below by $|G|$.   □

We have already described in Section 2.6 a pushdown automaton that solves the word problem in a virtually free group. In Chapter 11 we study Muller and Schupp's proof [198] that the word problem is context-free precisely when G is virtually free. We also study groups whose word problems are in other language families.

We have observed already that solubility of the word problem for a group G is independent of the choice of a finite generating set X for G. The following results show that, for most natural choices of a family C of formal languages, the property $\mathrm{WP}(G, X^{\pm}) \in C$ is also independent of the choice of X.

3.4.2 Independence of the choice of generating set Let X and Y be finite generating sets of the group G, and let $A := X^{\pm}$, $B := Y^{\pm}$. Suppose that $\mathrm{WP}(G, A) \in C$ for some family C of languages. For each $y \in Y$, we can choose a specific word $\sigma(y) \in A^*$ with $\sigma(y) =_G y$. We can then extend the domain of σ to B^* by putting $\sigma(\varepsilon) = \varepsilon$, $\sigma(y^{-1}) = \sigma(y)^{-1}$ for $y \in Y$, and $\sigma(b_1 \cdots b_n) = \sigma(b_1) \cdots \sigma(b_n)$ for words $b_1 \cdots b_n \in B^*$ with $n > 1$. In fact this extends σ to a *monoid homomorphism* $\sigma : B^* \to A^*$. Now, for $w \in Y^*$, we have $w \in \mathrm{WP}(G, B)$ if and only if $\sigma(w) \in \mathrm{WP}(G, A)$.

Recall from 2.2.24 that a family of languages C is *closed under inverse homomorphisms* if, for any monoid homomorphism $\tau : B^* \to A^*$ (with A and B finite) and any language $L \subseteq A^*$, we have $L \in C$ implies $\tau^{-1}(L) \in C$. Since $\mathrm{WP}(G, B) = \sigma^{-1}(\mathrm{WP}(G, A))$, we have the following result.

3.4.3 Proposition *If the language family C is closed under inverse homomorphisms, then the property of the word problem $\mathrm{WP}(G, X^{\pm})$ of a group G being in C is independent of the chosen finite generating set X of G.*

(More generally this independence extends to arbitrary finite monoid generating sets of G.)

We saw in Chapter 2 that the families of regular (Proposition 2.5.14), context-free (Proposition 2.6.27), context-sensitive (2.8.3), recursive and recursively enumerable (2.9.4) languages are all closed under inverse homomorphisms. We shall see later that the same applies to one-counter, indexed, and real-time languages. It is also true for growing context-sensitive languages, although that is less straightforward to show.

So, from now on we can safely talk about groups G having word problem in C without reference to the generating set. It is customary to refer to such groups as *C-groups*. In situations where we know that the property of WP($G, X^\pm$) in question is independent of the finite generating set X, then we may just write WP(G) rather than WP($G, X^\pm$). For example, the statement 'WP(G) is context-free' is unambiguous, and is equivalent to 'G is a context-free group'.

3.4.4 Finitely generated subgroups Again let $G = \langle X \rangle$ with X finite and let $A := X^\pm$. Let H be a subgroup of G generated by a finite set Y and let $B := Y^\pm$. Assuming that WP(G, A) is independent of the generating set, we can adjoin the elements of Y as new generators of G and thereby assume that $Y \subseteq X$. So now a word $w \in B^*$ is also an element of A^*, and we can use the same automaton that we use to test membership of WP(G, A) to test membership of WP(H, B). So we expect WP(H, B) to lie in the same language family C as WP(G, A).

Again we can formulate this in terms of properties of C. Assuming that $Y \subseteq X$, we have WP(H, B) = WP(G, A) $\cap B^*$. So it is sufficient to have closure under inverse homomorphism in order to allow change of the generating set followed by intersection with the regular language B^*. This justifies the following fundamental result.

3.4.5 Proposition *Suppose that the language family C is closed under inverse homomorphisms and intersection with regular languages. If the word problem of a group G lies in C, then the word problem of any finitely generated subgroup of G also lies in C.*

Once again, the property of closure under intersection with regular languages is satisfied by all of the familiar language families C.

3.4.6 Finite index overgroups Suppose that F is a finite index overgroup of G; that is G is a subgroup of F and $|F : G|$ is finite. We now assume that WP(G) lies in the language family C, and we would like to be able to say the same about the word problem of F.

Suppose that $G = \langle X \rangle$ with X finite and let $A := X^\pm$. Provided that C is closed under inverse homomorphisms, we are free to choose our preferred generating set for F. Let $F = \cup_{i=1}^n Gt_i$ be a decomposition of F into right cosets of G with $t_1 = 1$, and let $Y = X \cup \{t_2, t_3, \ldots, t_n\}$. Then it is easy to see that Y generates F. Let $B := Y^\pm$.

For an element $f \in F$ we use t_{if} to denote the coset representative of $Gt_i f$ for $1 \le i \le n$. So $Gt_i f = Gt_{i^f}$ and $i \mapsto i^f$ defines a permutation of $\{1, 2, \ldots, n\}$. Then for each i with $1 \le i \le n$ and each $b \in B$, we have an equation of the form $t_i b = w(i, b) t_{i^b}$ for some word $w(i, b) \in A^*$.

Hence, for $b_1 b_2 \cdots b_l \in B^*$, we have

$$b_1 b_2 \cdots b_l = t_1 b_1 b_2 \cdots b_l = w(1, b_1) t_{1^{b_1}} b_2 \cdots b_l$$
$$= w(1, b_1) w(1^{b_1}, b_2) \cdots w(1^{b_1 \cdots b_{l-2}}, b_{l-1}) w(1^{b_1 \cdots b_{l-1}}, b_l) t_{1^{b_1 \cdots b_l}}.$$

So $b_1 b_2 \cdots b_l \in \mathrm{WP}(F, B)$ if and only if $1^{b_1 \cdots b_l} = 1$ and

$$w(1, b_1) w(1^{b_1}, b_2) \cdots w(1^{b_1 \cdots b_{l-2}}, b_{l-1}) w(1^{b_1 \cdots b_{l-1}}, b_l) \in \mathrm{WP}(G, A).$$

3.4.7 Proposition *Suppose that the language family C is closed under inverse GSMs and under intersection with regular languages. Let $G \leq F$ be groups with $|F : G|$ finite and $\mathrm{WP}(G) \in C$. Then $\mathrm{WP}(F) \in C$.*

Proof Closure of C under inverse GSMs implies closure under inverse homomorphisms, so membership of $\mathrm{WP}(F, Y^\pm) \in C$ is independent of the choice of Y, and hence we may choose $Y = X \cup \{t_2, t_3, \ldots, t_n\}$ as above.

Using the above notation, we define a generalised sequential machine M (as defined in 2.2.25) with input alphabet $B = Y^\pm$, output alphabet $A = X^\pm$, with state set $\{1, 2, \ldots, n\}$, initial state 1, and transition function defined by $\tau(i, b) = (i^b, w(i, b))$. Let ϕ be the associated GSM.

Then, from the discussion above, we see that a word $w \in B^*$ lies in $\mathrm{WP}(F, B)$ if and only if M is in state 1 after reading w, and $\phi(w) \in \mathrm{WP}(G, A)$. The sets of words satisfying these two conditions are respectively the language of the dfsa associated to M with accepting states $\{1\}$, and $\phi^{-1}(\mathrm{WP}(H, A))$, so $\mathrm{WP}(F, B) \in C$ by the assumed closure properties of C. □

3.4.8 Syntactic monoid of the word problem In the next few results we allow A to be an arbitrary finite subset of a group G that generates G as a monoid, so that it need not be of the form $X^\pm$ for a group generating set X.

We recall from 2.4 the definitions of the syntactic congruence $\approx$ of a language $L \subseteq A^*$, and the syntactic monoid, $\mathrm{Syn}(L) = A^* / \approx$. Work of Parkes and Thomas [212] uses syntactic monoids to derive a characterisation of languages that are word problems of groups. We observe first the following elementary result.

3.4.9 Proposition *If G is a group generated as a monoid by a finite set A, then G is isomorphic to the syntactic monoid of $\mathrm{WP}(G, A)$ in A^*.*

Proof Consider the natural monoid homomorphism $\varphi: A^* \to G$, and let $\approx$ be the syntactic congruence of WP(G, A). Then, for any $w_1, w_2 \in A^*$, we have

$$w_1 \approx w_2 \iff (uw_1v \in \mathrm{WP}(G,A) \iff uw_2v \in \mathrm{WP}(G,A) \; \forall u, v \in A^*)$$
$$\iff (\varphi(uw_1v) = 1 \iff \varphi(uw_2v) = 1 \; \forall u, v \in A^*)$$
$$\iff (\varphi(u)\varphi(w_1)\varphi(v) = 1 \iff \varphi(u)\varphi(w_2)\varphi(v) = 1 \; \forall u, v \in A^*)$$
$$\iff \varphi(w_1) = \varphi(w_2),$$

and the result follows. $\qquad\qquad\qquad\qquad\qquad\qquad\qquad\qquad\qquad\qquad\quad\square$

A pair of conditions now characterises word problems of groups.

3.4.10 Proposition *A language $L \subseteq A^*$ is the word problem of a group that is generated as a monoid by A if and only if it satisfies the following two conditions:*

(W1) if $w \in A^*$ then there exists $w' \in A^*$ such that $ww' \in L$;
(W2) if $w, u, v \in A^*$ with $w \in L$ and $uwv \in L$, then $uv \in L$.

Proof First suppose that L is the word problem WP(G, A) of a group. Choose $w \in A^*$ and suppose that w represents $g \in G$. Choose w' to be any word in A^* that represents g^{-1}. Then $ww' \in L$, and so L satisfies (W1). Now if $w, u, v \in A^*$ with $w \in L$ and $uwv \in L$ then $w =_G uwv =_G 1$, and so $uv =_G 1$ and hence $uv \in L$. Therefore L satisfies (W2).

Conversely, suppose that L satisfies (W1) and (W2); we want to show that L is the word problem WP(G, A) of a group. The first step is to deduce a third condition from (W1) and (W2):

(W3) if $w \in L$ and $uwv \notin L$ then $uv \notin L$.

For suppose that $w, u, v \in A^*$ with $w \in L$, but $uwv \notin L$. By (W1), we can find $w' \in A^*$ with $uwvw' \in L$. Since also $w \in L$, we can apply (W2) to deduce that $uvw' \in L$. Now if $uv \in L$, we derive a contradiction from two applications of (W2). First, since $uvw' = \varepsilon uvw' \in L$, we deduce that $\varepsilon w' = w' \in L$. Then since $w' \in L$ and $uwvw'\varepsilon = uwvw' \in L$, we deduce that $uwv\varepsilon = uwv \in L$, which is a contradiction. Hence we have proved that (W3) holds.

Now let $\approx$ denote the syntactic congruence of L, and let φ be the natural homomorphism from A^* onto the syntactic monoid Syn(L). Then L is a union of congruence classes of $\approx$ (this is immediate from the definition).

But additionally, for any $w \in L$, the combination of (W2) and (W3) ensures that, for all $u, v \in A^*$,

$$uv \in L \iff uwv \in L.$$

So, in fact L (which is certainly nonempty, since (W1) holds) consists of a single congruence class of $\approx$.

Now, where $w = \varepsilon w \varepsilon$ is any element of L, (W2) implies also that $\varepsilon \in L$. So L is the congruence class of ε, that is $L = \varphi^{-1}(1)$.

If $m \in \text{Syn}(L)$, choose $w \in A^*$ with $\varphi(w) = m$, and then (using (W1)) choose $w' \in A^*$ such that $ww' \in L$. If $n = \varphi(w')$, then $mn = \varphi(w)\varphi(w') = \varphi(ww') = 1$. So every element of $\text{Syn}(L)$ has a right inverse; hence the monoid $\text{Syn}(L)$ is a group.   □

3.5 Dehn presentations and Dehn algorithms

Dehn solved the word problem in surface groups by exploiting features of their natural presentations. Presentations with these properties are now called Dehn presentations.

Recall from Section 1.9 that, for a set R of relators, the *symmetric closure* $\hat{R}$ of R is the closure of R under taking of inverses and cyclic conjugates. A finite presentation $G = \langle X \mid R \rangle$ is called a *Dehn presentation* if, for any reduced word w over X with $w =_G 1$ and $w \neq \varepsilon$, w contains a subword u such that for some word v with $|u| > |v|$, $uv \in \hat{R}$

3.5.1 Theorem   *If $G = \langle X \mid R \rangle$ is a Dehn presentation then the Dehn function of $\phi := \phi_{X,R}$ of G satisfies $\phi(n) \leq n$.*

Proof   We need to prove that any word of length at most n that represents the identity is equal to the free reduction of a product of at most n conjugates of elements of $\hat{R}$. The proof is by induction on n; the case $n = 0$ is clear.

So suppose that $w =_G 1$, with $|w| \leq n$. We may replace w by its free reduction and thereby assume that it is reduced. Then there are words w_1, w_2, u, v over X, with $w = w_1 u w_2$, $uv \in \hat{R}$, $|u| > |v|$, and w is equal to the free reduction of $w_1 u v w_1^{-1} w_1 v^{-1} w_2$. Now $w_1 u v w_1^{-1} =_G 1$ implies $1 =_G w =_G w_1 v^{-1} w_2$, and $|v| < |u|$ implies that $|w_1 v^{-1} w_2| < n$. So, by induction, we can write $w_1 v^{-1} w_2$ as a product of at most $n - 1$ conjugates of elements of $\hat{R}$; we precede this product by $w_1 u v w_1^{-1}$ to get an expression for w.   □

3.5.2 Dehn algorithms   Let $\langle X \mid R \rangle$ be a finite presentation for a group G, and let $\tilde{R} := \hat{R} \cup \{xx^{-1}, x^{-1}x : x \in X\}$. Then we can carry out the following algorithm on input words w over X.

Algorithm: While w contains a subword u for which there exists $uv \in \tilde{R}$ with $|u| > |v|$, replace u in w by v^{-1}.

If this algorithm reduces w to the empty word if and only if $w =_G 1$, then we call it a *Dehn algorithm* for the presentation and it is straightforward to see that this is the case if and only if $\langle X \mid R \rangle$ is a Dehn presentation for G. In that case, as we shall show in Proposition 3.5.5, the Dehn algorithm provides a linear time solution to the word problem in G.

We shall see in Chapter 6 that Dehn presentations exist not just for surface groups but for all (word-)hyperbolic groups, and that this property characterises this class of groups, which include virtually free groups and hyperbolic Coxeter groups.

3.5.3 Example We look at reduction in the group of the double torus.

$$G = \langle a, b, c, d \mid a^{-1}b^{-1}abc^{-1}d^{-1}cd \rangle.$$

We reduce $d^{-1}c^{-1}ddcb^{-1}a^{-1}bad^{-1}b^{-1}a^{-1}ba$ as follows

$$d^{-1}c^{-1}d\underline{dcb^{-1}a^{-1}b}ad^{-1}b^{-1}a^{-1}ba \rightarrow d^{-1}c^{-1}\underline{dc}\overline{d\mathbf{a}^{-1}\mathbf{a}}d^{-1}b^{-1}a^{-1}ba$$

$$\rightarrow \underline{d^{-1}c^{-1}dcb^{-1}a^{-1}}ba$$

$$\rightarrow \overline{\mathbf{a^{-1}b^{-1}aa}^{-1}\mathbf{ba}} \rightarrow \epsilon.$$

The application of a rule is marked using underlining and overlining, and subwords that freely reduce to the empty word are marked in bold. The second and final reductions are just free reductions. Notice that we need to backtrack; that is, the third reduction is further left in the word than the first two.

3.5.4 Rewriting systems in groups It is convenient at this point to briefly introduce rewriting systems, which play a major role in practical solutions to the word problems of their associated groups. This topic is studied in more detail in Chapter 4.

Let $G = \langle X \rangle$ be a group and $A = X^{\pm}$. A *rewriting system* for G is a set $\mathcal{S}$ of ordered pairs $(u, v) \in A^* \times A^*$ with $u =_G v$. The pairs (u, v) are called the *rewrite rules* of the system. We call u and v the *left- and right-hand sides* of the rule, and we usually choose the rules such that $u > v$ in some reduction ordering of A^* (as defined in 4.1.5).

A word $w \in A^*$ is called $\mathcal{S}$*-reduced* if it does not contain the left-hand side of any rule in $\mathcal{S}$ as a subword. If w is not $\mathcal{S}$-reduced, then it contains the left-hand side u of a rule (u, v) of $\mathcal{S}$, and we can substitute v for u in w, thereby replacing w by a G-equivalent word that precedes it in the ordering of A^*. This process is easiest to carry out when $\mathcal{S}$ is finite. The following result demonstrates that it leads to an efficient solution of the word problem in groups with a Dehn algorithm.

3.5.5 Proposition *If the group G has a Dehn presentation, then the word problem of G is soluble in linear time.*

Proof   This seems intuitively clear, because each substitution in the application of Dehn's algorithm to a word w reduces the length of w, and so there can be at most $|w|$ such reductions altogether. But we have to be careful with the implementation. A naive implementation would involve moving large subwords of w to fill gaps created by reductions and would then run only in quadratic time.

The process can however be carried out in linear time by using two stacks, which are initially empty. Given a Dehn presentation $\langle X \mid R \rangle$ of G, we start (as part of the pre-processing stage) by constructing a rewriting system S for G in which the rules are

$$\{(u, v) : uv^{-1} \in \hat{R}, |u| > |v|\} \cup \{(xx^{-1}, \varepsilon) : x \in A\}.$$

(For increased efficiency, we could use an fsa that accepts the set of left-hand sides of these rules and whose accept states provide pointers to the right-hand sides of the corresponding rules, as described in Section 4.2.)

The first of the two stacks is used to store an S-reduced word equal in G to a prefix of the word w read so far (but with the beginning of the word at the bottom of the stack), and the second stack is used as a queue of letters waiting to be processed. If the second stack is nonempty, then we read the next letter to be processed from that stack, and otherwise we read it from the input word w.

Let $m = \max\{|r| : r \in R\}$. The letters read for processing are put onto the top of the first stack as long as this does not contain the left side u of a rewrite rule (u, v). Whenever such a word u appears (in reversed order) on the top of the first stack, we first remove it. Next, we put the right-hand side v of that rule onto the second stack. Finally, we transfer $m - 1$ letters (or fewer if we reach the bottom of the stack) from the first stack to the second stack. We are then ready to carry on reading letters for processing.

Observe that, when we resume reading after a substitution, we start by putting the $m - 1$ transferred letters from the second stack back onto the first stack, which may seem strange. This is necessary because the substitution of v for u in w could result in a further substitution that starts up to $m - 1$ letters to the left of u in w, so we need to backtrack in order to recognise the left-hand side of such a substitution. Since the total work involved in each substitution is constant, and the total number of substitutions is bounded by $|w|$, the process runs in linear time.   □

3.5.6 Generalised Dehn algorithms   Goodman and Shapiro [108] introduce a generalisation of Dehn's algorithm, which they call a *Cannon algorithm*, but which is also referred to in the literature as a *generalised Dehn algorithm*

[151]. This generalisation is a rewrite system in which the rewrite rules are still length reducing, but use an alphabet that may be larger than the set $A = X^{\pm}$ of group generators. A group is said to have a Cannon algorithm (or generalised Dehn algorithm) if there exists such a rewriting system for which, for $w \in A^*$, w reduces to the empty word if and only if $w =_G 1$. (So the alphabet symbols outside of A may be used in the middle of such a reduction.) Theorem 3.5.5 still applies, so such groups have word problem soluble in linear time.

From now on, for consistency within this book, we shall call a rewrite system of the type we have just defined a generalised Dehn algorithm, although some of the results to which we refer (notably [108, 155]) are stated in their published form using the other, original terminology.

It is proved [108] that finitely generated nilpotent groups and many types of *relatively hyperbolic groups*, including the *geometrically finite hyperbolic groups*, which are also studied and proved automatic by Epstein et al. [84, Chapter 11], have generalised Dehn algorithms.

It is also proved that various types of groups do not have generalised Dehn algorithms, including direct products $F_m \times F_n$ of free groups with $m > 1$ and $n \geq 1$, Baumslag–Solitar groups $BS(m, n)$ with $|m| \neq |n|$, braid groups on three or more strands, Thompson's group F, and the fundamental groups of various types of closed 3-manifolds.

3.5.7 Non-deterministic generalised Dehn algorithms and growing context-sensitive languages

A word w may contain the left-hand side of more than one reduction rule. The generalised Dehn algorithms defined by Goodman and Shapiro in [108] are deterministic, in that they specify exactly which rule is to be applied to w in that situation. (Two different ways of doing this are considered.)

It is possible also to define a non-deterministic version of a generalised Dehn algorithm in which the rule to be applied is not specified. There may even be more than one reduction rule having the same left-hand side. However, when the restriction on choice of rules to be applied is removed, even if we forbid two rules to have the same left-hand side, then it is conceivable that a word in the group generators that is not equal to the identity element could reduce to the empty word, in which case the algorithm would be behaving incorrectly. Hence it is not obvious that a group that has a deterministic generalised Dehn algorithm also has a non-deterministic one. However it is true, and is proved by Holt, Rees and Shapiro [155] that the word problem WP(G) of a group is a growing context-sensitive language (see 2.8.1) if and only if G has a non-deterministic generalised Dehn algorithm. So, for example, if G is a nilpotent or a geometrically finite hyperbolic group, then WP(G) is in this language family. It is also proved that, for each of the groups G listed above that do *not* have generalised Dehn algorithms, the language WP(G) is not growing context-sensitive. As we shall see in Chapter 12, WP(G) $\in \mathcal{CS}$ for most of these examples.

3.6 Filling functions

The Dehn function $\phi \colon \mathbb{N}_0 \to \mathbb{N}_0$ that we defined in Section 3.3 is an example of a *filling function* defined on a finite presentation $\langle X \mid R \rangle$ of a group G. There are a number of other functions of this type, which were introduced by Gromov [126] in a more general context, some of which will arise later in the book, so we summarise them briefly here.

Let Δ be a van Kampen diagram for a word $w \in$ WP(G, A), where $A := X^{\pm}$. We defined area(Δ) to be the number of 2-cells in Δ. The *diameter* of Δ is the maximum distance of a vertex of Δ to its base point. The *radius* of Δ is the maximum distance of a vertex of Δ to its boundary. The *filling length* of Δ is the minimal bound on the length of the boundary loop among combinatorial null-homotopies of w down to the base point.

As with the area, we define the filling diameter, radius and length functions on words $w \in$ WP(G, A) to be the minumum value of these functions over all van Kampen diagrams for w, and then define their values on $n \in \mathbb{N}_0$ to be the maximum such value over all $w \in$ WP(G, A) with $|w| \leq n$.

It may be helpful to give an alternative and equivalent description of the filling length of $w \in$ WP(G, A). Consider reductions

$$w \to w_1 \to w_2 \to \ldots \to w_n = \varepsilon,$$

where each w_{i+1} results from w_i either by substituting r_{i+1} for a subword r_i of w_i, where $r_i^{-1} r_{i+1}$ is in the symmetric closure of R, or by inserting or deleting a subword xx^{-1} for some $x \in A$. The filling length of the reduction is the maximum of the word lengths $|w_i|$ and the filling length of w is the minimum among the filling lengths of all such reductions.

As with the area, it is straightforward to show that changing the finite presentation for G replaces any of these functions with a $\simeq$-equivalent function in the sense of 3.3.2. So we can, for example, unambiguously refer to groups with linear filling function.

PART TWO

FINITE STATE AUTOMATA AND GROUPS

4

Rewriting systems

4.1 Rewriting systems in monoids and groups

When working with an infinite finitely generated group $G = \langle X \rangle$, a principal goal is to find a *normal form* for the group elements. More precisely, for each element of G, we would like to find a representative word in A^* (where $A := X^{\pm 1}$) for that element, and we refer to these representative elements as words in normal form. This normal form should ideally satisfy two requirements:

(i) it should be easy to recognise whether a given word is in normal form;
(ii) for a given word $v \in A^*$, it should be possible to find the unique word w in normal form with $v =_G w$.

Of course, we can only hope to achieve these objectives if G has soluble word problem. As we will see in Section 10.1, it has been proved by Novikov and Boone [206, 31, 32] that this is not the case in general. Fortunately, many of the interesting classes of finitely presented groups that arise in practice, particularly those that come from topology or geometry, turn out to have soluble word problem. Furthermore, as we shall see in this chapter and in Chapter 5, many of them have convenient normal forms.

4.1.1 Word-acceptors In this chapter, we consider normal forms arising from rewriting systems, and in Chapters 5 and 6 we study automatic groups, and the important special case of hyperbolic groups. In each of these cases, the words in normal form constitute a regular set, and so there is an fsa with alphabet A, known as a *word-acceptor*, of which the accepted language is the set of normal form words. From this description, one might expect a word-acceptor to accept a unique word for each group element, but it turns out that the following more general definition is more convenient.

If $G = \langle X \rangle$ is a finitely generated group and $A := X^{\pm}$, then an fsa W with input alphabet A is called a *word-acceptor* for G if it accepts at least one word in A^* for each group element. It is called a *unique word-acceptor* for G if it accepts a unique word for each group element.

Of course, according to this definition, the fsa accepting A^* is a (not very useful) word-acceptor. Word-acceptors arising from rewriting systems are always unique word-acceptors.

4.1.2 Growth series of fsa We observe that the *growth series* $\sum_{n=0}^{\infty} a_n t^n$ of an fsa W with

$$a_n = |\{w \in L(W) : |w| \leq n\}|$$

is equal to the power series of a rational function of t with integral coefficients, and there are algorithms to compute the series from W. (One such is described in [87], and there is an implementation in Holt's KBMAG package [146].) In the important special case in which the words $w \in L(W)$ of a unique word-acceptor W for G are all geodesic (i.e. $|w| = |w|_G$), the growth series (see 1.6.5) of G is the same as that of W, so it can be effectively computed.

4.1.3 Further reading There is a large body of literature on the topic of rewriting systems, which has applications to many branches of algebra. For a historical account with an extensive bibliography, we refer the reader to the article by Buchberger [57]. One of the earliest papers describing the use of critical pairs and the completion process in a very general setting is that of Knuth and Bendix [172], after whom this process is usually named. Applications to group theory are described by Gilman [96, 97]. For those interested in the algorithmic aspect, there is an extensive treatment with details of algorithms and their implementation by Sims [240], and a briefer treatment in Chapter 12 of Holt et al. [144]. On account of the existing detailed literature on rewriting systems, we present only an outline account here.

4.1.4 Rewriting systems for groups and monoids It is convenient to study rewriting systems with respect to a finitely presented monoid $M = \text{Mon}\langle A \mid \mathcal{R} \rangle$ (see 1.2.3 for the definition of a monoid presentation). As we saw in Exercise 1.2.4, a finitely presented group $G = \langle X \mid R \rangle$ has a corresponding monoid presentation $M = \text{Mon}\langle A \mid \mathcal{I}_X \cup \mathcal{R} \rangle$, where

$$\mathcal{I}_X = \{ (xx^{-1}, \varepsilon) \mid x \in A \} \quad \text{and} \quad \mathcal{R} = \{ (w, \varepsilon) \mid w \in R \}.$$

A rewriting system for M (over A) is a set $\mathcal{S}$ of ordered pairs (w_1, w_2) with $w_1, w_2 \in A^*$, such that $\text{Mon}\langle A \mid \mathcal{S} \rangle$ is a presentation of M. If M is defined from G as described above, then we also call it a rewriting system for G (over X).

The elements of S are called *rewrite rules*, and w_1 and w_2 are called, respectively, the left- and right-hand sides of the rule.

For $u, v \in A^*$, we write $u \to_S v$ if there exist strings $x, y, w_1, w_2 \in A^*$ such that $u = x w_1 y$, $v = x w_2 y$ and $(w_1, w_2) \in S$; in other words, if v is obtained from u by making a single substitution using a rewrite rule. We omit the subscript S when there is no danger of ambiguity.

A word $u \in A^*$ is said to be *(S-)irreducible* or *(S-)reduced* if there is no word $v \in A^*$ with $u \to v$. Notice that, if S is finite, then it follows from Proposition 2.5.10 that the set of S-irreducible words is regular, and can be recognised by an fsa.

We denote the reflexive, transitive closure of $\to$ by $\to^*$. Hence $u \to^* v$ if and only if, for some $n \geq 0$, there exist $u = u_0, u_1, \ldots, u_n = v \in A^*$ with $u_i \to u_{i+1}$ for $0 \leq i < n$.

The symmetric closure of $\to^*$ is denoted by $\leftrightarrow^*$. Thus $u \leftrightarrow^* v$ if and only if, for some $n \geq 0$, there exist $u = u_0, u_1, \ldots, u_n = v$ with

From our assumption that $w_1 =_M w_2$ for all rewrite rules $(w_1, w_2) \in S$, it follows that $u \leftrightarrow^* v$ implies $u =_M v$. So we can repeatedly replace occurrences of w_1 in a string by w_2 without changing the corresponding element of M.

We see from the definition of a monoid presentation in 1.2.3 that our assumption that $M = \mathrm{Mon}\langle A \mid S \rangle$ means that the relation $\leftrightarrow^*$ is equal to the congruence on A^* generated by $\mathcal{R}$, and so $A^* / \leftrightarrow^*$ is equal to M. In other words, for $u, v \in A^*$, $u =_M v$ if and only if $u \leftrightarrow^* v$.

4.1.5 Reduction orderings It would be desirable if the replacement of w_1 by w_2 with $(w_1, w_2) \in S$ resulted in some kind of simplification of the string, containing w_1.

To make this idea more precise, we suppose that a well-ordering $\leq$ is defined on A^* that has the property that $u \leq v$ implies $uw \leq vw$ and $wu \leq wv$ for all $w \in A^*$. Such an ordering is known as a *reduction ordering*. Then we assume that $w_1 > w_2$ for all $(w_1, w_2) \in S$. The assumptions on the ordering ensure that $u \to^* v$ implies $u \geq v$. So application of the rewrite rules to strings results in the words decreasing with respect to this ordering. In this situation, we say that S is *compatible* with the reduction ordering $\leq$.

There are several different examples of reduction orderings that have been used in this context, and the most appropriate one depends to some extent on the nature of the group or monoid in question. Probably the most commonly used and simplest to define are the *shortlex* (also called *lenlex*) orderings, which we defined in 1.1.4.

A more complicated class of reduction orderings, appropriate for working with polycyclic groups, is the class of *wreath product* orderings; these include the *recursive path* orderings. See Sims' book [240, Section 2.1] for more details.

4.1.6 Exercise Show that ε is the least element in A^* for any reduction ordering.

4.1.7 Confluence and completeness A rewriting system S is called *Noetherian* (or terminating) if there is no infinite chain of strings u_i ($i > 0$) with $u_i \to u_{i+1}$ for all i. This implies that for any $u \in A^*$ there exists an irreducible $v \in A^*$ with $u \to^* v$. Since reduction orderings are by definition well-orderings, any rewriting system that is compatible with a reduction ordering is automatically Noetherian.

We call S *confluent* if, whenever $u, v_1, v_2 \in A^*$ with $u \to^* v_1$ and $u \to^* v_2$, there exists $w \in A^*$ with $v_1 \to^* w$ and $v_2 \to^* w$, and we call S *complete* (or *convergent*) if it is Noetherian and confluent.

Furthermore, S is called *locally confluent* if, whenever $u, v_1, v_2 \in A^*$ with $u \to v_1$ and $u \to v_2$, there exists $w \in A^*$ with $v_1 \to^* w$ and $v_2 \to^* w$. It is proved in [144, Lemmas 12.15, 12.16] that a Noetherian locally confluent rewriting system is complete, in which case each equivalence class under $\leftrightarrow^*$ contains a unique S-irreducible element.

These results depend on the assumption that no two rules in S have the same left-hand side, so we assume that to be the case for the remainder of this chapter. If the rewriting system is derived from a finite group presentation as described above then, to achieve this condition, we may need to adjust the presentation slightly.

For testing for local confluence (and hence for completeness), the following result [144, Lemma 12.17] is important.

4.1.8 Proposition *The rewriting system S is locally confluent if and only if the following conditions are satisfied for all pairs of rules $(u_1, t_1), (u_2, t_2) \in S$.*

(i) *If $u_1 = rs$ and $u_2 = st$ with $r, s, t \in A^*$ and $s \neq \varepsilon$, then there exists $w \in A^*$ with $t_1 t \to^* w$ and $r t_2 \to^* w$.*

(ii) *If $u_1 = rst$ and $u_2 = s$ with $r, s, t \in A^*$ and $s \neq \varepsilon$, then there exists $w \in A^*$ with $t_1 \to^* w$ and $r t_2 t \to^* w$.*

4.1.9 Knuth–Bendix completion A pair of rules satisfying either of the two conditions of the proposition is called a *critical pair*. If one of these conditions fail (so S is not confluent), then we can find two distinct irreducible strings w_1 and w_2 that are equivalent under $\leftrightarrow^*$. We can resolve this instance of incompleteness by adjoining either (w_1, w_2) or (w_2, w_1) to S as a new rule. We can then continue with the check for local confluence. The procedure of examining critical pairs and adjoining new rules to S if necessary is known as the *Knuth–Bendix completion process*.

We assume from now on that S is finite unless explicitly stated otherwise. We also assume that S is compatible with a suitable reduction ordering $\leq$. So when we need to form a new rule out of a relation $w_1 = w_2$, we make the left-hand side of the rule the larger of the two words w_1, w_2 with respect to $\leq$.

In the case that our rewriting system is derived from a monoid presentation $M = \text{Mon}\langle A \mid \mathcal{R} \rangle$, as described in 4.1.4 above, we initialise S to $\mathcal{R}$, but with an element $(w_1, w_2) \in \mathcal{R}$ replaced by (w_2, w_1) if $w_2 > w_1$. (As we observed earlier, we may also need to do a little extra massaging if some of the rewrite rules turn out to have the same left-hand sides.)

Since the Knuth–Bendix completion process results in adjoining new rules (u, v) with $u =_M v$, $\text{Mon}\langle A \mid S \rangle$ remains a presentation of M after adjoining these new rules, and we still have $M = A^* / \leftrightarrow^*$. (In practice, some of the rewrite rules may become redundant during the completion process, and we may choose to remove them from S, but $\text{Mon}\langle A \mid S \rangle$ still remains a presentation of M, and we shall not concern ourselves with these details here.)

It can happen that the completion process eventually terminates with a finite complete S. In this situation, the S-irreducible words define a normal form for M, which is the language of an fsa. Furthermore, as we explain in more detail in Section 4.2, the rewrite rules in S can be used to reduce an arbitrary word in A^* to its equivalent normal form, and so we have achieved the objectives (i) and (ii) specified at the beginning of this chapter.

It is not hard to show (see, for example, [144, Corollary 12.21]) that the completion process always completes if the monoid M is finite.

4.1.10 Example Let $G := \langle x, y \mid x^3 = 1, y^2 = 1, (xy)^2 = 1 \rangle$ be the dihedral group of order 6. Putting $X := x^{-1}$, $Y := y^{-1}$, we have $A = \{x, X, y, Y\}$ and, using the order $x < X < y < Y$ on A, we can start with

$$S := \{(xX, \varepsilon), (Xx, \varepsilon), (yY, \varepsilon), (Yy, \varepsilon), (xx, X), (Y, y), (yX, xy)\}.$$

(When each generator has an inverse generator and we are using a shortlex ordering, then we can always rewrite our rules so that the length of the left-hand side is greater than the right-hand side by at most 2, and it is generally efficient to do this.)

One critical pair is $(xX, \varepsilon), (xx, X)$ with the overlap x. This leads to the word xxX reducing in two different ways to x and XX, and to resolve this we adjoin the rule (XX, x) to S. Another critical pair is $(Yy, \varepsilon), (Y, y)$ with overlap Y, and Yy reduces to both ε and y^2, so we add (yy, ε) to S. It can be seen that the rules (yY, ε) and (Yy, ε) are now both redundant, and can be removed from S. The overlap X between (yX, xy) and (Xx, ε) leads to the new rule (xyx, y) and then the overlap x between this and (Xx, ε) leads to the rule (yx, Xy). The rule (xyx, y) is now redundant and can be removed. There are several more overlaps to check, but no further new rules arise, so

$$S = \{(xX, \varepsilon), (Xx, \varepsilon), (xx, X), (XX, x), (Y, y), (yy, \varepsilon), (yX, xy), (yx, Xy)\}$$

is now complete. The reader will now have realised that the completion process is tedious to carry out by hand, but it is straightforward for a computer!

4.1.11 Example of an infinite monoid Even when M is infinite, the Knuth–Bendix process may complete. For example, let $G := \langle x, y \mid yx = xy \rangle$ be the free abelian group on two generators. The corresponding monoid is

$$M = \text{Mon}\langle x, y, X, Y \mid (xX, \varepsilon), (Xx, \varepsilon), (yY, \varepsilon), (Yy, \varepsilon), (yx, xy) \rangle.$$

If we use the ordering $x < X < y < Y$ on A with the associated shortlex ordering on A^*, then the Knuth–Bendix completion process rapidly terminates with the finite complete rewriting system

$$S = \{(xX, \varepsilon), (Xx, \varepsilon), (yY, \varepsilon), (Yy, \varepsilon), (yx, xy), (yX, Xy), (Yx, xY), (YX, XY)\}.$$

4.1.12 Example of an infinite rewriting system As a general rule, however, when the monoid is infinite, the completion process does not terminate, but it does, in principle, generate an infinite complete rewriting system. In a few examples, such an infinite system can have a very transparent structure, which can make it almost as useful as a finite complete set of rewrite rules.

For example, with the monoid defined from the free abelian group in the previous example, if instead we use the ordering $x < y < X < Y$ of A, then the completion process generates the infinite complete set of rewrite rules

$$\{(xX, \varepsilon), (Xx, \varepsilon), (yY, \varepsilon), (Yy, \varepsilon), (yx, xy), (Xy, yX), (Yx, xY),$$
$$(YX, XY), (xy^n X, y^n), (yX^n Y, X^n) : \text{for all } n > 0\}.$$

Since the set of left-hand sides of these rules still forms a regular set, the set of S-reduced words is again regular, and we can still achieve our objectives (i) and (ii).

4.2 The use of fsa in the reduction process

In computer implementations of the Knuth–Bendix completion process, there are two principal problems to be solved. The first is to locate the critical pairs, which involves finding overlaps between left-hand sides of rewrite rules. The second is the reduction of words to equivalent irreducible words using the rewrite rules. The first of these concerns string matching algorithms, about which there is plentiful literature.

In practice, it appears that the bulk of the computer time is taken by word reduction. The reduction process can be rendered significantly more efficient by the use of a certain dfsa M_S, which we describe briefly in this section.

4.2.1 Index automata We remove redundant rules from S, and may assume that no left-hand side of any rule of S is a proper substring of any other left-hand side.

The alphabet of M_S is A. The state set $\Sigma = \Sigma_S$ is in one-one correspondence with the set of all prefixes of all left-hand sides of rules in S, with the subset F of accepting states corresponding to the proper prefixes; we denote by σ_u the accepting state corresponding to a proper prefix u of a left-hand side, but by $\sigma_{(u,v)}$ (rather than simply σ_u) the non-accepting state corresponding to the word u that is the left-hand side of the rule (u, v). The start state corresponds to the empty word ε and so is denoted by σ_ε. For $\sigma_u \in F$ and $a \in A$, σ_u^a is defined to be the state corresponding to the longest suffix of ua that corresponds to a state in Σ. For any $a \in A$, the target σ^a of a non-accepting state $\sigma \in \Sigma \setminus F$ is undefined (that is, leads to failure). So the non-accepting states are dead states.

It is straightforward to show by induction on $|w|$ that, when M_S reads a word $w \in A^*$, then σ_ε^w is either undefined or equal to the state corresponding to the longest suffix v of w that defines an element of Σ.

Note that, if $w = w_1 w_2 w_3$, where w_2 is the first occurrence of the left-hand side of a rule (w_2, v_2) in w, then after reading the prefix $w_1 w_2$ of w, M_S is in the dead state $\sigma_{(w_2, v_2)}$. So $L(M_S)$ is equal to the set of S-irreducible words.

To use S to reduce words, we read w into M_S and if, for some prefix $w_1 w_2$ of w, we find that $\sigma_\varepsilon^{w_1 w_2} = \sigma_{(w_2, v_2)}$ is a dead state, then we replace w_2 by v_2 in w, and restart the process. (Or, for greater efficiency, we can remember the 'state history' of M_S reading $w_1 w_2$, and then restart reading the modified word w into M_S after w_1 and before v_2.) This type of fsa is sometimes called an *index automaton* because the dead states provide an index into the rewrite rules of S.

4.2.2 Example Consider the complete rewriting system

$$S = \{(xX, \varepsilon), (Xx, \varepsilon), (xx, X), (XX, x), (Y, y), (yy, \varepsilon), (yX, xy), (yx, Xy)\}$$

for the dihedral group of order 6 with $A = \{x, X, y, Y\}$ that we obtained in Example 4.1.10. Let us number the states Σ of M_S by integers (positive integers $1, \ldots, 4$ for states in F, negative integers $1, \ldots, 8$ otherwise) as follows:

$$
\begin{aligned}
1 &= \sigma_\varepsilon, & 2 &= \sigma_x, & 3 &= \sigma_X, & 4 &= \sigma_y, \\
-1 &= \sigma_{(xX,\varepsilon)}, & -2 &= \sigma_{(Xx,\varepsilon)}, & -3 &= \sigma_{(xx,X)}, & -4 &= \sigma_{(XX,x)}, \\
-5 &= \sigma_{(Y,y)}, & -6 &= \sigma_{(yy,\varepsilon)}, & -7 &= \sigma_{(yX,xy)}, & -8 &= \sigma_{(yx,Xy)}.
\end{aligned}
$$

The transitions are as follows:

	x	X	y	Y
1	2	3	4	-5
2	-3	-1	4	-5
3	-2	-4	4	-5
4	-8	-7	-6	-5

For example, let us use this fsa to reduce the word $w = Yxxyx$. On reading the first letter Y of w, M moves from state 1 to $1^Y = -5$, which represents the left-hand side of the rule (Y, y), so Y is replaced by y in w. Restarting, we have $1^y = 4$, $4^x = -8$, so we have read the left-hand side of (yx, Xy) and we replace yx by Xy in w. Restarting again, $1^X = 3$, $3^y = 4$, $4^x = -8$ and we replace yx by Xy once more. So now $w = XXyyx$, and we have $1^X = 3$, $3^X = -4$, and XX is replaced by x. Then $1^x = 2$, $2^y = 4$, $4^y = -6$, and yy is replaced by ε, giving $w = xx$. Finally, xx is replaced by X to yield the reduced word X for w.

4.2.3 Remarks If, at some stage during the completion process, we have not reduced S by removing redundant rules, then M_S may fail to notice a substring w_2 of w which is a left-hand side of a rule but is also a proper substring of the left-hand side of another rule. Hence it may accept, and fail to reduce, some S-reducible words. This is not disastrous provided that S is reduced at regular intervals during the completion process.

We shall not go into the details of the algorithm for constructing M_S from S here, and refer the reader to Section 3.5 of Sims' book [240] for the code for doing this. During the Knuth–Bendix completion process, the set S of rules is continually changing, and so M_S needs to be reconstructed very frequently. It is therefore vital to use an efficient implementation of the construction of M_S. The fsa M_S can also be used to speed up the search for critical pairs; this usage is also discussed in detail in Section 3.5 of [240].

5

Automatic groups

In this chapter, we consider the class of *automatic groups* and some related classes, including the important subclass of *(word-)hyperbolic groups*, which are studied in more detail in Chapter 6. In the class of automatic groups, the two requirements for a normal form specified at the beginning of Chapter 4 are met. There is a normal form for group elements with the property that the set of words in normal form constitutes a regular set, and for which an arbitrary word can be reduced to normal form in quadratic time. The most comprehensive reference on automatic groups is the six-author book [84], by D.B.A. Epstein and co-authors and we refer the reader there for any necessary further information and references. For an alternative, more rapid, treatment, see the account by Baumslag, Gersten, Shapiro and Short [19, Part II].

A procedure is described by Epstein, Holt and Rees [86], and by Holt [145] and Holt et al. [144, Chapter 16], that starts with a finite presentation of a group and, if successful, outputs a set of fsa that between them define an automatic structure for that group; one fsa defines the normal form and the others enable the definition (and implementation) of an algorithm that reduces any input word to normal form. The most recent implementation of the procedure forms part of Holt's KBMAG package [146]. There is an alternative implementation by Alun Williams [252] which performs better on many examples,

In view of the multitude of existing descriptions, we present only a summary of this procedure here, and we concentrate on the fact that it makes heavy use of logical operations applied to regular languages, such as those described in Proposition 2.5.10, 2.10.5 and 2.10.8. It involves the use of the Knuth–Bendix completion process described in Chapter 4, but can be successful even in cases where this process fails to terminate.

Throughout this chapter, $G = \langle X \rangle$ denotes a group with finite generating set X, and $A := X^{\pm}$ is the alphabet of the fsa involved. But much of the theory can be generalised to allow arbitrary sets A that generate G as a monoid.

125

It may be helpful at this stage to recall some notation from Chapter 1. For a word $w \in A^*$ representing $g \in G$, $|w|$ denotes its length, and $|w|_G = |g|$ denotes the length of a shortest word representing g. The word w is geodesic if $|w| = |w|_G$, and the set of all geodesic words is denoted by $\mathcal{G}(G, X)$. For $i \geq 0$, $w(i)$ denotes the prefix of w of length i for $i < |w|$ and $w(i) = w$ for $i \geq |w|$.

The Cayley graph of G with respect to X is denoted by Γ or $\Gamma(G, X)$, and is a geodesic metric space with all edges having length 1. A word $w \in A^*$ can represent an element of G or a path in $\Gamma(G, X)$; this path is based at 1 by default, and $_g w$ denotes the path based at the vertex $g \in \Gamma(G, X)$. So geodesic words label geodesic paths in $\Gamma(G, X)$. For points v, w of $\Gamma(G, X)$, each of which may be either a vertex or a point on an edge, $d(v, w) = d_{\Gamma(G,X)}(v, w)$ denotes the distance between points v and w of $\Gamma(G, X)$. For the vertices g, h of $\Gamma(G, X)$, we have $d(g, h) = |g^{-1}h|$.

A language for G is a subset of A^* containing at least one representative of each group element, and a normal form is a language containing a unique such representative. Important examples of normal forms are languages that contain the least representative of each group element under a suitable well-ordering of A^*, such as a shortlex ordering with respect to some ordering of A.

5.1 Definition of automatic groups

The definition of the class of automatic groups is due to Thurston; it evolved in the mid-1980s following the appearance of a paper by Jim Cannon on groups acting discretely and cocompactly on a hyperbolic space [60]. For such a group the Cayley graph embeds naturally in the hyperbolic space, and the cocompactness of the action ensures that the embedding is a quasi-isometry (see Section 1.7); hence group geodesics embed as quasigeodesics of the space, and some properties of the space are inherited by the Cayley graph.

In particular, since a quasigeodesic in hyperbolic space must stay within a neighbourhood of the geodesic between its end points, and hence also of any other quasigeodesic, we see that geodesics in the Cayley graph 'fellow travel' (we provide a precise definition of this property below).

Cannon also observed (and we shall see later that this is actually a consequence of the fellow traveller property for geodesics) that the geodesics in such a group lie in finitely many equivalence classes, known as 'cone types'. Thurston noticed that both of these properties of the geodesics in these groups could be described in terms of fsa, and his definition of automatic groups was an attempt to abstract these properties and generalise them to a larger class of

Figure 5.1 Word-differences.

groups. That class turns out to include a large proportion of the fundamental groups of compact 3-manifolds.

The definition that we give at this stage is not Thurston's original definition, but an equivalent more geometric version, as we shall see in Theorem 5.2.3.

5.1.1 Definition Let $G = \langle X \rangle$ with X finite and put $A := X^{\pm}$. Then G is said to be *automatic* (with respect to X), if there exists a regular set L of words over A and a constant K, such that:

  (i) L contains at least one word representing each element of G;

  (ii) for all $v, w \in L$ for which $d(v, w) \leq 1$, we have $d((v(i), w(i)) \leq K$ for all
    $i \geq 0$.

The property described in Condition (*ii*) is an instance of the (synchronous) *fellow traveller* property, and two words v, w with $d(v(i), w(i)) \leq K$ for all $i \geq 0$ are said to *K-fellow travel*: see Figure 5.1. The elements $v(i)^{-1}w(i)$ of G, which correspond to the paths of length at most K that join $v(i)$ to $w(i)$ in $\Gamma(G, X)$, are known as *word-differences*. One such word-difference is labelled u in Figure 5.1.

Let W be an fsa that accepts L. We call L the *language*, W the *word-acceptor*, and K the *fellow traveller constant* of an *automatic structure* for G. Note that W satisfies the criteria defining a word-acceptor in 4.1.1. If L contains a unique representative of each group element, we say that it is a part of an *automatic structure with uniqueness*.

We shall now introduce variations of this concept, including an asynchronous version, which leads in turn to other classes of groups that are closely related to automatic groups.

5.1.2 Fellow traveller properties Two paths α and β in a metric space, which are parametrised by arc length, are said to *(synchronously) K-fellow travel* if $d(\alpha(t), \beta(t)) \leq K$ for all $t \geq 0$ (where $\alpha(t) = \alpha(\text{len}(\alpha))$ for all $t \geq \text{len}(\alpha)$).

More generally, we say that the paths α and β *asynchronously K-fellow travel* if there exist non-decreasing surjective functions $f, g \colon \mathbb{R}_{\geq 0} \to \mathbb{R}_{\geq 0}$ such

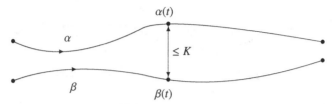

Figure 5.2 Fellow travelling paths

that $d(\alpha(f(t)), \beta(g(t))) \leq K$ for all $t \geq 0$. Of course paths that synchronously K-fellow travel also K-fellow travel asynchronously.

5.1.3 Combable and bicombable groups The group $G = \langle X \rangle$ is said to be *(asynchronously) combable* (with respect to X) if there is a language $L \subseteq A^*$ for G and $K \geq 0$ such that, if $v, w \in L$ satisfy $d(v, w) = |v^{-1}w|_G \leq 1$, then the paths v and w (asynchronously) K-fellow travel. In this situation L is called a *combing* with *fellow traveller constant K*.

We call G *(asynchronously) bicombable* (with respect to X) if there is a combing L with fellow traveller constant K and the additional property that, if $v, w \in L$ with $wv^{-1} =_G a \in A$, then av and w (asynchronously) K-fellow travel (or, equivalently, the paths $_1 w$ and $_a v$ fellow travel in $\Gamma(G, X)$).

We warn the reader that some authors define combings slightly differently.

5.1.4 Automatic and biautomatic groups A group G with finite generating set X is said to be *(asynchronously) automatic* (with respect to X) if it has a regular (asynchronous) combing $L \subseteq A^*$. Such a combing is also called an *(asynchronous) automatic structure* for G.

Similarly, we call G *(asynchronously) biautomatic* if it admits a regular (asynchronous) bicombing. Note that (asynchronous) biautomaticity implies (asynchronous) automaticity and hence (asynchronous) combability.

If L is a normal form for G, then we speak of an *(asynchronous) (bi)combing/ automatic structure with uniqueness*.

5.1.5 Exercise Verify that for F_n in its standard presentation the set of reduced words constitutes the language of a biautomatic structure with fellow traveller constant 1, while for $\mathbb{Z}^n$ in its standard presentation the set of words of the form $x_1^{i_1} x_2^{i_2} \cdots x_n^{i_n}$ (that is, the shortlex minimal words) is the language of a biautomatic structure with fellow traveller constant 2.

5.1.6 Examples The groups studied by Cannon in [60] are all biautomatic, as is proved in [84, Theorem 3.2.1]. It is proved by Bridson and Gilman [42]

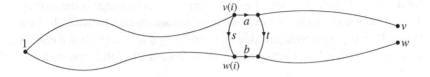

Figure 5.3 Transition in the word-difference automaton

that the fundamental groups of all compact geometrisable 3-manifolds are asynchronously combable; for compact 3-manifolds based on six of the eight geometries described by Thurston, these groups are automatic [84, Theorem 12.4.7]. Many more examples of automatic groups are described later in this chapter. Examples of combable groups that are not automatic are provided by Bridson [41], and also in unpublished work of Bestvina and Brady.

5.1.7 Exercise Show that, if L is the language of an automatic structure for G and $L^{-1} = L$, then L is the language of a biautomatic structure for G.

5.1.8 Open problem Do there exist automatic groups that are not biautomatic? See also 5.5.8.

5.2 Properties of automatic groups

5.2.1 Word-difference automaton Thurston recognised that the fellow traveller condition could also be expressed in terms of a 2-variable fsa; see Section 2.10. Suppose that G has an automatic structure with fellow traveller constant K.

The *word-difference automaton* $\mathrm{WD}(\Gamma_K)$ is defined to be the 2-variable fsa over A with state set $\Gamma_K := \{g \in G : |g| \le K\}$ (that is the vertices in the ball of radius K about the origin in the Cayley graph), start state 1, and transitions $g \xrightarrow{(a,b)} h$ if and only if $g, h \in \Gamma_K$, $a, b \in A^\mathrm{p} = A \cup \{\$\}$, and $h =_G a^{-1}gb$; see Figure 5.3. (Within the group product $a^{-1}gb$ we interpret the padding symbol $\$$ as a representative of the identity element.) We define accept states variously depending on our needs, and we denote the word-difference automaton with accept states $U \subseteq \Gamma_K$ by $\mathrm{WD}(\Gamma_K, U)$ (or $\mathrm{WD}(\Gamma_K, h)$ when $U = \{h\}$).

5.2.2 Multiplier automata Suppose that G is automatic with word-acceptor W. The language

$$\{(v, w)^\mathrm{p} : v, w \in L(W)\} \cap L(\mathrm{WD}(\Gamma_K, a))$$

for $a \in A \cup \{1\}$ is regular, and the dfsa accepting this language is denoted by M_a and known as a *multiplier automaton* for the automatic structure. The idea is that M_a recognises multiplication on the right by a in $A \cup \{1\}$. It is common to use the notation M_ε in preference to M_1, and we shall do that from now on.

5.2.3 Theorem *The group G is automatic if and only if there is a regular language $L = L(W)$ over A that maps onto G and, for each $a \in A \cup \{\varepsilon\}$, there exists a 2-variable* fsa M_a *such that $L(M_a) = \{(v, w)^\mathsf{p} : v, w \in L,\ va =_G w\}$.*

Proof (In brief: see [84, Section 2.3] for more details.) First suppose that G is automatic, and define the multiplier automata M_a for $a \in A \cup \{\varepsilon\}$ as above. Then, by construction, if $(v, w)^\mathsf{p} \in L(M_a)$, then $v, w \in L$ and $v^{-1}w =_G a$, so $va =_G w$. Conversely, if $v, w \in L$ and $va =_G w$ then, by the definition of an automatic group, v and w K-fellow travel, so $(v, w)^\mathsf{p} \in L(\mathrm{WD}(\Gamma_K, a))$ and hence $(v, w)^\mathsf{p} \in L(M_a)$.

Conversely, suppose that M_a exists for each $a \in A \cup \{\varepsilon\}$ with

$$L(M_a) = \{(v, w)^\mathsf{p} : v, w \in L,\ va =_G w\}.$$

Let l be the maximum of the numbers of states in the fsa M_a and put $K = 2l+1$.

Suppose that $v, w \in L$ and $d(v, w) \le 1$. Then $va =_G w$ for some $a \in A \cup \{\varepsilon\}$, so $(v, w)^\mathsf{p} \in L(M_a)$. Now, for any i with $0 \le i \le \max(|v|, |w|)$, there exist words $v', w' \in A^*$ with $|v'|, |w'| \le l$ (where v' is empty if $v(i)$ ends in \$, and similarly for w'), such that $(v(i)v', w(i)w')^\mathsf{p} \in L(M_a)$. Hence

$$|v(i)^{-1}w(i)|_G = |v'aw'^{-1}|_G \le K.$$

In other words, $d(v(i), w(i)) \le K$ and so G is automatic. □

From now on we shall use whichever definition of automaticity is more convenient. Since the language, the fellow traveller constant and the word-difference automata of an automatic structure can all be recovered from the word acceptor, together with the set of multipliers $M_a : a \in A \cup \{\varepsilon\}$, that set of fsa is frequently what is used to specify an automatic structure. Indeed it is this set of fsa that is constructed by the automatic group procedures referred to at the start of this chapter.

5.2.4 Biautomatic and asynchronously automatic structures Biautomaticity can be equivalently defined using automata in a similar way: a biautomatic structure consists of a collection $\{W, M_a, {}_aM\ (a \in A \cup \{\varepsilon\})\}$ of fsa, where $\{W, M_a\ (a \in A \cup \{\varepsilon\})\}$ is an automatic structure, and

$$L({}_aM) = \{(v, w)^\mathsf{p} : v, w \in L(W),\ av =_G w\}.$$

There is also an equivalent definition of asynchronously automatic groups involving asynchronous multiplier fsa M_a ($a \in A$) (see 2.10.2). For more details see [84, Chapter 7].

5.2.5 Exercise If G has automatic structure $\{L, M_a\}$ then, for any $g \in G$, the language $\{(v, w)^p : v, w \in L, \ vg =_G w\}$ is regular. (Use composite fsa as defined in 2.10.8.)

Similarly, if G has biautomatic structure $\{L, M_a, {}_aM\}$ then, for any $g \in G$, the language $\{(v, w)^p : v, w \in L, \ gv =_G w\}$ is regular.

5.2.6 Unique word-acceptors Note that we have not insisted in our definition that L is a normal form for G; this is not necessary in general, because the fsa M_ε can be used to decide whether two words in L represent the same element of G. In fact L is a normal form if and only if $L(M_\varepsilon) = \{(w, w): w \in L\}$. In any case, the following result, whose proof is a straightforward application of the logical operations on fsa, shows that we can easily replace a given (bi)automatic structure by one with uniqueness.

5.2.7 Theorem *If $\{L, M_a\}$ is a (bi)automatic structure for G, then there is a (bi)automatic structure $\{L', M'_a\}$ for G with uniqueness in which $L' \subseteq L$.*

Proof Let $\leq$ be any shortlex ordering on A^*. Then

$$\{(v, w)^p : v, w \in A^*, \ v < w\}$$

is a regular language by Proposition 2.10.4. Hence, by using the logical operations on fsa discussed in Section 2.10, we see that

$$L' = \{w \in L : \not\exists v \in A^* \ v < w \text{ and } (v, w)^p \in L(M_\varepsilon)\}$$

is a normal form for G. Since $L' \leq L$, it has the same fellow travelling properties as L and so L is a (bi)automatic structure for G. $\qquad\square$

This result is also true for asynchronous automatic structures, but the proof is less straightforward. See [84, Theorem 7.3.2].

5.2.8 Exercise Show that an infinite torsion group G cannot be automatic. *Hint*: Exploit the Pumping Lemma (2.5.17) [84, Example 2.5.12].

5.2.9 Independence of generating set It is proved in [84, Theorem 2.4.1] that automaticity is a property of G and does not depend on the (finite) generating set X; that is, if G is automatic with respect to one generating set, then it is automatic with respect to any other. We can therefore unambiguously say that

G is an *automatic group*. The same applies to biautomaticity and asynchronous automaticity (see [84, Theorem 7.33] for the latter).

Here is a summary of the proof for automaticity. Suppose that G has the automatic structure $L \subseteq A^*$ with respect to X, and let K be the fellow travelling constant of the structure. Let Y be another finite generating set of G and put $B := Y \cup Y^{-1}$. Let l and m be respectively the maximum lengths of elements of A and B when represented in G as words in B^* and A^*.

For each $x \in X$, we fix $w_x \in B^*$ with $|w_x| \leq l$ and $w_x =_G x$. We could then define $\sigma \colon A^* \to B^*$ as the monoid homomorphism with $\sigma(x^{\pm 1}) = w_x^{\pm 1}$, and $\sigma(L)$ would be a regular subset of B^* mapping onto G.

The problem is that elements of A^* of the same length can have images of different lengths and so the (synchronous) fellow travelling property of L does not carry over to $\sigma(L)$. We can remedy this as follows.

Fix some $y \in Y$. Let $w = x_1^{\epsilon_1} x_2^{\epsilon_2} \cdots x_n^{\epsilon_n} \in L$. Then we define $\sigma(w) := w_{x_1}^{\epsilon_1} \rho_1 w_{x_2}^{\epsilon_2} \rho_2 \cdots w_{x_n}^{\epsilon_n} \rho_n$, where ρ_i is a word consisting of repetitions of yy^{-1} such that, for each i, $|w_{x_1}^{\epsilon_1} \rho_1 w_{x_2}^{\epsilon_2} \rho_2 \cdots w_{x_i}^{\epsilon_i} \rho_i|$ is equal to li or $li + 1$.

Then $\sigma(L)$ is still regular, and σ preserves length up to the additive constant 1. If $v, w \in L$ with $|\sigma(v)^{-1}\sigma(w)|$ equal in G to an element of $B \cup \{\varepsilon\}$, then the group element $v^{-1}w$ has length at most m with respect to X. So v and w fellow travel with constant Km (cf. Exercise 5.2.5) and it is not hard to see that $\sigma(v)$ and $\sigma(w)$ fellow travel with constant $Klm + l + 1$. So $\sigma(L)$ is an automatic structure for G with respect to Y.

5.2.10 The word problem The following result from [84, Theorem 2.3.10], which implies an efficient solution of the word problem in automatic groups, is an immediate consequence of Definition 5.1.1 and the properties of 2-variable fsa. Note that the solution requires knowledge of a representative $w_\varepsilon \in L$ of the identity element; in all examples that the authors have encountered we can choose $w_\varepsilon = \varepsilon$, but this is not assured by the definition.

5.2.11 Theorem *Suppose that G has an automatic structure $\{L, M_a, w_\varepsilon\}$, and let C be the maximum number of states in any of the fsa M_a. Then for any $w \in A^*$, there is a word $w' \in L$ with $w =_G w'$ and $|w'| \leq C|w| + |w_\varepsilon|$. Furthermore, we can find such a word w' in time $O(|w|^2)$ and in space $O(|w|)$.*

Proof We prove the first statement by induction on $|w|$. Our assumption that w_ε is given ensures that it is true for $|w| = 0$. Hence, if $w = va$ with $a \in A$, then we can assume that there exists $v' \in L$ with $|v'| \leq C|v| + |w_\varepsilon| = C(|w| - 1) + |w_\varepsilon|$. Let $w' \in L$ with $w' =_G w$. Then $(v', w')^\rho \in L(M_a)$ and hence, by Proposition 2.10.6, we can choose w' with $|w'| \leq |v'| + C \leq C|w| + |w_\varepsilon|$, as claimed. Furthermore, by Theorem 2.10.7, we can find w' from v' in time $O(|v'|) = O(|v|)$

and hence also in space $O(|v|)$. So we can find w' from w and w_ε in time $O(|w|^2)$. Since the computation of w' from w consists of at most $|w|$ applications of Theorem 2.10.7, each of which can be done in space $O(|w|)$, so can the complete computation. □

5.2.12 Corollary *Let G be an automatic group. Then the word problem of G is soluble deterministically in quadratic time and in linear space.*

In particular, if G is automatic, then the word problem $\mathrm{WP}(G)$ of G lies in the class $\mathcal{DCS}$ of deterministic context-sensitive languages; this result is due to Shapiro [235].

By using the fellow traveller property, we can easily prove the following result [84, Theorem 2.3.12]. It was observed first by Thurston.

5.2.13 Theorem *An automatic group is finitely presented and has a quadratic Dehn function.*

Proof Let $\{L, M_a\}$ be an automatic structure for G. Fix $w_\varepsilon \in L$ with $w_\varepsilon =_G 1$, and let $w = a_1 a_2 \cdots a_n \in A^*$ with $w =_G 1$. By Theorem 5.2.11, there exist words $w_i \in L$ for $0 \le i \le n$, with $w_i =_G w(i)$ and $|w_i| \le Ci + |w_\varepsilon|$ (we can take $w_0 = w_\varepsilon$), where C is the maximum number of states in the fsa M_a. By the fellow traveller property, there is a constant K (we could take $K = 2C + 1$) for which $d(w_i(j), w_{i+1}(j)) \le K$ for $0 \le i < n$ and all $j \ge 0$.

Define R to be the subset of A^* consisting of w_ε together with all reduced words $w \in A^*$ with $w =_G 1$ and $|w| \le 2K + 2$. Then, by inserting paths labelled w_i from 1 to $w(i)$ and paths of length at most K from $w_i(j)$ to $w_{i+1}(j)$ for all relevant i and j, we can construct a van Kampen diagram for the relator w with all interior regions labelled by words in R; see Figure 5.4. So $G = \langle X \mid R \rangle$ is a finite presentation of G. Furthermore, since $|w_i| \le Cn + |w_0|$ for $1 \le i \le n$, this diagram has at most $(Cn + |w_0|)n$ interior regions and hence G has quadratic Dehn function. □

It is proved in [84, Theorem 7.3.4] that the Dehn function of an asynchronously automatic group is bounded above by an exponential function, and it can be shown similarly that its word problem is soluble in exponential time.

5.2.14 Biautomatic groups and the conjugacy problem In the same way as in Exercise 5.2.5, we can show that, if G has biautomatic structure $\{W, M_a, {}_aM\}$ then, for any $g, h \in G$, the language $\{(v, w)^p : v, w \in L(W), vg =_G hw\}$ is regular.

By intersecting this language with the regular language $\{(v, v) : v \in A^*\}$ we find that, for any $g, h \in G$, $\{(v, v) : v \in L(W), vgv^{-1} =_G h\}$ and hence

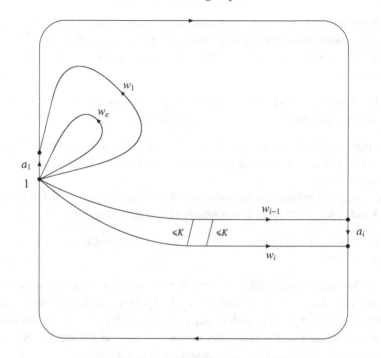

Figure 5.4 Van Kampen diagram for an automatic group

$\{v \in L(W) : vgv^{-1} =_G h\}$ is also regular. But since $L(W)$ maps onto G and we can determine whether a regular language is empty, we can decide whether there is an element $v \in G$ with $vgv^{-1} =_G h$. In other words:

5.2.15 Theorem   *If G is biautomatic, then the conjugacy problem is soluble in G.*

However it seems likely that the number of states in the fsa recognising the language $\{(v, v) : v \in L(W), vgv^{-1} =_G h\}$ is exponential in $|g| + |h|$, in which case any algorithm to construct that fsa must have no better than exponential complexity, and similarly any algorithm for the solution to the conjugacy problem based on this approach.

The above proof is due to Gersten and Short [92, Theorem 3]. An alternative proof is provided by Short [236], and shows that in any bicombable group two conjugate elements of length at most n are conjugate by an element of length bounded by an exponential function in n.

In the more restricted class of word-hyperbolic groups, it is proved by Epstein and Holt [85] that the conjugacy problem is soluble in linear time.

5.2.16 Open problem Is the conjugacy problem soluble in every automatic group?

5.3 Shortlex and geodesic structures

A *shortlex automatic structure* for a group G with respect to a finite generating set X is an automatic structure with uniqueness in which the word-acceptor W accepts precisely the minimal words for the group elements under some shortlex ordering of A^*, as defined in 1.1.4.

More generally, a *geodesic automatic structure* for G with respect to X is an automatic structure (not necessarily with uniqueness) in which the language L consists of geodesic words. The group G is called *shortlex* or *geodesically automatic* with respect to X if it has a structure of one of these types.

Most of the implementations of algorithms to construct automatic structures work only for shortlex automatic structures, so they are important in practice. We present a brief description of these algorithms in the next section. Unfortunately, for a given A, a group can be automatic, but not shortlex or even geodesically automatic with respect to any ordering of A; see [84, Section 3.5] for an example.

5.3.1 Exercise [84, Example 4.4.1] Let

$$G = \langle x, y, t \mid t^2 = 1, xy = yx, txt = y \rangle.$$

So $G = \mathbb{Z} \wr C_2$ is the standard wreath product of an infinite cyclic group with a group of order 2.

Show that G is shortlex automatic using the ordering $x < x^{-1} < y < y^{-1} < t$ of A but not using the opposite ordering $t < y^{-1} < y < x^{-1} < x$.

5.3.2 Open problem Is every automatic group G geodesically, or even shortlex, automatic with respect to some ordering of some monoid generating set of G?

5.4 The construction of shortlex automatic structures

We present only an outline of the procedure here, without any discussion of implementation details and performance. See [86], [145] or [144, Chapter 16] for a more detailed description. The process can be conveniently split into three steps.

5.4.1 Step 1: Using Knuth–Bendix to construct a word-difference automaton We start by running the Knuth–Bendix completion process (see 4.1.9) on the group presentation using the specified shortlex ordering $<$ on A^*. For each rewrite rule (w_2, w_1) that is found, we compute the associated word-differences $w_2(i)^{-1}v_2(i)$, for $0 \le i \le |w_2|$, and store the set D of all such word-differences coming from all rewrite rules.

The elements of D are represented as words in A^* that are reduced with respect to the rewrite rules found so far. (So it is not guaranteed that words in D represent distinct group elements, which can occasionally cause technical difficulties, but we shall not discuss that further here.)

Let $\mathcal{R}$ be the set of all rewrite rules that would be generated by the Knuth–Bendix process if it were run for a sufficiently long time. Then $\mathcal{R}$ is infinite in general, and the process itself does not always halt. So it is interrupted (either automatically using input parameters, or by manual intervention) as soon as the set D appears to have stabilised and new rewrite rules are not producing any new word-differences. If D fails to stabilise in this way, then the complete algorithm fails.

If G really is shortlex automatic, then the set of word-differences coming from the rules in $\mathcal{R}$ is finite and, if D does appear to stabilise, then we assume provisionally that D is equal to this set. In practice, this assumption may turn out to be wrong but, as we shall see, it is often possible to detect the error and extend the set D in Step 2.

If either the Knuth–Bendix process halts with a confluent rewriting system or it is interrupted because D has (apparently) stabilised, then we construct a 2-variable fsa $\mathrm{WD}(D)$ with state set D, in which ε is the start state and the single accepting state, and the transitions are $s \xrightarrow{(a,b)} t$ whenever $d_1, d_2 \in D$, $a, b \in A \cup \{\varepsilon\}$ and $a^{-1}d_1b$ reduces to d_2. Our intention is that $\mathrm{WD}(D)$ should be equal to the accessible portion of $\mathrm{WD}(\Gamma_k)$, but there may well be missing word-differences and/or missing transitions between word-differences, and/or more than one representative in D of the same element of G; so D and $\mathrm{WD}(D)$ are often modified later in the algorithm.

During this first step of the algorithm, we have used the Knuth–Bendix reduction rules to reduce words. Of the data computed in Step 1, we only keep the fsa $\mathrm{WD}(D)$, but we can use this fsa in subsequent steps to carry out word reduction, as described in the proof of Theorem 5.2.11.

5.4.2 Step 2: Constructing the word-acceptor and multiplier automata We now make repeated attempts to construct the word-acceptor W and the multiplier automata M_a, followed by partial correctness tests on the results. If the automata fail these tests, then we try again. If and when they pass them, we proceed to Step 3 for the complete correctness verification.

The proof of Proposition 2.5.10 shows how to construct fsa accepting $L \cup L'$, $L \cap L'$, $A^* \setminus L$, LL' from fsa accepting the regular languages L and L'. In 2.10.5 we explained how to perform quantification by constructing the fsa $M_\exists$ from the 2-variable fsa M. We use various combinations of these operations during this step of the algorithm. After each such operation, we carry out the minimisation process described in Theorem 2.5.4 to construct an fsa with the same language and a minimal number of states.

Let $L \subset A^*$ be the set of shortlex minimal representatives of the elements of G. So, if G is shortlex automatic, then L is the language of its word-acceptor.

We start by computing an fsa W with language

$$L' := \{w \in A^* : \nexists v \in A^* \text{ with } (w, v)^\rho \in L(\mathrm{WD}(D)) \text{ and } v < w\}.$$

This is motivated by the following lemma.

5.4.3 Lemma *We have $L \subseteq L'$ and, if D contains all word-differences from all rules in $\mathcal{R}$, then $L = L'$.*

Proof Let $w \in L$. Then, by definition of L, there is no $v \in A^*$ with $v =_G w$ and $v < w$. Since $(w, v)^\rho \in L(\mathrm{WD}(D))$ implies $v =_G w$, we have $w \in L'$, and hence $L \subseteq L'$.

Suppose that $w \in A^* \setminus L$. Then w has a minimal subword v with $v \notin L$, and there is a unique $v' \in L$ with $v =_G v'$. Let w' be the word obtained by substituting v' for the occurrence of v in w. So $v' < v$ and $w' < w$. The minimality of v ensures that $(v, v') \in \mathcal{R}$ and hence, if D contains all word-differences from $\mathcal{R}$, then $(v, v') \in L(\mathrm{WD}(D))$ and then (omitting some technical details) $(w, w') \in L(\mathrm{WD}(D))$, so $w \notin L'$. Hence $L = L'$ under this assumption. $\square$

We now construct the fsa M_a from $\mathrm{WD}(D)$ and W exactly as described in their definition in 5.2.2, but using $\mathrm{WD}(D, a)$ rather than $\mathrm{WD}(\Gamma_K, a)$.

Since L' should contain a unique representative of each group element, we should have $L(M_\varepsilon) = \{(w, w) : w \in L'\}$. It may happen that, during the construction of the M_a, we find distinct words v, w with $(v, w)^\rho \in L(M_\varepsilon)$. This indicates that D is incomplete, and we interrupt the process, adjoin the word-differences from (v, w) to D, reconstruct $\mathrm{WD}(D)$ and restart Step 2.

If, on the other hand, we complete the construction of the M_a, then we perform a partial correctness verification by constructing the fsa E_a accepting

$$\{w \in L' : \exists v \in L' \text{ with } (w, v) \in L(M_a)\}$$

for each $a \in A$. If W and the M_a are correct, then $L(E_a) = L'$ for all a. If not, then we can find some word w and $a \in A$ such that there is no $v \in L'$ with $(w, v) \in L(M_a)$. The missing v is the reduction of the word wa, so we use the existing (incorrect) automatic structure and the algorithm described in the proof of Theorem 5.2.11 to reduce wa to a word v, and then adjoin the word-differences arising from the pair (wa, v) to D, reconstruct WD(D) and (attempt to) reconstruct the M_a.

Note that, even if D contains all word-differences from rules in $\mathcal{R}$, there may be other word-differences from the equations (wa, v), so it is inevitable that we need to repeat the construction of the M_a in some examples.

5.4.4 Step 3: Verifying correctness Let $R' = R \cup \{aa^{-1} : a \in A\}$. (Hence Mon$\langle A \mid \mathcal{R} \rangle$ with $\mathcal{R} = \{(r, 1) : r \in R'\}$ presents G as a monoid.) In Step 3, for each $r = a_1 a_2 \cdots a_n \in R'$, with $a_i \in A$, we construct the composite M_r of the fsa M_{a_i}, as defined in 2.10.8 which, by Exercise 5.2.5, should have the language

$$\{(v, w)^\rho : v, w \in L', \ vr =_G w\} = \{(v, w)^\rho : v, w \in L', \ v =_G w\},$$

which should be equal to $L(M_\varepsilon)$.

If $L(M_r) \neq L(M_\varepsilon)$ for some $r \in R'$, then the verification process fails. (However, in practice, we have not yet encountered an example in which we have got as far as Step 3, and then verification has failed.) It is proved by Epstein, Holt and Rees [86, Theorem 2.5] that, if $L(M_r) = L(M_\varepsilon)$ for all $r \in R'$, then the automatic structure (W, M_a) is correct.

After checking that $L(M_{aa^{-1}}) = L(M_\varepsilon)$ for each $a \in A$, rather than constructing M_r for each $r \in R$, we can replace r by the relation $r_1 = r_2$ with $\|r_1\| - |r_2\| \leq 1$, and verify that $L(M_{r_1}) = L(M_{r_2})$. Since the complexity of constructing M_v for $v \in A^*$ is potentially exponential in $|v|$, this is more efficient than checking that $L(M_r) = L(M_\varepsilon)$.

5.4.5 Examples The implementations of the automatic groups programs have been used to prove infiniteness of a number of finitely presented groups for which this property had been previously unknown. One such example is the group defined by the presentation

$$G = \langle x, y, z \mid [x, [x, y]] = z, \ [y, [y, z]] = x, \ [z, [z, x]] = y \rangle.$$

This was proposed by Heineken in the 1980s as a new candidate for a finite group defined by a presentation with three generators and three relations. (For $n \geq 4$, no finite groups are known that can be defined by presentations with n generators and n relators, and are not $(n - 1)$-generated. It is a consequence of the Golod–Shafarevich Theorem [107] that there are no finite nilpotent groups with this property for $n \geq 4$. The first examples with $n = 3$ were found by Mennicke in 1959 [190], and several families of examples have been found since then.)

Preliminary investigation of the group G revealed it to have a finite quotient of order $2^{24}.60$ (and hence, if it were finite, its large size would make this property hard to verify), but the long-standing approach of proving infiniteness by finding a finite index subgroup with infinite abelianisation could not be made to work. However, using KBMAG [146], it turned out not to be difficult to prove that the group is automatic and infinite; in fact it is hyperbolic.

Two other such examples are described by Havas and Holt [133]. The more difficult of these was the group

$$\langle x, y \mid x^3, y^5, (xy)^7, [x,y]^2 \rangle.$$

This was first proved to be automatic and infinite by Alun Williams, using his MAF package [252], and later using KBMAG [146]. The minimised word-acceptor has 47 613 states, and the multiplier has 277 371 states. More straight-forward (and earlier) examples of applications of KBMAG of this type are described by Edjvet and Juhàsz [80].

KBMAG [146] has also been used to compute the growth series (defined in 1.6.5) of a number of automatic groups including, for example, some 3-manifold groups. It was pointed out in 4.1.2 that, if the group has a unique word-acceptor W whose language consists of geodesic words, then this series is a rational function that can be computed directly from W.

5.5 Examples of automatic groups

5.5.1 Hyperbolic and geometrically finite hyperbolic groups
Groups that are *word-hyperbolic*, as defined by Gromov [125], are biautomatic on the set of all geodesic words, and also shortlex biautomatic with respect to any ordering of any finite generating set. These are studied in more detail later in this book, in Section 5.8 and in Chapter 6. The multi-author paper [5] and the book by Ghys and de la Harpe [94] are the standard references.

This class includes all groups that act discretely and cocompactly on hyperbolic space, as considered by Cannon [60], and various small cancellation groups – see below.

Geometrically finite hyperbolic groups are proved biautomatic in [84, Theorem 11.4.3]. It is proved by Neumann and Shapiro [204, Theorem 5.6] that they have a geodesic biautomatic structure. This result is generalised by Antolín and Ciobanu [8, Theorem 1.5] to all groups that are hyperbolic relative to a set of subgroups with geodesic biautomatic structures. This applies, in particular, to geometrically finite hyperbolic groups; these are hyperbolic relative to a set of virtually abelian subgroups. Rebecchi proved in his PhD thesis [216] that a group that is hyperbolic relative to a subgroup having a prefix-closed biautomatic structure is biautomatic.

It is proved in [8, Theorem 7.7] that a group that is hyperbolic relative to a collection of groups with shortlex biautomatic structures is itself shortlex biautomatic. This applies, in particular, if the parabolic subgroups are all abelian [84, Theorem 4.3.1], which is the case for many examples of geometrically finite hyperbolic groups.

5.5.2 Small cancellation groups It is proved in [183, Theorem 4.5, Chapter V] and in Strebel's article [243] that groups satisfying either of the metric small cancellation conditions $C'(1/6)$ or $C'(1/4) + T(4)$ have a Dehn algorithm which, as we prove in Chapter 6, is equivalent to their being hyperbolic. Results of this type are often referred to as *Greendlinger's Lemma*, due to earlier proofs of related results by Greendlinger.

Groups satisfying $C(7)$, or else $T(p)$ and $T(q)$ for (p, q) equal to any one of $(3, 7)$, $(4, 5)$, $(5, 4)$, are proved to be hyperbolic by Gersten and Short [91, Corollary 4.1]. The same authors prove in [92, Section 8] that groups satisfying $C(6)$, or $C(4)$ and $T(4)$, or $C(3)$ and $T(6)$ are biautomatic.

5.5.3 Virtually abelian, nilpotent and soluble groups Euclidean groups (that is, groups with torsion-free finitely generated abelian subgroups of finite index or, equivalently, finitely generated virtually abelian groups) are biautomatic [84, Section 4.2], but non-abelian torsion-free nilpotent groups are not automatic [84, Section 8.2]. Thurston conjectured that soluble groups cannot be automatic unless they are virtually abelian. That conjecture has been verified by Harkins [131] for torsion-free polycyclic groups.

5.5.4 Braid groups, Artin groups, Garside groups and Coxeter groups Braid groups are automatic [84, Chapter 9]. That result is generalised by Charney [65] who establishes biautomaticity of all Artin groups of finite type, and then generalised further by Dehornoy and Paris [78] who show that Garside groups are biautomatic.

Artin groups of extra-large type (or 3-generator of large type) and right-angled Artin groups are also biautomatic; this is due to Peifer [213], Brady and McCammond [38], and Hermiller and Meier [139]. Furthermore, those of large type are proved shortlex automatic over their natural generating set by

Holt and Rees [152]. Brink and Howlett [49] proved that Coxeter groups are shortlex automatic using the natural generating set.

It is proved by Caprace and Mühlherr [64, Corollary 1.6] that, if the Coxeter group G has no affine standard subgroup (see 1.10.2) of rank greater than 2, then G is biautomatic. A similar condition for biautomaticity was proved independently by Bahls [11]. The proofs of both of these results make use of an earlier result of Niblo and Reeves [205] that a Coxeter group with only finitely many conjugacy classes of subgroups isomorphic to triangle groups is biautomatic. Moussong proves in [197] that a Coxeter group is word-hyperbolic (see Chapter 6) if and only if this condition holds, together with the condition that there are no pairs of infinite commuting standard subgroups.

5.5.5 Open problem Are all Artin groups of large type and all Coxeter groups biautomatic?

5.5.6 Mapping class groups Mapping class groups are proved automatic by Mosher [195] and biautomatic by Hamenstädt [130].

5.5.7 Baumslag–Solitar groups The Baumslag–Solitar groups

$$BS(m, n) = \langle x, y \mid y^{-1} x^m y = x^n \rangle$$

for nonzero integers m, n are examples of asynchronously automatic groups, and they are automatic if and only if $|m| = |n|$; see [84, Example 7.4.1].

5.5.8 Automatic but not biautomatic? Although a number of candidates have been proposed at various times, there is as yet no proven example of a group that is automatic but not biautomatic. There are some examples of groups that are automatic but have not yet been proved to be biautomatic; for example, it has not yet been proved that all Coxeter groups are biautomatic, although this has been proved under various additional hypotheses (see 5.5.4). So far as we are aware, the question of automaticity is still open for the groups $SL_2(O)$, where O is the ring of integers of a real quadratic number field $\mathbb{Q}(\sqrt{d})$; these groups, and others, were proved by Gersten and Short not even to be subgroups of biautomatic groups [93, section 6]. The groups $Aut(F_n)$ and $Out(F_n)$ for $n \geq 3$ were proved not to be biautomatic by Bridson and Vogtmann [44]. Subsequently these groups were proposed as candidates to be automatic, non-biautomatic examples [45], but they were eventually proved to be non-automatic with exponential Dehn functions [46].

Even in groups that are biautomatic, not every automatic structure need be a biautomatic structure. For example, let $G = \mathbb{Z} \wr C_2$ be the group of Exercise 5.3.1. Since G is virtually abelian, it is biautomatic. The language

$$L(W) = \{x^m y^n t^i \mid m, n \in \mathbb{Z}, i \in \{0, 1\}\},$$

is the shortlex automatic structure provided by the ordering

$$x < x^{-1} < y < y^{-1} < t$$

of A, which is referred to in the exercise. However, this is not a biautomatic structure, since, for any $n \geq 0$, the words $v = x^n y^n t$ and $w = x^n y^n$ of $L(W)$ satisfy $tv =_G w$, but $d(tv(n), w(n)) = |x^{-n} t x^n|_G = |x^{-n} y^n t|_G = 2n + 1$.

5.5.9 Combable but not automatic The existence of non-automatic combable groups was mentioned in 5.1.6. But it is easier to find examples of asynchronously combable groups that are not asynchronously automatic; such examples (all nilpotent or soluble) are described by Bridson and Gilman [42] and by Gilman, Holt and Rees [100]. The Heisenberg group (the free 2-generator nilpotent group of class 2) is an example. In these examples, the combings are indexed or real-time languages. By [84, Theorem 8.2.8] none of these groups can be asynchronously automatic; it is also proved in [42] that a nilpotent group with a context-free asynchronous combing with uniqueness must be virtually abelian.

5.6 Closure properties

- If H has finite index in G, then H is automatic if and only if G is automatic. Further if G is biautomatic, then H is biautomatic [84, Theorem 4.1.4], but the converse is not known.
- If N is a finite normal subgroup of G, then G is (bi)automatic if and only if G/N is (bi)automatic. (This can safely be left as an exercise to the reader.)
- If G and H are (bi)automatic, then so is $G \times H$ [84, Theorem 4.1.1]. But it is not known whether direct factors of automatic groups must be automatic. However, direct factors and central quotients of biautomatic groups are biautomatic. This was proved by Mosher [196].
- If G and H are (bi)automatic, then $G * H$ is (bi)automatic, and more generally so is $G *_A H$, for any finite subgroup A embedding in both G and H [84, Theorem 12.1.4]. So also is any HNN-extension $G*_{A,t}$ of G with finite associated subgroup A [84, Theorem 12.1.9].

  Conversely, if $G *_A H$ is automatic and A is finite, then G and H are automatic [19, Theorem H] or [84, Theorem 12.1.8]. The same proofs work for biautomatic groups, but the corresponding result for asynchronously automatic groups has been proved only when $A = 1$ [19, Theorem F].

Various other results on the automaticity of amalgamated products are proved in [19]. For example, any amalgamated product of two finitely generated abelian groups is automatic; the amalgamated product of two hyperbolic groups is automatic under some extra technical hypotheses, which include all cases when the amalgamated subgroup is cyclic; the product of two finitely generated free groups with a finitely generated amalgamated subgroup is asynchronously automatic.

- A graph product of finitely many (asynchronously) (bi)automatic groups is also (asynchronously) (bi)automatic. This was proved by Hermiller and Meier [139].
- Groups having a central finitely generated abelian subgroup with hyperbolic quotient group are biautomatic. This was proved by Neumann and Reeves [202].
- Rational subgroups (that is, subgroups corresponding to rational subsets of L) of automatic (or biautomatic, synchronously automatic, hyperbolic) groups are automatic (or biautomatic, synchronously automatic, hyperbolic). This was proved by Gersten and Short [93]; see Theorem 8.1.3.
- Centres and element centralisers in biautomatic groups are rational subgroups, and hence biautomatic, as proved by Gersten and Short [93]; see Proposition 8.1.6.

5.7 The falsification by fellow traveller property

The property FFTP to be discussed in this section was introduced by Neumann and Shapiro [204]. Although there is no proven connection with automaticity in general, there seems to be a strong correlation among known examples. As we shall see below, unlike automaticity, FFTP depends on the generating set X of G. But we know of no examples of groups that are known to be automatic but also not to satisfy FFTP with respect to any generating set. Conversely, all examples that we know of groups with FFTP for some X are geodesically automatic with respect to that X.

5.7.1 The property FFTP The group $G = \langle X \rangle$ has the *falsification by fellow traveller property (FFTP)* with respect to X if there is a constant $K \geq 0$ such that, for any non-geodesic word $v \in A^*$, there exists a word $w \in A^*$ with $v =_G w$ and $|w| < |v|$ such that v and w K-fellow travel.

In fact the definition given in [204] stipulates the apparently weaker condition that v and w *asynchronously* K-fellow travel.

5.7.2 Exercise Show that the two definitions are equivalent. (Or see Lemma 1 in Elder's paper [81].)

Several of the properties proved for automatic groups can be proved in a similar fashion for groups with FFTP, as we now demonstrate.

5.7.3 Theorem *Suppose that G has FFTP with respect to X and $w \in A^*$ is a non-geodesic word. Then we can find a word $v \in A^*$ with $v =_G w$ and $|v| < |w|$ in time $O(|w|)$. The word problem for G can be solved in quadratic time.*

Proof Suppose that G has FFTP with fellow travelling constant K, let Γ_K be the set of all elements of G with X-length at most K, and let M be an fsa with language equal to the intersection of $L(\text{WD}(\Gamma_K))$ (as defined in 5.2.1) and the regular set $\{(u_1, u_2)^p : u_1, u_2 \in A^*, |u_1| < |u_2|\}$. Then, since w is not geodesic, there exists a word v with $(v, w)^p \in L(M)$ and, by Theorem 2.10.7, we can find such a word in time $O(|w|)$. It follows as in the proof of Theorem 5.2.11 that the word problem for G is soluble in quadratic time. $\square$

5.7.4 Theorem *A group with FFTP is finitely presented and has a quadratic Dehn function.*

Proof The proof is similar to that of Theorem 5.2.13. Let $w = a_1 a_2 \cdots a_n \in A^*$ with $w =_G 1$. We first show how to define geodesic words w_i with $w_i =_G w(i)$ such that w_i and w_{i+1} $2K$-fellow travel for $0 \le i < n$, and then the result follows as in the proof of Theorem 5.2.13. Let $w_0 = \varepsilon$ and suppose by induction that w_j satisfy the required condition for $j < i$. Since w_{i-1} is geodesic, we have $|w_{i-1}| + 1 \ge |w_{i-1} a_i|_G \ge |w_{i-1}| - 1$, so there are just three possibilities for $|w_{i-1} a_i|_G$.

(i) If $|w_{i-1} a_i|_G = |w_{i-1}| + 1$, then put $w_i = w_{i-1} a_i$.

(ii) If $|w_{i-1} a_i|_G = |w_{i-1}|$, then there is a word w_i for which $w_i =_G w_{i-1} a_i$ and $|w_i| < |w_{i-1} a_i|$ such that w_i and $w_{i-1} a_i$ K-fellow travel. Then $|w_i| = |w_{i-1}|$, so w_i is geodesic.

(iii) If $|w_{i-1} a_i|_G = |w_{i-1}| - 1$, then there is a word v for which $v =_G w_{i-1} a_i$ and $|v| < |w_{i-1} a_i|$ such that v and $w_{i-1} a_i$ K-fellow travel. If $|v| = |w_{i-1}| - 1$, then we put $w_i = v$. Otherwise $|v| = |w_{i-1}|$, and there is a word w_i of length $|w_{i-1}| - 1$ with $w_i =_G v =_G w_{i-1} a_i =_G w(i)$ that K-fellow travels with v. So w_i is a geodesic word that $2K$-fellow travels with w_{i-1}, and we are done. $\square$

5.7.5 Theorem *Let G have FFTP with respect to X. Then the set $\mathcal{G}(G, X)$ of geodesic words over A is regular.*

Proof Suppose that G has FFTP with fellow travelling constant K, and let Γ_K be the set of all elements of G with X-length at most K. Then

$$\mathcal{G}(G, X) = \{w \in A^* : \nexists v \in A^* \text{ with } |v| < |w| \text{ and } (v, w) \in L(\text{WD}(\Gamma_K))\}$$

(cf. proof of Theorem 5.2.7). □

But the converse of Theorem 5.7.5 does not hold. Elder [82] proved that the group $\mathbb{Z} \wr C_2$ in Exercise 5.3.1, with the presentation $\langle x, t \mid t^2, [x, txt] \rangle$ obtained by eliminating the generator $y = txt$, does not satisfy FFTP but has regular geodesics with respect to $X = \{x, t\}$. However, it is a virtually abelian group, and it is proved by Neumann and Shapiro [204, Proposition 4.4] that all such groups have FFTP with respect to suitable generating sets. In fact it has FFTP with respect to the original generating set $\{x, y, t\}$.

This example also shows that, unlike automaticity, FFTP depends on the generating set. Even before the Neumann and Shapiro paper, Jim Cannon had observed that this same example does not have regular geodesics with respect to $\{a, c, d, t\}$, where $a = x$, $c = a^2$ and $d = ab = atat$, so regularity of geodesics also depends on the generating set of the group. We return to these topics in Chapter 7.

5.8 Strongly geodesically automatic groups

5.8.1 Definition [84, Page 56] A group $G = \langle X \rangle$ is said to be *strongly geodesically automatic* with respect to X if the set $\mathcal{G}(G, X)$ of all geodesic words forms the language of an automatic structure for G.

The results proved by Cannon [60] show that groups that act discretely and cocompactly on a hyperbolic space have this property and, more generally, it is not difficult to show that (word-)hyperbolic groups, which are the subject of the next chapter, are strongly geodesically automatic. More surprisingly, the converse result, that strongly geodesically automatic groups are hyperbolic is proved by Papasoglu [210].

The following result follows immediately from the proof of Theorem 5.2.7.

5.8.2 Proposition *A strongly geodesically automatic group G is shortlex automatic with respect to any ordering of A.*

5.8.3 Proposition *The group G is strongly geodesically automatic with* *respect to X if and only if there is a constant K such that v and w K-fellow travel for any $v, w \in \mathcal{G}(G, X)$ with $|v^{-1}w|_G \leq 1$.*

Proof By Theorem 5.2.3, the necessity of the fellow travelling condition follows immediately and, to prove the sufficiency, it is enough to show that the fellow travelling condition implies that $\mathcal{G}(G, X)$ is regular. We shall show that the condition implies FFTP, and then apply Theorem 5.7.5.

Let $w \in A^* \setminus \mathcal{G}(G, X)$, and let $w = uv$, where u is the shortest prefix of w with $u \notin \mathcal{G}(G, X)$. Then $u = tx$ with $t \in \mathcal{G}(G, X)$ and $x \in A$. Let $s \in \mathcal{G}(G, X)$ with $u =_G s$. Then, by hypothesis, s and t K-fellow travel and hence, since $\|u| - |s\| \leq 2$, $w = uv$ and sv K-fellow travel (provided that $K \geq 2$), and $|sv| < |w|$, so G has FFTP and the result follows.   □

5.8.4 Proposition *The language $\mathcal{G}(G, X)$ of geodesic words in a strongly geodesically automatic structure of G with respect to X forms a biautomatic structure.*

Proof We have to show that, if $v, w \in \mathcal{G}(G, X)$ with $wv^{-1} =_G a \in A$, then w and v K-fellow travel for some constant K. But $v, w \in \mathcal{G}(G, X)$ implies $v^{-1}, w^{-1} \in \mathcal{G}(G, X)$, and $w^{-1}a =_G v^{-1}$, so w^{-1} and v^{-1} K-fellow travel by the previous result. Hence so do v and w.   □

5.8.5 Example Let $G \cong \mathbb{Z}^n$ be free abelian on the finite generating set $X = \{x_1, \ldots, x_n\}$. Then, for any $m \in \mathbb{N}$, $x^m y^m$ and $y^m x^m$ are geodesic words representing the same group element that do not K-fellow travel for $K < 2m$. So G is not strongly geodesically automatic with respect to X.

We continue our investigation of strongly geodesically automatic groups in Section 6.3 of the next chapter, where we link this concept with the hyperbolicity of the group G.

5.9 Generalisations of automaticity

There have been a number of attempts to generalise the class of automatic groups with the aim of including more groups. We have already discussed asynchronously automatic groups, and (asynchronously) combable groups. It is shown by Bridson and Gilman [42] that the fundamental groups of all compact 3-manifolds that satisfy Thurston's Geometrisation Conjecture have asynchronous combings in the class $\mathcal{I}$ of indexed languages, which was introduced in 2.2.21. This includes the manifolds in the classes Nil and Sol, for which the fundamental groups are not automatic.

Remark: As a result of the combined efforts of Perelman and others, the Geometrisation Conjecture has now been proved; see [250].

The class of *parallel poly-pushdown groups* is defined by Baumslag, Shapiro and Short [20]. The definition is similar to the automata-theoretical characterisation of asynchronously automatic groups mentioned in 5.2.4 and described in detail in [84, Chapter 7]. For this class the regularity conditions for both the word-acceptor and multiplier languages are replaced by the requirement that they lie in the class $\mathcal{DPCF}$ of deterministic poly-context-free languages (or, equivalently, languages that are the intersection of finitely deterministic context-free languages; see 2.2.22).

This class of groups includes the fundamental groups of all 3-manifolds that obey Thurston's Geometrisation Conjecture. It also includes nilpotent groups of arbitrary class and polynomial Dehn function. It is closed under wreath products, and so contains many groups that are not finitely presented. The word problem is soluble but there are no useful upper bounds on its complexity. It includes the class of asynchronously automatic groups, and the best known upper bound for the word problem there is exponential.

5.9.1 Cayley graph automaticity Kharlampovich, Khoussainov and Miasnikov [166] introduce the notion of *Cayley graph automatic groups*. These include the Baumslag–Solitar groups $BS(1, n)$, nilpotent groups of class 2, and some higher class nilpotent groups, such as upper unitriangular integral matrices. The word problem is still soluble in quadratic time and there are many closure properties. Perhaps the main difference from standard automatic groups is that the alphabet over which the word-acceptor language is defined may contain symbols that do not correspond to group elements.

Miasnikov and Šunić [191] give examples of groups that are Cayley graph automatic but not Cayley graph biautomatic.

This class has been generalized further by Elder and Taback [83] to the class of *C-graph automatic groups*, in which the language of the word-acceptor may belong to a language class other than $\mathcal{Reg}$. This further extends the range of examples, and includes some groups that are not finitely generated. In general, the solubility of the word problem in quadratic time is no longer guaranteed, but for some language classes, such as the quasi-realtime counter-graph automatic groups, it is soluble in polynomial time.

5.9.2 Stackable and autostackable groups A generalisation of automaticity in a different direction is introduced by Brittenham, Hermiller and Holt [50] and [51], where the classes of stackable and autostackable groups are defined.

Let G be a group with inverse closed generating set A. Let $\Gamma = \Gamma(G, A)$ be the Cayley graph of G, and suppose that E, P are the sets of directed edges and directed paths of Γ. The group G is defined to be *stackable* with respect to A if we can associate with Γ a maximal tree $\mathcal{T}$ in the undirected graph defined by Γ, and a *bounded flow function* $\Phi : E \to P$, such that the following properties hold.

(F1) For each directed edge $e \in E$ the directed path $\Phi(e)$ has the same initial and terminal vertices as e.

(F2) If the edge e is within $\mathcal{T}$ then $\Phi(e) = e$.

(F3) Let $<_\Phi$ be the transitive closure of the relation on directed edges of the tree that is defined by $e' < e$ whenever e' is on $\Phi(e)$. Then $<_\Phi$ is a strict partial ordering that is well-founded (i.e. it has no infinite descending chains).

(F4) There is a constant k that bounds the length of the paths $\Phi(e)$ with $e \in E$.

The conditions (F1)–(F3) are the defining properties of a flow function, while (F4) renders it bounded.

Suppose that G is stackable with respect to A, as above. For $\pi \in P$, define $\lambda(\pi)$ to be the word over A^* that labels the path π. Now define L to be the set of labels of non-backtracking paths in $\mathcal{T}$ that start at the base point of Γ, and note that L is a prefix closed normal form for G. For $g \in G$, we define y_g to be the unique representative of g in L. Now we can define a function $\phi \colon L \times A \to A^*$ by the rule

$$\phi(y_g, a) = \lambda(\Phi(e_{g,a})),$$

where $e_{g,a}$ is the edge from the vertex g of Γ to the vertex ga.

Let graph(ϕ) be the set

$$\text{graph}(\phi) = \{(y_g, a, \phi(y_g, a)) : g \in G, a \in A\}.$$

The stackable group G is defined to be *autostackable* if graph(ϕ) is the language of a synchronous 3-variable fsa, following the definition of 2.10.1. In the following, we shall say 'graph(ϕ) is regular' when we mean the above.

Note that we can extend the flow function $\Phi \colon E \to P$ to a map from P to P by defining the image of the path $e_1 \cdots e_n$ to be the path $\Phi(e_1) \cdots \Phi(e_n)$. Iterating this function starting from any path $e_1 \cdots e_n$, the resulting path must lie within the tree T after finitely many steps.

Examples of autostackable groups include groups with asynchronously automatic structures with uniqueness in which the accepted language is prefix closed [51, Theorem 4.1], groups with finite complete rewrite systems [51, Corollary 5.4], and the fundamental groups of all closed 3-manifolds [51, Corollary 1.5] and [53]. (Note that it is a long-standing unresolved conjecture that all automatic groups have a prefix closed automatic structure with uniqueness, and so it has not been proved that all automatic groups are autostackable.)

The class of autostackable groups is closed under extensions, finite index super-groups, and graph products, and hence it includes all virtually polycyclic groups. Moreover, in contrast to automatic groups and groups with finite complete rewriting systems, all of which are of homological type FP_∞, this class includes groups that are not of type FP_3 [52]. It also includes groups with non-elementary primitive recursive Dehn function, namely the Baumslag–Gersten group $\langle a, b \mid a^{a^b} = a^2 \rangle$, and there are stackable groups with insoluble word problem [138].

5.9.3 Characterisation We define a *prefix rewrite system* for a group G with respect to a finite inversed-closed generating set A to be a set R of rewrite rules over A that may only be applied to words containing a left-hand side of a rule as a prefix, rather than as an arbitrary subword; then R is said to be *bounded* if for some k the two sides of any rule differ in suffixes of length at most k.

It is proved in [51, Theorem 5.3] that a group G is

 (i) stackable if and only if it admits a bounded complete prefix rewriting system R with respect to some finite inverse-closed generating set A;
 (ii) autostackable if and only if it admits such a rewrite system R that is synchronously regular.

5.9.4 Solving the word problem in stackable groups We can solve the word problem in a stackable group G provided that graph(ϕ) is recursive. Suppose that a freely reduced word w is input with $w \notin L$ (where, as above, L is defined to be equal to the set of labels of non-backtracking paths in $\mathcal{T}$ that start at the base point of Γ). Let y be the maximal prefix of w that is in L, and let a be the next letter of w. Let w' be the word formed from w by replacing a by $\phi(y, a)$ and then freely reducing. An iteration of this process reduces w to a word in L, which is the normal form representative of the group element represented by w. That the process terminates follows from the well-foundedness of the ordering $<_\Phi$ on edges.

6

Hyperbolic groups

The class of hyperbolic groups was introduced and studied in Gromov's paper [125]. In that paper a not necessarily finitely generated group is called *hyperbolic* if it is hyperbolic as a metric space, in the sense that the space has a hyperbolic inner product, which we define in 6.2.1 below. The group is called *word-hyperbolic* if it is finitely generated and hyperbolic with respect to the word metric. Since we are only concerned with finitely generated groups in this book, we shall from now on refer to such groups simply as *hyperbolic groups*.

The fact that there is a large variety of apparently different conditions on a finitely generated group that turn out to be equivalent to hyperbolicity (we present a list of several such conditions in Section 6.6) is itself a strong indication of the fundamental position that these groups occupy in geometric group theory. We have already encountered two of these conditions: groups having a Dehn presentation (or algorithm) in Section 3.5, and strongly geodesically automatic groups in Section 5.8.

Gromov's paper is generally agreed to be difficult to read, but there are several accessible accounts of the basic properties of hyperbolic groups, including those by Alonso et al. [5], Ghys and de la Harpe [94], Bridson and Haefliger [39, Part II, Section Γ] and Neumann and Shapiro [203].

6.1 Hyperbolicity conditions

We begin by comparing various notions of hyperbolicity. The definitions that we consider here apply to an arbitrary geodesic metric space, as defined in Section 1.6, but we are mainly interested in the case when the space is the Cayley graph of a finitely generated group.

Let (Γ, d) be a geodesic metric space. A *geodesic triangle xyz* in Γ consists

150

of three points x, y, z together with geodesic paths $[xy]$, $[yz]$ and $[zx]$. We can define hyperbolicity of Γ in terms of 'thinness' properties of geodesic triangles.

6.1.1 Slim triangles For $\delta > 0$ ($\delta \in \mathbb{R}$), the triangle xyz is called *δ-slim* if, for any point p on one of the sides, there is a point q in the union of the other two sides with $d(p, q) \leq \delta$. In other words, each side of the triangle is contained in the δ-neighbourhood of the union of the other two sides.

6.1.2 Thin triangles In order to define thinness of a geodesic triangle xyz, we first define

$$d_x := \frac{d(x, y) + d(x, z) - d(y, z)}{2}$$
$$d_y := \frac{d(y, x) + d(y, z) - d(x, z)}{2}$$
$$d_z := \frac{d(z, x) + d(z, y) - d(x, y)}{2}.$$

Then $d_x + d_y = d(x, y)$, etc.

Let c_x, c_y, c_z be the points on $[yz]$, $[zx]$, $[xy]$ with

$$d(c_x, y) = d_y \quad \text{and} \quad d(c_x, z) = d_z,$$
$$d(c_y, x) = d_x \quad \text{and} \quad d(c_y, z) = d_z,$$
$$d(c_z, x) = d_x \quad \text{and} \quad d(c_z, y) = d_y.$$

We call these the *interior points* of the triangle; see Figure 6.1.

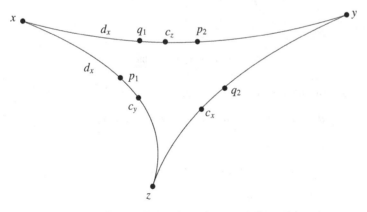

Figure 6.1 Interior and corresponding points

We say that a point on $[xc_y]$ *corresponds* to a point on $[xc_z]$ if the two points are at the same distance from x, and similarly for two points on $[yc_z]$ and $[yc_x]$, and for points on $[zc_x]$ and $[zc_y]$. In Figure 6.1, the point p_1 corresponds to q_1, and p_2 corresponds to q_2. Note also that any two of the interior points correspond.

The triangle xyz is called δ-*thin* if $d(p,q) \le \delta$ whenever p and q correspond.

6.1.3 Theorem *Let xyz be a geodesic triangle. Then*

(i) *if xyz is δ-thin then it is δ-slim;*
(ii) *if xyz is δ-slim then it is 4δ-thin.*

Proof (i) is clear from the definition.

To prove (ii), we suppose, without loss, that $p \in [c_zy] \subseteq [xy]$ and let q correspond to p on $[c_xy] \subseteq [zy]$; see Figure 6.2. By δ-slimness of xyz, there exists $r \in [yz] \cup [xz]$ with $d(p,r) \le \delta$. Let $[pr]$ be a geodesic from p to r.

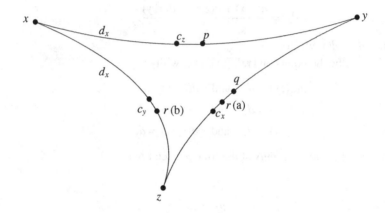

Figure 6.2 Slim triangles are thin

Suppose first (a) that $r \in [yz]$. So

$$d(r,q) = |d(r,y) - d(q,y)| = |d(r,y) - d(p,y)|.$$

The triangle inequality gives

$$d(p,y) - d(r,p) \le d(r,y) \le d(p,y) + d(r,p),$$

and hence $d(r,q) \le \delta$. So $d(p,q) \le 2\delta$.

So now suppose (b) that $r \in [xz]$. Now the triangle inequality gives

$$d(p,x) - d(r,p) \le d(r,x) \le d(p,x) + d(r,p)$$

and hence that $|d(r,x) - d(p,x)| \le \delta$.

Now let $\delta_1 := d(p, x) - d_x$ and $\delta_2 := d(r, x) - d_x$. Then $\delta_1 = d(p, c_z) \geq 0$, and $|\delta_1 - \delta_2| \leq \delta$. Further $\delta_1 = d_y - d(p, y)$, $\delta_2 = d_z - d(r, z)$, and the triangle inequality gives $d(y, z) \leq d(p, y) + d(p, r) + d(r, z)$. We rewrite this as

$$d_y + d_z \leq d_y - \delta_1 + d(p, r) + d_z - \delta_2$$

and deduce that $\delta_1 + \delta_2 \leq d(p, r) \leq \delta$. This, together with $|\delta_1 - \delta_2| \leq \delta$ implies $\delta_1 = d(p, c_z) \leq \delta$.

If $d_y = d(c_z, y) \leq \delta$, then $d(p, q) \leq 2\delta$, so assume that $d_y > \delta$. Choose $\epsilon > 0$ such that $d_y \geq \delta + \epsilon$. Let p' be the point on $[c_z y]$ at distance $\delta + \epsilon$ from c_z and let q' be the point corresponding to p' on $[c_x y]$. Since $d(p', c_z) > \delta$, Case (a) must apply to p', and hence $d(p', q') \leq 2\delta$. Then also

$$d(q, q') = d(p, p') \leq d(c_z, p') = \delta + \epsilon,$$

and so $d(p, q) \leq 4\delta + 2\epsilon$. Since ϵ can be chosen arbitrarily small, we deduce that $d(p, q) \leq 4\delta$. $\qquad\qquad\qquad\qquad\qquad\qquad\qquad\qquad\qquad\qquad\qquad\square$

6.2 Hyperbolicity for geodesic metric spaces and groups

We say that the geodesic metric space (Γ, d) is *hyperbolic* if there exists δ such that all geodesic triangles in Γ are δ-slim. The above theorem shows that we could alternatively express the definition in terms of δ-thin triangles (for a different choice of δ). We say that a finitely generated group $G = \langle X \rangle$ is *hyperbolic* if $\Gamma(G, X)$ is hyperbolic when considered as a metric space with respect to the *word metric*, in which all edges of $\Gamma(G, X)$ have length 1.

This definition turns out to be invariant under change of generating set and, more generally, under quasi-isometry as defined in 1.7.2. This will follow from Corollary 3.3.4 once we have established the equivalence of hyperbolicity with having a linear Dehn function, which we do in Sections 6.4 and 6.5.

Since by Lemma 1.7.6 the image of a geodesic under a quasi-isometry is at a bounded Hausdorff distance from a quasigeodesic, it will also follow from the definition in terms of slim triangles, once we have proved that quasigeodesics lie in a bounded neighbourhood of a geodesic with the same endpoints, which we do in Section 6.7.

In fact this result about quasigeodesics is true in any hyperbolic metric space [5, Proposition 3.3]. It follows that, if two geodesic metric spaces are quasi-isometric and one is hyperbolic, then so is the other. (This is also proved in [94, Chapter 5, Section 2, Theorem 12].)

It is proved by Milnor in [193, Lemma 2], and also in [39, Proposition I.8.19] and [84, Theorem 3.3.6], that if G acts cocompactly and discretely by isometries on a geodesic metric space Γ (or, more generally, on a length space), then there is a quasi-isometric embedding of the Cayley graph of G into Γ. (This result was first proved by Russian mathematicians in the 1950s, and is known as the Švarc–Milnor Lemma.) So, if Γ is hyperbolic, then so is the group G. In particular [84, Theorem 3.4.1], since the fundamental group of a path connected, locally path connected and locally simply connected topological space X acts freely by deck transformations on the universal covering space of X with quotient space X, it follows that the fundamental group of a negatively curved compact manifold or orbifold or convex manifold with boundary is hyperbolic.

6.2.1 The inner product Given a base point w and points x, y in a geodesic metric space Γ, we define the *inner product*

$$(x.y)_w = \frac{1}{2}(d(x, w) + d(w, y) - d(x, y)).$$

The inner product (with respect to w) is δ-hyperbolic if

$$\forall x, y, z \in \Gamma, \ (x.y)_w \geq \min\{(x.z)_w, (z.y)_w\} - \delta.$$

The existence of a fixed $\delta > 0$ such that the inner product is δ-hyperbolic for every choice of base point is used as the basic definition of hyperbolicity by Gromov [125]. It is proved in [5, Prop 2.1] that this definition is equivalent to thinness and slimness of triangles. Specifically, δ-thinness of triangles with vertex w implies that the inner product with base point w is δ-hyperbolic, while δ-hyperbolicity of the inner product with respect to w implies 2δ-slimness of any geodesic triangle with w as one of its vertices.

6.3 Thin bigons, biautomaticity and divergence

A geodesic *bigon* in a geodesic metric space (Γ, d) consists of two points $x, y \in \Gamma$ together with two geodesic paths $[xy]$ and $[xy]'$ joining them. Note that by reversing the second path, we see that a geodesic bigon is equivalent to a geodesic triangle in which two of the three corner points are equal.

For a constant $\delta > 0$, we say that Γ has δ-*thin geodesic bigons* if, for any geodesic bigon as defined above, the paths $[xy]$ and $[xy]'$ synchronously δ-fellow travel. Since a δ-thin geodesic bigon is a δ-thin geodesic triangle with two equal corner points, the following result is immediate.

6.3.1 Proposition *A hyperbolic metric space has δ-thin geodesic bigons for some constant δ.*

We can now easily deduce the following result, which was effectively proved by Cannon in [60].

6.3.2 Theorem *If G is hyperbolic, then it is strongly geodesically automatic with respect to any generating set X.*

Proof We prove the result by verifying the hypotheses of Proposition 5.8.3.

We observed in Section 6.2 that hyperbolicity does not depend on the choice of generating set so, by the previous result, the Cayley graph $\Gamma(G, X)$ has δ-thin geodesic bigons for some constant δ.

Now let $v, w \in \mathcal{G}(G, X)$ with $|v^{-1}w|_G \leq 1$. If $v =_G w$, then the paths based at 1 in $\Gamma(G, X)$ labelled by v and w form the sides of a geodesic bigon in $\Gamma(G, X)$, and hence v and w δ-fellow travel.

Otherwise $va =_G w$ for some $a \in A$. If $|w| = |v| \pm 1$, then the paths based at 1 labelled va and w when $|w| = |v| + 1$, or v and wa^{-1} when $|w| = |v| - 1$, δ-fellow travel. Otherwise $|v| = |w|$, in which case the paths based at 1 labelled by v and w, each followed by half of the edge labelled a from vertex v to vertex w in $\Gamma(G, X)$, form the sides of a geodesic bigon in $\Gamma(G, X)$, so again v and w δ-fellow travel. □

6.3.3 Corollary *Hyperbolic groups are biautomatic with respect to any finite generating set X, with language equal to all geodesic words.*

Proof This follows from Proposition 5.8.4. □

In fact, Papasoglu [210] proved the converse of Theorem 6.3.2, and hence the following result.

6.3.4 Theorem *For a finite generating set X, the group G is strongly geodesically automatic with respect to X if and only if it is hyperbolic with respect to X.*

This is perhaps surprising, because this result does not extend to arbitrary geodesic metric spaces: $\mathbb{R}^n$ has 0-thin geodesic bigons but is not hyperbolic for $n \geq 2$. The proof of Papasoglu's theorem, which we shall outline in 6.3.7, relies on an examination of the divergence of geodesics in the space, which needs a little introduction.

6.3.5 Divergence of geodesics A divergence function for a geodesic metric space is a function $e\colon \mathbb{R}^+ \to \mathbb{R}^+$, with $e(0) > 0$ that satisfies the following condition.

Let γ and γ' be any pair of geodesic paths with $\gamma(0) = \gamma'(0)$ and $d(\gamma(R), \gamma'(R)) > e(0)$ for some $R > 0$. Then, for any $r \geq 0$, the length of any path between $\gamma(R + r)$ and

$\gamma'(R + r)$ that is outside the ball of radius $R + r$ about $\gamma(0)$, except at its end points, is greater than $e(r)$.

Geodesics in the space are said to diverge if there is a divergence function e with $\lim_{r\to\infty} e(r) = \infty$. (This is a uniform property for the space; there is, for example, no divergence function for the Euclidean plane, although any two intersecting geodesics diverge linearly.)

The following is an important component of the proof of Theorem 6.3.4, but we shall not give any details of its proof here.

6.3.6 Theorem   *For a finitely generated group G with Cayley graph $\Gamma :=$ $\Gamma(G, X)$ the following are equivalent, and hence each provides a definition of the hyperbolicity of G.*

  (i)   *There exists δ such that all geodesic triangles in Γ are δ-thin.*
 (ii)   *There exists δ such that all geodesic triangles in Γ are δ-slim.*
(iii)   *Geodesics in Γ exhibit superlinear divergence.*
(iv)   *Geodesics in Γ exhibit exponential divergence.*

Proof   We have already shown the equivalence of the first two. That thin triangles imply exponential divergence of geodesics is [5, Theorem 2.19], and that superlinear divergence of geodesic implies slimness of triangles is [5, Proposition 2.20].   □

6.3.7 Sketch of proof of Theorem 6.3.4   Theorem 6.3.2 deals with the 'if' part of the statement; we need only deal with the converse of Theorem 6.3.2. We sketch the proof of that below, using the notation of [210] wherever possible. So we assume that all geodesic bigons in Γ are δ-thin for some δ, and aim to deduce hyperbolicity of Γ.

The first step is to prove that, if geodesics in Γ (or, more generally, in any geodesic metric space) diverge, then they diverge exponentially, and hence Γ is hyperbolic by Theorem 6.3.6. We will not attempt to summarise the details of that step here, and refer the reader to [210, Theorem 1.1].

So we suppose, for a contradiction, that geodesics do not diverge. We shall show that we can construct an M-thick bigon in Γ for any given $M \in \mathbb{N}$; that is, a pair of geodesic paths between two points of Γ which at some stage are distance at least M apart. Of course, for $M > \delta$ such a bigon is not δ-thin, and so we have a contradiction. So let $M \in \mathbb{N}$ be fixed.

For $r \in \mathbb{R}_{>0}$, define $f(r)$ to be the infimum of $d(\gamma(R + r), \gamma'(R + r))$ over all $R \in \mathbb{N}$ and all geodesic paths γ, γ' in Γ with $\gamma(0) = \gamma'(0) = 1$ (the base point of Γ) and $d(\gamma(R), \gamma'(R)) \geq 2M^2$. Then our assumption that geodesics do not diverge implies that $\liminf_{r\to\infty} f(r)$ is a finite number K, and the fact that Γ is a graph implies that $K \in \mathbb{N}$ and hence that $f(r) = K$ for infinitely many values of r.

The proof now splits into two cases. In the first case, there exist geodesic paths γ, γ' in Γ with $\gamma(0) = \gamma'(0) = 1$, $d(\gamma(R), \gamma'(R)) \geq 2M^2$, and $t_1, t_2 \in \mathbb{N}$ with $t_1 < t_2$ and

$$d(\gamma(R + t_1), \gamma'(R + t_1)) \geq Md(\gamma(R + t_2), \gamma'(R + t_2)) > 0.$$

We choose a geodesic path β from $\gamma(R + t_2)$ to $\gamma'(R + t_2)$, and a sequence $\gamma = \gamma_0, \gamma_1, \ldots, \gamma_l = \gamma'$ of geodesic paths from 1 to the successive vertices on β, where $|\beta| = l$. We see then that there must be two neighbouring paths γ_i, γ_{i+1} with $d(\gamma_i(R+t_1), \gamma_{i+1}(R+t_1)) \geq M$, and then γ_i and γ_{i+1} together with the edge joining their endpoints on β form an M-thick bigon in Γ.

In the second (and more difficult) case, there are no such pairs of paths γ, γ', and hence we can choose two geodesic paths γ, γ' in Γ with $\gamma(0) = \gamma'(0) = 1$, $d(\gamma(R), \gamma'(R)) \geq 2M^2$ and $d(\gamma(R + r), \gamma'(R + r)) = K$ for some large value of r; in fact we choose $r \geq (2^{KM} - 1)(KM)^2$. Then, since we are not in Case 1, we have

$$2M < d(\gamma(R + t), \gamma'(R + t)) < KM$$

for all $t \in \mathbb{N}$ with $0 \leq t \leq r$. So there must exist some J with $2M < J < KM$ for which $d(\gamma(R + t), \gamma'(R + t)) = J$ for at least $(2^{KM} - 1)KM$ distinct values of t. We call points $\gamma(R + t)$ or $\gamma'(R + t)$ J-*points* if $d(\gamma(R + t), \gamma'(R + t)) = J$ and define the J-*distance* d_J between two J-points on γ or on γ' by making the J-distance between successive J-points 1.

Since $J > 2M$, it is enough to prove that there is a $J/2$-thick bigon in Γ, so assume not. The crucial lemma [210, Lemma 1.5] is that, if $P := \gamma(t)$ is a J-point on γ, if $n \in \mathbb{N}_0$, and if $Q := \gamma'(t')$ is a J-point on γ' with $t' \geq t$ and $d_J(Q, \gamma'(t)) = (2^n - 1)J$, then $d(P, Q) \leq t' - t + J - n$. This is proved by induction on n. The case $n = 0$ is easy. For the inductive step, suppose that the result is proved for $n = m$, and let $Q := \gamma'(t')$ be a J-point with $t' \geq t$ and $d_J(Q, \gamma'(t)) = (2^{m+1} - 1)J$.

Now consider the J-points $Q_1 := \gamma'(t_1)$ and $Q_2 := \gamma'(t_2)$ with $t_1 \geq t$, $t_2 > t$, $d_J(Q_1, \gamma'(t)) = (2^m - 1)J$ and $d_J(Q_2, \gamma'(t)) = 2^m J$, and let $P_2 := \gamma(t_2)$. Note that $t' \geq t_1, t_2$. Then the inductive hypothesis gives $d(P, Q_1) \leq t_1 - t + J - m$ and $d(P_2, Q) \leq t' - t_2 + J - m$, from either of which we deduce, using the triangle inequality, that $d(P, Q) \leq t' - t + J - m$. Since all of the quantities are integers, we can replace m by $m + 1$ in this inequality, except when the two concatenations of geodesic paths $[PQ_1] \cup [Q_1 Q]$ and $[PP_2] \cup [P_2 Q]$ from P to Q are both geodesics, in which case it can be shown that they define a $J/2$-thick bigon, in which P_2 is at distance at least $J/2$ from any point on $[PQ_1] \cup [Q_1 Q]$, contrary to hypothesis. This completes the proof of the lemma.

Now, since the number of J-points is at least $(2^{KM} - 1)KM > (2^{J+1} - 1)J$, we can apply the lemma with $n = J + 1$ and find points $P = \gamma(t)$ and $Q = \gamma'(t')$ with $d(P, Q) \leq t' - t - 1$, which contradicts the fact that γ' is a geodesic, and hence completes the proof.

6.4 Hyperbolic groups have Dehn presentations

We shall need the concept of a k-local geodesic. For $k \in \mathbb{N}$, we call a word $w \in A^*$ a *k-local geodesic* if every subword u of length at most k is geodesic. The proof of the following result comes from [5, Theorem 2.12].

6.4.1 Theorem *Hyperbolic groups have Dehn presentations.*

Proof Let G be hyperbolic. So geodesic triangles in the Cayley graph $\Gamma :=$ $\Gamma(G, X)$ are δ-thin, for some $\delta > 0$, and we assume that $\delta \in \mathbb{N}$. We let

$$R = \{w \in A^* \mid w =_G 1, w \text{ reduced}, |w| < 8\delta\},$$

and we shall prove that $\langle X \mid R \rangle$ is a Dehn presentation of G.

Put $k = 4\delta$. Then the k-local geodesics are exactly those reduced words that do not contain more than half of any word in R as a subword. Let $w \in A^*$ with $w =_G 1$ and $|w| > 0$. If w is not a k-local geodesic, then $w = w_1 u w_2$ where $u =_G v$ with $|v| < |u| \leq 4\delta$. Then $uv^{-1} =_G 1$ with $|uv^{-1}| < 8\delta$, so $uv^{-1} \in R$. Hence, to prove the theorem, it is enough to prove that every k-local geodesic w with $w =_G 1$ has length 0.

We first need to prove some technical results about local geodesics (we use the first lemma in the proof of the second).

6.4.2 Lemma *Let w be a k-local geodesic with $k = 4\delta$, and let v be a geodesic with $w =_G v$, and assume that $|v| \geq 2\delta$. Let $g \in G$ be the endpoint of the paths based at the origin and labelled w and v in Γ. Let r and s be the vertices of Γ on v and w at distance 2δ from g. Then $d(r, s) \leq \delta$.*

6.4.3 Lemma *Let w be a k-local geodesic with $k = 4\delta$, and let v be a geodesic with $w =_G v$. Then $d(z, v) \leq 3\delta$ for all vertices z on w.*

Using these we can complete the proof of the theorem. For let w be a k-local geodesic with $w =_G 1$. Then we can take v to be the empty word in Lemma 6.4.3. So $d(z, 1) \leq 3\delta$ for all vertices z on w. Now if $|w| > 3\delta$ then, by definition of k-local geodesics, the vertex z at distance $3\delta + 1$ from 1 along w

satisfies $d(z, 1) = 3\delta + 1$, a contradiction. So $|w| \leq 3\delta \leq 4\delta$ and hence w is a geodesic, and must have length 0. This completes the proof of the theorem. □

Proof of Lemma 6.4.2 The proof is by induction on $|w|$. If $|w| \leq 4\delta$, then w is a geodesic by hypothesis, and then the result follows from the thin-triangles condition applied to the geodesic bigon with sides w and v. Hence we can assume that $N \geq 4\delta$. So now suppose that the result is true when $|w| \leq N$ for some $N \geq 4\delta \in \mathbb{N}$.

Suppose that $N < |w| \leq N + k$, and let p be the vertex of Γ at distance $k = 4\delta$ from g along w (recall $k \in \mathbb{N}$). Let w_{pg} be the final segment of w of length k, joining p to g. So w_{pg} is a geodesic by hypothesis.

Let u be a geodesic from 1 to p. If $|u| < 2\delta$, then $d(s, p) > |u|$ and so s and r must be corresponding points on the geodesic triangle with sides u, w_{pg}, v, and then $d(s, r) \leq \delta$. So assume that $|u| \geq 2\delta$, and let q, t be the points on w, u at distance 2δ from p, as shown in the diagram below. The segment of w from q to s has length k and so is a geodesic, and $d(q, s) = k = 4\delta$. By the induction hypothesis, $d(q, t) \leq \delta$. Hence $d(t, s) \geq 3\delta$. But then once again, s and r must be corresponding points on the geodesic triangle with sides u, w_{pg}, v, so $d(s, r) \leq \delta$ as required. □

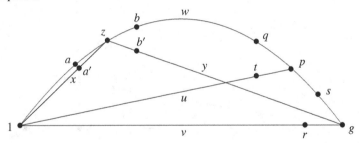

Proof of Lemma 6.4.3 Again let g be the common endpoint of w and v in Γ. Let z be a vertex on w. We have to prove that $d(z, v) \leq 3\delta$. We may clearly assume that $d(z, 1) \geq 2\delta$ and $d(z, g) \geq 2\delta$, since otherwise we are done. Let x and y be geodesics from 1 to z and from z to g. Let a, b be the points at distance 2δ from z along w in the directions of 1 and g, respectively, and let a' and b' be the points at distance 2δ from z along x and y, respectively, as shown in the diagram above. Then by Lemma 6.4.2 we have $d(a, a') \leq \delta$ and $d(b, b') \leq \delta$. Since w is a k-local geodesic, $d(a, b) = 4\delta$, so $d(a', b') \geq 2\delta$. But, then, since x, y, v are the sides of a geodesic triangle, a' and b' are at distance at most δ from v, and hence $d(z, v) \leq 3\delta$, as required. □

The following result also uses Lemma 6.4.2, and is used in the proof that hyperbolic groups have real-time word problem by Holt [147] (see 13.1.6).

It implies that 4δ-local geodesics in groups with δ-thin geodesic triangles are $(2, 0)$-quasigeodesics (see 1.7.5). An analogue of this implied result in the more general context of hyperbolic geodesic metric spaces is proved in [5, Theorem 3.11].

6.4.4 Proposition *Let $G = \langle X \rangle$ be a hyperbolic group in which geodesic triangles are δ-thin, with $0 < \delta \in \mathbb{N}$, and let $k = 4\delta$. Let w be a k-local geodesic from 1 to g in $\Gamma := \Gamma(G, X)$ with $|w| > 1$, and let v be a geodesic with $w =_G v$. Then $|v| \ge |w|/2 + 1$.*

Proof The proof is by induction on $|w|$. If $|w| \le k$ then w is a geodesic and the result is clear (note that $k \ge 4$), so we may assume that $|w| > k$. Our argument is illustrated by Figure 6.3.

Let p be the point on w at distance 2δ from the vertex g, and let u be a geodesic path from 1 to p. Let w_{1p} denote the subpath of w from 1 to p.

Case 1: $|v| < 2\delta$. Since $|w| > k$, we certainly have $|w_{1p}| > 2\delta$. If $|w_{1p}| \le 4\delta$ then w_{1p} is a geodesic and so $|w_{1p}| = |u| > 2\delta$, and otherwise, by the induction hypothesis, $|u| \ge |w_{1p}|/2 + 1 > 2\delta$.

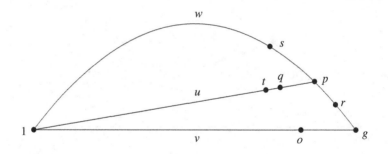

Figure 6.3

So we have $|u| > 2\delta > |v|$, $|w_{pg}| = 2\delta$, and hence (since $|u|, |v|$ and δ are all integers) $(|w_{pg}| + |u| - |v|)/2 \ge \delta + 1$. Hence, in the δ-thin geodesic triangle with vertices $1, p, g$ and edges u, w_{pg}, v, the points q and r on u and w_{pg} at distance $\delta + 1$ from p must correspond, and so $d(q, r) \le \delta$. Now let s and t be the points on w_{1p} and u at distance 2δ from p. By Lemma 6.4.2 applied with p, w_{1p}, u in place of g, w, v, we have $d(s, t) \le \delta$. Now $|w_{sr}| = 3\delta + 1 \le 4\delta$ and so w_{sr} is a geodesic by hypothesis. But $d(s, r) \le d(s, t) + d(t, q) + d(q, r) \le 3\delta - 1$, which is a contradiction.

Case 2: $|v| \ge 2\delta$. Let o be the point on v at distance 2δ from g, and let v_{1o} be the subpath of v from 1 to o. Then $d(p, o) \le \delta$ by Lemma 6.4.2. We can apply our inductive hypothesis to w_{1p} to get $|w_{1p}| = |w| - 2\delta \leq 2|u| - 2$. Also, $|u| \leq |v_{1o}| + d(p, o) \leq |v| - \delta$, and combining these inequalities yields $|w| \leq 2|v| - 2$, as claimed.   □

6.5 Groups with linear Dehn functions are hyperbolic

We proved in the previous section that hyperbolic groups have Dehn presentations which, by Theorem 3.5.1, implies that their Dehn functions are linear.

The converse of this result is more difficult. The proof that we present here is similar to that of [5, Theorem 2.5]. In fact we prove the stronger result that the Dehn function of a group that is not hyperbolic is at least quadratic. The proof makes use of the following result on van Kampen diagrams. For the remainder of this section, it is convenient to use the same symbol w to represent both a path in a Cayley graph or van Kampen diagram and a word that labels that path, and we leave the reader to choose the appropriate interpretation.

6.5.1   Lemma   *Let Δ be a van Kampen diagram over a finite group presentation $\langle X \mid R \rangle$, and suppose that there are integers $m, n \geq 0$, such that the boundary path of Δ has the form $w_1 w_2 w_3 w_4$, where $d_\Delta(\rho, \rho') \geq m$ for all vertices $\rho \in w_1$, $\rho' \in w_3$, and $d_\Delta(\rho, \rho') \geq n$ for all vertices $\rho \in w_2$, $\rho' \in w_4$. Then $\mathrm{area}(\Delta) \geq mn/l^2$, where $l = \max\{|r| : r \in R\}$.*

Proof   We use induction on mn. In the base case $mn = 0$, there is nothing to prove, so suppose that $mn > 0$.

We wish to reduce to the situation in which each edge of the boundary path $w := w_1 w_2 w_3 w_4$ is traversed only once. Since $m, n > 0$, w_1 and w_3 are disjoint, as are w_2 and w_4. So if a boundary edge is traversed twice, then this must happen within two adjacent components of w, such as $w_1 w_2$. In that case $w_1 w_2$ has a subword/subpath of the form $aw'a^{-1}$, that starts and finishes at a vertex p, runs along a from p to a vertex p', along w' from p' back to p' and then along a^{-1} back to p; so a labels the edge (from p to p') that is traversed twice. Then the word $aw'a^{-1}$ is the boundary label of a subdiagram of Δ through p and p' and we can delete the edges and 2-cells of that subdiagram from Δ (together with all vertices not also in Δ) and get a subdiagram of Δ that still satisfies the hypotheses of the lemma with the same m and n, whose boundary has fewer edges traversed twice. By repeating this process as necessary, we can achieve the desired reduction. So each boundary edge of w is adjacent to exactly one internal region.

We now remove from Δ all internal regions that have an edge in common with w_4 and also all vertices and edges that are incident only with the removed regions. The hypothesis on w_1 and w_3 ensures that $|w_4| \geq m$, and so we have removed at least m/l regions from Δ. Hence $\text{area}(\Delta) \geq m/l$. If $n \leq l$, we deduce that $\text{area}(\Delta) \geq mn/l^2$ and we are done.

So we may assume that $n > l$, and then the hypothesis on w_2 and w_4 ensures that there is a subpath $w_1' w_2 w_3'$ of the original boundary that has not been removed, where w_1' is a nonempty suffix of w_1 and w_3' is a nonempty prefix of w_3. The remaining part of Δ may be disconnected, but it has a connected component Δ' with boundary $w_1' w_2 w_3' w_4'$, where each vertex $\rho' \in w_4'$ was on one of the removed regions, and hence satisfies $d_\Delta(\rho', w_4) \leq l$. So we have $d_{\Delta'}(\rho, \rho') \geq n - l$ for all $\rho \in w_2$, $\rho' \in w_4'$ and, since we still have $d_{\Delta'}(\rho, \rho') \geq m$ for all $\rho \in w_1'$, $\rho' \in w_3'$, we may apply our inductive hypothesis to Δ' to get $\text{area}(\Delta') \geq m(n-l)/l^2$ and hence $\text{area}(\Delta) \geq \text{area}(\Delta') + m/l \geq mn/l^2$, as required.   □

6.5.2 Remark By arguing more carefully it can be proved that $\text{area}(\Delta) \geq 4mn/l^2$ in the result above.

6.5.3 Theorem *Suppose that the group $G = \langle X \mid R \rangle$ with X and R finite is not hyperbolic. Then the Dehn function ϕ of G is at least quadratic; that is, $\phi(n) = \Theta(n^2)$.*

Proof Suppose that G is not hyperbolic. Choose any $n \in \mathbb{N}$. Then there is a geodesic triangle $[xyz]$ in $\Gamma := \Gamma(G, X)$ that is not n-slim. Let $\pi(n)$ be the smallest perimeter of any such geodesic triangles and, amongst all geodesic triangles of perimeter $\pi(n)$ that are not n-slim, choose a triangle $[xyz]$ and a vertex $p \in [xy]$ for which $m := d_\Gamma(p, [xz] \cup [yz]) > n$ is as large as possible. Our aim is to construct a word $w = w_1 w_2 w_3 w_4$ that labels a closed path in Γ, to which we can apply Lemma 6.5.1, and thereby conclude that G cannot have a linear Dehn function. The distances we compute within the proof are all distances between vertices of Γ, computed using d_Γ, but we can derive from these the lower bounds on distances in a van Kampen diagram that we need to apply the lemma, since $d_\Delta(\rho, \rho') \geq d_\Gamma(\rho, \rho')$ for any vertices ρ, ρ' in a van Kampen diagram Δ over the presentation.

Let q' be the vertex on $[px]$ closest to p for which $d_\Gamma(q', [yz] \cup [zx]) \leq m/4$. Then choose a vertex q on $[pq']$ at an appropriate distance from p, as below.

Case 1: if $d_\Gamma(p, q') > 3m/2$, then $d_\Gamma(p, q) = \lfloor 3m/2 \rfloor$;
Case 2: if $d_\Gamma(p, q') \leq 3m/2$, then $d_\Gamma(q, q') = 1$.

In either case, $d_\Gamma(p, q) \leq 3m/2$.

Choose a vertex $r \in [yz] \cup [zx]$ with $d_\Gamma(q, r)$ minimal. So $d_\Gamma(q, r) \leq m$ by choice of p. (Note that Figure 6.4 shows r on $[zx]$, but it is possible that

$r \in [yz]$.) Note that in Case 2 above, the choice of q' gives $d_\Gamma(q, r) \leq m/4 + 1$, so $d_\Gamma(p, q) \geq 3m/4 - 1$ (since otherwise we would have $d_\Gamma(p, r) < m$).

Choose vertices s', s on $[py]$ similarly, and $t \in [yz] \cup [zx]$ with $d_\Gamma(s, t)$ minimal. There are now four possibilities for Cases 1 and 2, two of which are symmetrical, so we have to consider three situations.

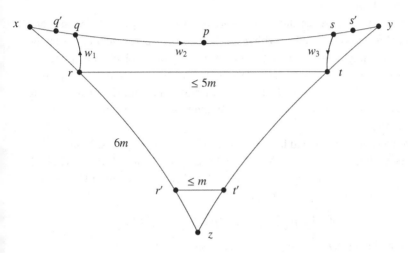

Figure 6.4 Constructing a word with large area

Case 11: both q' and s' in Case 1. Then

$$d_\Gamma(q, r), \ d_\Gamma(s, t) \leq m \ \text{and} \ 3m - 1 \leq d_\Gamma(q, s) \leq 3m.$$

Case 22: both q' and s' in Case 2. Then

$$d_\Gamma(q, r), \ d_\Gamma(s, t) \leq m/4 + 1 \ \text{and} \ 3m/2 - 2 \leq d_\Gamma(q, s) \leq 3m.$$

Case 12: q' in Case 1 and s' in Case 2. Then

$$d_\Gamma(q, r) \leq m, \ d_\Gamma(s, t) \leq m/4 + 1 \ \text{and} \ 9m/4 - 2 \leq d_\Gamma(q, s) \leq 3m.$$

We are now ready to define the word/path $w = w_1 w_2 w_3 w_4$ to which we can apply Lemma 6.5.1. Let w_1, w_2 and w_3 be (words labelling) geodesic paths $[rq]$, $[qs]$ and $[st]$, where w_2 is a subpath of $[xy]$. So $|w_1|, |w_3| \leq m$ and $|w_2| \leq 3m$, and hence $d_\Gamma(r, t) \leq 5m$. We observe that, for vertices $\rho \in w_1$ and $\rho' \in w_3$, we have $d_\Gamma(\rho, \rho') \geq m - 4$, since otherwise, in each of the above three cases, we would get $|[q\rho] \cup [\rho\rho'] \cup [\rho's]| < d_\Gamma(q, s)$.

To define w_4, we consider three different cases.

Case A: If r and t are both on the same edge $[zx]$, say, then we choose w_4 to be the subpath $[tr]$ of $[zx]$. So $|w_4| \leq 5m$.

Otherwise, we assume, as in Figure 6.4, that $r \in [zx]$ and $t \in [yz]$. (The other possibility is that $r \in [yz]$ and $t \in [zx]$, in which case the diagram would look different and involve crossing lines, but the arguments do not depend on the diagram and are essentially the same.)

Case B: If $d_\Gamma(r, z) \leq 6m$ and $d_\Gamma(z, t) \leq 6m$, then we choose $w_4 := [tz] \cup [zr]$. So $|w_4| \leq 12m$.

Case C: Otherwise, as shown in Figure 6.4, we have $d_\Gamma(r, z) > 6m$, say, and we let r' be the vertex on $[rz]$ with $d_\Gamma(r, r') = 6m$. The perimeter of the triangle $[rtz]$ is certainly no longer than that of $[xyz]$ so, by our choice of $[xyz]$, there is a vertex $t' \in [rt] \cup [tz]$ with $d_\Gamma(r', t') \leq m$. In fact we must have $t' \in [tz]$, since otherwise $[r't'] \cup [t'r]$ would be a path from r' to r of length less than $6m$. Note that $d_\Gamma(t', t) \leq m + 6m + 5m = 12m$. We define $w_4 := [tt'] \cup [t'r'] \cup [r'r]$, so $|w_4| \leq 19m$.

We have $|w_4| \leq 19m$ and hence $|w| \leq 24m$ in each of Cases A, B and C. Let ρ, ρ' be vertices on w_2 and w_4. We claim that $d_\Gamma(\rho, \rho') \geq m/4$. If $\rho' \in [zx] \cup [yz]$, then this is true by our choice of q' and s'. Otherwise, we are in Case C and $\rho' \in [r't']$. But if $d_\Gamma(\rho, \rho') < m/4$, then we get

$$|[rq] \cup [q\rho] \cup [\rho\rho'] \cup [\rho'r']| \leq 5m + m/4,$$

contradicting $d_\Gamma(r, r') = 6m$.

So, putting $l := \max\{|r| : r \in R\}$, Lemma 6.5.1 gives $\mathsf{area}(w) \geq m(m-4)/4l^2$. But $|w| \leq 24m$, where m can be arbitrarily large, so the Dehn function ϕ of G satisfies $\phi(n) = \Theta(n^2)$ as claimed.   □

Now that we have proved that G is hyperbolic if and only if it has a linear Dehn function, the following result follows from Proposition 3.3.3 and Corollary 3.3.4.

6.5.4 Proposition *If G and H are finitely generated quasi-isometric groups, then G is hyperbolic if and only if H is. In particular, hyperbolicity of G does not depend on the choice of finite generating set, and subgroups and overgroups of hyperbolic groups of finite index are hyperbolic.*

6.6 Equivalent definitions of hyperbolicity

Let us now summarise all of the equivalent definitions of hyperbolicity that we have encountered so far.

6.6.1 Theorem *For a finitely generated group G with Cayley graph $\Gamma := \Gamma(G, X)$ the following are equivalent, and hence each provides a definition of word-hyperbolicity.*

(i) *There exists δ such that all geodesic triangles in Γ are δ-thin.*

(ii) *There exists δ such that all geodesic triangles in Γ are δ-slim.*

(iii) *There exists δ such that the inner product on Γ with respect to its base point is δ-hyperbolic.*

(iv) *There exists δ such that all geodesic bigons in Γ are δ-thin.*

(v) *The set of all geodesic words for G over A is an automatic structure.*

(vi) *Geodesic paths in Γ exhibit super-linear divergence.*

(vii) *Geodesic paths in Γ exhibit exponential divergence.*

(viii) *G has a Dehn presentation.*

(ix) *G has a linear Dehn function.*

(x) *G has a subquadratic Dehn function.*

(xi) *G has a language L over A for which the set*

$$\{u\#v\#w : u, v, w \in L, uvw =_G 1\}$$

is a context-free language [98].

We have proved, or at least discussed, most of these. The equivalence of (ix) and (x), which follows from Theorems 6.4.1, 3.5.1 and 6.5.3, is attributed to Gromov. There are published proofs by Ol'shanskii [207], Bowditch [35] and Papasoglu [209].

6.7 Quasigeodesics

We defined quasigeodesics in 1.7.5. The following result, which says that quasigeodesics in the Cayley graphs of hyperbolic groups lie within a bounded Hausdorff distance of geodesics between the same endpoints, is fundamental. The corresponding result is valid in the more general context of hyperbolic geodesic metric spaces, and is proved in [5, Proposition 3.3].

6.7.1 Theorem *Let $G = \langle X \mid R \rangle$ be a hyperbolic group in which geodesic triangles in the Cayley graph $\Gamma := \Gamma(G, X)$ are δ-thin. Let w be a (λ, ϵ)-quasigeodesic in Γ for some $\lambda \geq 1$, $\epsilon \geq 1$, and let v be a geodesic joining the endpoints of w. Then there are numbers $L, M \geq 0$ that depend only on $\delta, \lambda, \epsilon$ and the presentation, such that v lies in the L-neighbourhood of w and w lies in the M-neighbourhood of v.*

Proof By Theorems 6.4.1 and 3.5.1, G has a linear Dehn function; that is, there is a constant K depending only on δ and the presentation, such that area$(w) \leq Kl$ for any word w over X with $w =_G 1$.

The proof of the existence of the constant L is similar to that of Theorem

6.5.3, but less involved. If there is no such L, then we can find w and v as in the statement of the theorem such that some vertex p on v is at distance m from w, where m can be arbitrarily large, and we choose p to maximise m. We define q', q, r, s', s, t as in the proof of Theorem 6.5.3, but using distance from w in place of distance from $[yz] \cup [zx]$. We define geodesic paths w_1, w_2, w_3 as in that proof, and for w_4 we just take the subpath w_{tr} of w. Then, as before, we have $d(\rho, \rho') \geq m - 4$ for all vertices $\rho \in w_1, \rho' \in w_3$ and, by construction, we have $d(\rho, \rho') \geq m/4$ for all $\rho \in w_2, \rho' \in w_4$.

Let $w' = w_1 w_2 w_3 w_4$. (This was called w in Theorem 6.5.3.) The fact that w is a (λ, ϵ)-quasigeodesic implies that $|w'| \leq 5m(\lambda + 1)/2 + \epsilon$. So we have an upper bound on $|w|$ that is linear in m, but Lemma 6.5.1 gives us a lower bound on $\text{area}(w')$ that is quadratic in m, so we contradict the linear Dehn function of G. So L exists as claimed.

To prove the existence of the number M, let $1 = p_0, p_1, \ldots, p_n = g$, where $|g| = |v| = n$, be the successive vertices on the path v and, for $0 \leq i \leq n$, let q_i be a vertex on w with $d(p_i, q_i) \leq L$, where $q_0 = p_0$ and $q_n = p_n$. Then the length of the subpath of w from q_{i-1} to q_i is at most $\lambda(2L + 1) + \epsilon$ for $1 \leq i \leq n$ and, since every point on w must lie within at least one of the intervals $[q_{i-1}, q_i]$, we can take $M = (\lambda(2L + 1) + \epsilon)/2 + L$. □

6.8 Further properties of hyperbolic groups

6.8.1 Subgroups of hyperbolic groups This topic is handled in more detail in Chapter 8. According to an elementary construction of Rips [219], hyperbolic groups satisfying arbitrarily stringent small cancellation properties can contain non-hyperbolic finitely generated subgroups. There is, however, a class of subgroups, called *quasiconvex* or *rational*, which are guaranteed to be hyperbolic. By considering subgroups of this type, we shall show in Proposition 8.1.7 that certain types of groups, such as non-cyclic free abelian groups, cannot occur as subgroups of hyperbolic groups.

6.8.2 Decision problems Since hyperbolic groups have a Dehn presentation, their word problem is soluble in linear time by Proposition 3.5.5.

Since hyperbolic groups are biautomatic, their conjugacy problem is soluble in exponential time. But we can do much better than that. It is proved by Epstein and Holt in [85] that, assuming arithmetic operations on integers take constant time, the conjugacy problem is soluble in linear time.

A quadratic time algorithm for conjugacy between finite ordered lists of non-torsion elements is presented by Bridson and Howie in [43]. This was improved by Buckley and Holt in [58] to a linear time algorithm, without the restriction to non-torsion elements.

We already mentioned in 6.8.1 Rips' construction of hyperbolic groups to satisfy various conditions; that construction can be used to give hyperbolic groups for which the generalised word problem is not soluble.

The isomorphism problem for hyperbolic groups has been shown to be soluble. A solution in torsion-free hyperbolic groups was given by Sela in [234]. A solution for hyperbolic groups in general has now been given by Dahmani and Guirardel in [72]. Their approach to the solution also provides an algorithm to solve Whitehead's problem, namely whether two tuples of elements of a hyperbolic group G are in the same orbit under the action of $\mathrm{Aut}(G)$.

6.8.3 Theorem [5, Corollary 2.17] *A hyperbolic group has only finitely many conjugacy classes of torsion elements.*

Proof Let $\langle X \mid R \rangle$ be a Dehn presentation of G. Now let $g \in G$ have finite order n, and let $[g]$ be its conjugacy class. We choose a shortest word w with $w \in [g]$. Now $w^n = 1$, and so w^n contains more than half of some $r \in \hat{R}$. But certainly neither w nor any of its cyclic conjugates can be shortened, so r must be longer than w. Hence the number of conjugacy classes of elements of finite order is bounded by the number of words of length at most $\max\{|r| : r \in R\}$. $\square$

Our proof of the following stronger result is taken from Bridson and Haefliger's book [39, Theorem III.Γ.3.2].

6.8.4 Theorem *A hyperbolic group has only finitely many conjugacy classes of finite subgroups.*

The proof uses the following lemma about hyperbolic geodesic metric spaces. In the lemma $B(x, r)$ denotes the closed ball in Γ of radius r and centre x.

6.8.5 Lemma *Let (Γ, d) be a hyperbolic geodesic metric space in which all geodesic triangles are δ-slim, and let Δ be a nonempty bounded subspace. Let*

$$r_\Delta := \inf\{r > 0 : \Delta \subseteq B(x, r) \text{ for some } x \in \Gamma\}.$$

Then, for any $\epsilon > 0$, the set $C_\epsilon(\Delta) := \{x \in \Gamma : \Delta \subseteq B(x, r_\Delta + \epsilon)\}$ has diameter at most $4\delta + 2\epsilon$.

Proof Let $x, x' \in C_\epsilon(\Delta)$, and let m be the midpoint of a geodesic segment $[xx']$. For each $y \in \Delta$ we consider a geodesic triangle with vertices x, x', y and with $[xx']$ as one of its sides. Then, since this triangle is δ-slim, we have $d(m, p) \leq \delta$ for some $p \in [xy] \cup [yx']$; suppose without loss that $p \in [xy]$.

Then, since $d(x, m) = d(x, x')/2$ and $d(p, x) \geq d(x, m) - \delta$, we get

$$d(y, p) = d(y, x) - d(p, x) \leq d(y, x) + \delta - d(x, x')/2.$$

By definition of $C_\epsilon(\Delta)$, we have $d(y, x) \leq r_\Delta + \epsilon$, and so

$$d(y, m) \leq d(y, p) + d(p, m) \leq r_\Delta + \epsilon + 2\delta - d(x, x')/2.$$

But, from the definition of r_Δ, for any $\epsilon' > 0$ there exists $y \in \Delta$ with $d(y, m) > r_\Delta - \epsilon'$, and so this inequality gives $d(x, x') \leq 4\delta + 2\epsilon + 2\epsilon'$. Since this is true for all $\epsilon' > 0$ we have $d(x, x') \leq 4\delta + 2\epsilon$, as claimed.   □

Proof of Theorem 6.8.4   Let Γ be the Cayley graph of a hyperbolic group G over some finite generating set. Then Γ has δ-slim geodesic triangles for some $\delta > 0$. Let $H \leq G$ be a finite subgroup, and let Δ be the set of vertices of Γ that correspond to elements of H. Then Δ is finite and hence bounded, so we can define r_Δ as in the lemma. We apply the lemma to $C_1(\Delta)$. Since any point of Γ is at distance at most $1/2$ from some vertex of Γ, we see that $C_1(\Delta)$ must contain some vertex of Γ, which corresponds to some $g \in G$. We know from the lemma that $C_1(\Delta)$ has diameter at most $4\delta + 2$, and hence so does $g^{-1}C_1(\Delta)$, which contains the base point of Γ.

Now the elements of the finite set Δ are permuted under left multiplication by the elements of H. So $C_1(\Delta)$ is left invariant under the action of H, and hence $g^{-1}C_1(\Delta)$ is left invariant under the action of $g^{-1}Hg$. So the image of the base point of Γ under each element $g^{-1}Hg$ lies in $g^{-1}C_1(\Delta)$. But these images correspond to the elements of $g^{-1}Hg$, and hence these elements all have length at most $4\delta + 2$. Since there are only finitely many elements of G with this property, there are only finitely many possibilities for the subgroup $g^{-1}Hg$, which proves the theorem.   □

7

Geodesics

7.1 Introduction

Recall that, for a group G with finite generating set X, we use $\mathcal{G}(G, X)$ to denote the set of geodesic words for G over A, where $A := X^{\pm 1}$; that is, the words $w \in A^*$ for which no shorter word represents the same group element as w. We are interested in the formal language class of $\mathcal{G}(G, X)$, with particular emphasis on the pairs (G, X) for which $\mathcal{G}(G, X)$ is a regular language.

The following example from Neumann and Shapiro's paper [204, p. 268], attributed there to Jim Cannon, shows that for a given G the regularity of $\mathcal{G}(G, X)$ can depend on the choice of X. This group arises frequently as a counterexample: it was considered in Exercise 5.3.1 and in 5.5.8.

7.1.1 Proposition *Let*

$$G = \langle a, b, t \mid ab = ba, \ t^2 = 1, \ tat = b \rangle \cong \mathbb{Z} \wr C_2 \cong \mathbb{Z}^2 \rtimes C_2,$$

and let X and Y be the generating sets $\{a, b, t\}$ and $\{a, c, d, t\}$ of G, respectively, where $c = a^2$ and $d = ab$. Then $\mathcal{G}(G, X)$ is regular but $\mathcal{G}(G, Y)$ is not.

Summary of proof It can be verified that

$$\mathcal{G}(G, X) = \bigcup_{\zeta, \eta \in \{\pm 1\}} \{a^\eta, b^\zeta\}^* \cup \{a^\eta, b^\zeta\}^* t \{a^\zeta, b^\eta\}^*,$$
$$\mathcal{G}(G, Y) \cap c^* t c^* t = \{c^m t c^n t : m < n\}.$$

The result then follows from the fact that intersections and unions of finitely many regular sets are regular, together with the fact that the set $\{c^m t c^n t \mid m < n\}$ is not regular. □

Indeed, the only classes of groups for which $\mathcal{G}(G, X)$ is known to be regular for all finite inverse-closed generating sets A are hyperbolic groups (Proposition 5.8.3 and Theorem 6.3.2) and finitely generated abelian groups (Proposition 7.2.2 below).

There appears to be no known counterexample to the 'conjecture' that a group is automatic if and only if $\mathcal{G}(G, X)$ is regular for some generating set X, although there is no particular reason to believe that this should be true. Indeed, it does not appear to be possible to infer much at all about G from the assumption that $\mathcal{G}(G, X)$ is regular. We can however deduce the solubility of the word problem from the following much more general result, which we leave as an exercise.

7.1.2 Exercise If G is a recursively presented group and $\mathcal{G}(G, X)$ is recursive, then G has soluble word problem.

7.2 Virtually abelian groups and relatively hyperbolic groups

As we saw in Theorem 5.7.5, if the group G satisfies the falsification by fellow traveller property (FFTP) with respect to X, then $\mathcal{G}(G, X)$ is regular. In this section, we establish this property for virtually abelian groups. The following two sections will do the same for Coxeter groups and Garside groups, for suitable generating sets X. Another class of groups for which this is true is the class of Artin groups of large type, studied by Holt and Rees [152].

The proofs of the following results are due to Neumann and Shapiro [204, Proposition 4.4].

7.2.1 Lemma *Order the set* $\mathbb{N}_0^r$ *via* $(m_1, \ldots, m_r) \leq (n_1, \ldots, n_r)$ *if and only if* $m_i \leq n_i$ *for* $1 \leq i \leq r$. *Then any subset R of* $\mathbb{N}_0^r$ *contains only finitely many elements that are minimal in R.*

Proof We use induction on r. The claim is true for $r = 1$ since $\mathbb{N}_0$ is well-ordered. Suppose that $R \subseteq \mathbb{N}_0^r$, let S be the set of minimal elements of R, and suppose that S is infinite. Let $s = (s_1, \ldots, s_r) \in S$. Then, for any $t = (t_1, \ldots, t_r) \in S$, we must have $t_k \leq s_k$ for some k with $1 \leq k \leq r$, since otherwise we would have $s < t$ and t would not be minimal. Hence there is a k such that $t_k \leq s_k$ for infinitely many $t \in S$, and hence there is some fixed $m \in \mathbb{N}_0^r$ such that the subset $T := \{t \in S \mid t_k = m\}$ is infinite. But then our inductive hypothesis applied to the subset

$$T' := \{(t_1, \ldots, t_{k-1}, t_{k+1}, \ldots t_r) : (t_1, \ldots, m, \ldots t_r) \in T\}$$

of $\mathbb{N}_0^{r-1}$ implies that T is finite, which is a contradiction. $\qquad\square$

7.2.2 Proposition *Let G be a finitely generated abelian group, and let A be any monoid generating set for G. Then G has FFTP with respect to A.*

Proof Let $A = \{a_1, \ldots, a_r\}$. For $w \in A^*$, define $e(w) := (n_1, \ldots, n_r)$, where n_i is the number of occurrences of a_i in w for $1 \le i \le r$. Then, since G is abelian, $w =_G a_1^{n_1} \cdots a_r^{n_r}$. Let $\mathcal{G}'$ be the set of non-geodesic words over A. Then, by Lemma 7.2.1 applied to the set $\{e(w) \mid w \in \mathcal{G}'\}$, there is a finite subset $\mathcal{G}'_0$ of $\mathcal{G}'$ with the property that, for any $w \in \mathcal{G}'$, there exists $v \in \mathcal{G}'_0$ with $e(v) \preceq e(w)$.

Let K be the maximum value of $|v|$ for $v \in \mathcal{G}'_0$, and let $w \in \mathcal{G}'$. We show now that w $2K$-fellow travels with a G-equivalent shorter word. Let $v \in \mathcal{G}'_0$ with $e(v) \preceq e(w)$. Then we can write $w = w_0 x_1 w_1 x_2 w_2 \cdots x_r w_r$ with each $x_i \in A$ and $w_i \in A^*$, where $v' := x_1 x_2 \cdots x_r$ is a permutation of the word v, so $r = |v| \le K$. Since G is abelian, $\mathcal{G}'_0$ is closed under taking permutations, so $v' \in \mathcal{G}'_0$, and hence there exists $u \in A^*$ with $u =_G v'$ and $|u| < |v'|$. Now, again using G abelian, the word w K-fellow travels with $w_0 w_1 \cdots w_r v'$ and, since $|u| < |v'| \le K$, $w_0 w_1 \cdots w_r v'$ K-fellow travels with $w_0 w_1 \cdots w_r u$, so w $2K$-fellow travels with the G-equivalent shorter word $w_0 w_1 \cdots w_r u$. $\qquad\square$

7.2.3 Theorem *Let G be a finitely generated virtually abelian group with monoid generating set B. Then there exists a monoid generating set A containing B such that G has FFTP with respect to A. We can choose A to be inverse-closed.*

Proof Let $N \trianglelefteq G$ with N abelian and G/N finite, and let $B_N = B \cap N$. We extend B to a monoid generating set A with the following properties, where $A_N := A \cap N$.

(i) A contains at least one representative of each non-identity coset of N in G.

(ii) If $v \in (A \setminus A_N)^*$ is a word of length at most 3 representing an element of Na, where $a \in (A \setminus A_N) \cup \{1\}$, then $va^{-1} \in A_N \cup \{1\}$.

(iii) A_N is closed under conjugation in G.

(iv) (If required) A is inverse-closed.

First we extend B to a monoid generating set that satisfies (i), (iv) (if required), and then (ii). Since G/N is finite, the generators that lie in N have only finitely many conjugates under G, so we can extend to a finite set satisfying (iii) and then close under inversion again if required. Adding new generators that lie in N does not affect properties (i) or (ii), so the resulting set A satisfies each of the properties (i)–(iv).

For a word $w \in A^*$, let $o(w)$ be the number of generators in w that lie in $A \setminus A_N$. We note the following.

(a) The closure of A_N under conjugation in G implies that, for any $w \in A^*$ and any $k \leq o(w)$, the word w is G-equivalent to, and $2k$-fellow travels with, a word of the same length for which k of the generators in $A \setminus A_N$ are at the end of the word.

(b) Properties (i) and (ii) together ensure that, if w is a geodesic word over A, then

(b1) $o(w) < 3$;

(b2) if $o(w) = 2$, then there is a word v with $|v| = |w|$, $v =_G w$ and $o(v) = 1$.

Let w be a non-geodesic word in A^*. If $o(w) = 0$, then $w \in A_N^*$. In this case we apply FFTP in N, using the constant K of Proposition 7.2.2, to deduce that w K-fellow travels with a shorter G-equivalent word.

If $o(w) \geq 3$ then, by (a) above, w 6-fellow travels with a G-equivalent word uv of the same length with $v \in (A \setminus A_N)^*$ and $|v| = 3$. By Properties (i) and (ii) above, the suffix v is G-equivalent to a shorter word, with which v must 2-fellow travel. Hence w 8-fellow travels with a G-equivalent shorter word.

If $o(w) = 2$ then, by (a), w 4-fellow travels with a G-equivalent word uv of the same length with $u \in A_N^*$, $v \in (A \setminus A_N)^*$ and $|v| = 2$. If $w \in N$, then $v \in N$, and Property (ii) (applied with $a = 1$) implies that w 6-fellow travels with a G-equivalent shorter word ux, where $x \in A_N \cup \{\varepsilon\}$. Otherwise, let w' be a geodesic word with $w' =_G w$; so $|w'| < |w|$. By (b1) above, we can choose w' with $o(w') < 3$ and then, by (a) above, we can choose w' such that $o(w') \leq 1$, and any generator from $A \setminus A_N$ within w' is at its right-hand end.

Since $w \notin N$, we have $o(w') = 1$ and hence $w' = u'y$ with $u' \in A_N^*$ and $y \in A \setminus A_N$. Then, since $w =_G w'$, $vy^{-1} \in N$, so property (ii) applies and $vy^{-1} \in A_N \cup \{1\}$. Then either $v =_G y$, in which case w 5-fellow travels with the G-equivalent shorter word uy, or $v =_G xy$ for some $x \in A_N$, and then w 6-fellow travels with the G-equivalent word uxy of the same length. But in that case, $ux =_N u'$ with $u', ux \in A_N^*$ and $|u'| < |ux|$ so ux is non-geodesic and we can apply FFTP in N to the non-geodesic word ux and deduce that ux K-fellow travels with an N-equivalent (and so G-equivalent) shorter word in A_N^*. Hence w $(6 + K)$-fellow travels with a G-equivalent shorter word.

If $o(w) = 1$ (and so $w \notin N$) then, by (a), w 2-fellow travels with a G-equivalent word uz of the same length, with $u \in A_N^*$ and $z \in A \setminus A_N$. Again we let w' be a geodesic word with $w' =_G w$; so $|w'| < |w|$. Just as above, we can choose w' with $o(w') = 1$, so that $w' = u'y$, with $u' \in A_N^*$ and $y \in A \setminus A_N$ and, since $|w'| < |w|$, we have $|u'| < |u|$. As above we deduce that either $z = y$, or $z =_G xy$ for some $x \in A_N$. If $z = y$, then $u =_G u'$ with $|u| > |u'|$ and so u is non-geodesic, and again the FFTP in N implies that u K-fellow travels with a G-equivalent shorter word in A_N^*, and w $(2 + K)$-fellow travels with a G-equivalent shorter word. If $z =_G xy$, then $ux =_G u'$ with $|ux| \geq |u'| + 2$. Now we apply FFTP in N, possibly twice, to see that ux $2K$-fellow travels with a G-equivalent word u'' of length at most $|ux| - 2$. So the word w $(2K + 3)$-fellow travels with the G-equivalent shorter word $u''y$. □

7.2.4 Relatively hyperbolic groups It is proved by Neumann and Shapiro [204, Theorem 4.3] that geometrically finite hyperbolic groups have FFTP and hence regular geodesics relative to suitable generating sets. This result is generalised by Antolín and Ciobanu [8, Theorem 1.3] to all groups that are hyperbolic relative to a set of subgroups with FFTP. This applies to geometrically finite hyperbolic groups; these groups are hyperbolic relative to a set of virtually abelian subgroups.

7.3 Coxeter groups

We recall from 1.10.2 that a finitely generated *Coxeter group* is a group defined by a presentation of the form $\langle S \mid R \rangle$, where $S = \{s_1, s_2, \ldots, s_m\}$ is finite, and $R = \{(s_i s_j)^{m_{ij}} \mid 1 \leq i \leq j \leq m\}$, with each $m_{ii} = 1$ and $2 \leq m_{ij} \leq \infty$ for $i \neq j$. We interpret $m_{ij} = \infty$ as meaning that there is no relator $(s_i s_j)^{m_{ij}}$. The relators s_i^2 imply that the generating set S is inverse-closed.

In 1991 Davis and Shapiro circulated a preprint claiming a proof of the result that Coxeter groups are shortlex automatic with respect to S. The proof turned out to be incomplete, since it relied on an erroneous proof of a result known as the *Parallel Wall Theorem*, which also implied that Coxeter groups have FFTP with respect to S. Both theorems were subsequently proved by Brink and Howlett [49]; their proof of shortlex automaticity of Coxeter groups relied on a proof of the finiteness of a certain set Δ' to be defined below, and they pointed out that this fact is equivalent to the Parallel Wall Theorem. A stronger version of the Parallel Wall Theorem was conjectured by Niblo and Reeves [205] and proved by Caprace and Mühlherr [64].

In this section, we present a proof that Coxeter groups have FFTP with respect to S, but we refer the reader to [49, Theorem 2.8] for a proof that the set Δ' is finite. We also make use of some of the basic theory of Coxeter groups, which we summarise in the following three paragraphs. Proofs of these statements can all be found, for example, in Sections 5.1–5.4 of Humphrey's book [162].

For a Coxeter group $G = \langle S \rangle$ defined, as above, by the numbers m_{ij} (with $1 \le i \le j \le m$), we let $\Pi := \{a_1, \ldots, a_m\}$ be a basis of $\mathbb{R}^m$, and define a symmetric bilinear form on $\mathbb{R}^m$ by $\langle a_i, a_j \rangle = -\cos(\pi/m_{ij})$ if $m_{ij} \ne \infty$ and $\langle a_i, a_j \rangle = -1$ if $m_{ij} = \infty$. Now let $\overline{G}$ be the group generated by the reflections r_α ($\alpha \in \Pi$) of $\mathbb{R}^m$ where, for $\beta \in \mathbb{R}^m$, $r_\alpha \cdot \beta := \beta - 2\langle \beta, \alpha \rangle \alpha$. So r_α is a reflection in the hyperplane orthogonal to α, and the elements of G preserve the bilinear form. Then there is an isomorphism $G \to \overline{G}$ induced by $s_i \mapsto r_{a_i}$; furthermore each $r_{a_i} r_{a_j}$ and hence also $s_i s_j$ has order exactly m_{ij}. We use this isomorphism to identify G with $\overline{G}$ and regard G as being generated by $S = \{r_{a_1}, \ldots, r_{a_m}\}$ from now on.

The set of *roots of G* is defined to be $\Phi := \{w \cdot \alpha \mid w \in G, \ \alpha \in \Pi\}$, and the subset Π of Φ is the set of *fundamental roots*. Then every root has the form $\sum_{i=1}^m x_i a_i$ for $x_i \in \mathbb{R}$, and it turns out that for each such root we either have $x_i \ge 0$ for $1 \le i \le m$ or $x_i \le 0$ for $1 \le i \le m$. Hence we have a partition of Φ into the *positive roots* Φ^+, which include the fundamental roots, and the *negative roots* Φ^-. For each root $\beta = w \cdot \alpha \in \Phi \setminus \Pi$ we denote by r_β the reflection in the hyperplane perpendicular to β; we observe that $r_\beta = w r_\alpha w^{-1}$.

As usual $\mathcal{G} = \mathcal{G}(G, S)$ denotes the set of geodesic words in S^*. These are generally known as *reduced* words in the theory of Coxeter groups. We note the following basic result.

7.3.1 Lemma *If $w = r_{\alpha_1} \cdots r_{\alpha_n} \in \mathcal{G}$ with $\alpha_i \in \Pi$, and $\alpha \in \Pi$, then $r_{\alpha_1} \cdots r_{\alpha_n} r_\alpha \in \mathcal{G}$ if and only if $(r_{\alpha_1} \cdots r_{\alpha_n}) \cdot \alpha \in \Phi^+$.*

For $w \in G$, we define
$$N(w) := \{\alpha \in \Phi^+ : w \cdot \alpha \in \Phi^-\}.$$

Now, for $\beta \in \Phi^+$ and $a_i \in \Pi$, we have $\beta - r_{a_i} \cdot \beta = 2\langle \beta, a_i \rangle a_i$, so β and $r_{a_i} \cdot \beta$ have the same coefficients of all basis elements except for that of a_i. Hence, provided that β is not a scalar multiple of a_i, we must have $r_{a_i} \cdot \beta \in \Phi^+$. Since G preserves the bilinear form, the only scalar multiple of a_i in Φ^+ is a_i itself, so $N(r_{a_i}) = \{a_i\}$ for $1 \le i \le n$.

For positive roots α, β, we say that α *dominates* β if
$$\alpha \ne \beta \quad \text{and} \quad (w \cdot \alpha \in \Phi^- \Rightarrow w \cdot \beta \in \Phi^-), \text{ for all } w \in G.$$

We define $\Delta := \{\alpha \in \Phi^+ : \alpha \text{ dominates some } \beta \in \Phi^+\}$ and Δ' to be the complement of Δ' in Φ^+; that is, Δ' is the set of those positive roots that do not dominate any other positive roots. Note that, since $N(r_\alpha) = \{\alpha\}$ for $\alpha \in \Pi$, we have $\Pi \subseteq \Delta'$.

The key result for Coxeter groups that Δ' is finite, already mentioned above, is Theorem 2.8 of [49]; the proof is also summarised at the end of [161]. The following easy lemma is proved in [161].

7.3.2 Lemma *If $\beta \in \Delta'$ and $r_\alpha \cdot \beta \in \Delta$ for some $\alpha \in \Pi$, then $r_\alpha \cdot \beta$ dominates α.*

Proof Since $r_\alpha \cdot \beta \in \Delta$, the root $r_\alpha \cdot \beta$ dominates some $\gamma \in \Phi^+$. Let $w \in G$ and suppose that $w \cdot \beta \in \Phi^-$. Then $wr_\alpha \cdot (r_\alpha \cdot \beta) = w \cdot \beta \in \Phi^-$, and so we must have $wr_\alpha \cdot \gamma \in \Phi^-$, that is $w \cdot (r_\alpha \cdot \gamma) \in \Phi^-$. Hence, if $r_\alpha \cdot \gamma$ were in Φ^+, then we would have shown that β dominates $r_\alpha \cdot \gamma$. But since $\beta \in \Delta'$, this is not possible, so we must have $r_\alpha \cdot \gamma \in \Phi^-$. But $N(r_\alpha) = \{\alpha\}$, so $\gamma = \alpha$, and $r_\alpha \cdot \beta$ dominates α, as claimed. $\square$

7.3.3 Theorem *The Coxeter group G has FFTP with respect to S.*

Proof Suppose that $w = r_1 \cdots r_n \notin G$, where each $r_i = r_{\alpha_i}$ for some $\alpha_i \in \Pi$. We recall from 1.1.2 that, for any word w and $i \in \mathbb{N}_0$, $w(i)$ denotes the prefix of w of length i, or w when $i > |w|$. In addition, for $1 \le i \le i' \le n = |w|$, we define $w_{i,i'} := r_i \cdots r_{i'}$ (so that $w(i) = w_{1,i}$), and we define $w_{i,i-1} := \varepsilon$.

Let k be minimal with $w(k) \notin G$. Then $w(k-1) \in G$ and, by Lemma 7.3.1, $w(k-1) \cdot \alpha_k \in \Phi^-$. Now let j be maximal such that $w_{j,k-1} \cdot \alpha_k \in \Phi^-$. Then $w_{j+1,k-1} \cdot \alpha_k \in \Phi^+$, and is mapped by r_j to a negative root. So, since $N(r_j) = \{\alpha_j\}$, we have $w_{j+1,k-1} \cdot \alpha_k = \alpha_j$. It follows that $w_{j+1,k-1} r_k w_{j+1,k-1}^{-1} =_G r_j$ and hence that $w_{j,k} =_G w_{j+1,k-1}$.

For $1 \le i \le k$, let $\beta_i := w_{i,k-1} \cdot \alpha_k$, so $\beta_k = \alpha_k$ and $\beta_{j+1} = \alpha_j$ and, by choice of j, $\beta_i \in \Phi^+$ for $j + 1 \le i \le k$.

We claim that $\beta_i \in \Delta'$ for $j+1 \le i \le k$. If not, then choose i maximal such that this is false. Since $\beta_k = \alpha_k \in \Delta'$, we have $i < k$. So $\beta_{i+1} \in \Delta'$ but $\beta_i = r_i \cdot \beta_{i+1} \in \Delta$ and so, by Lemma 7.3.2, β_i dominates α_i. But $w_{j,i-1} \cdot \beta_i = -\alpha_j \in \Phi^-$ so, by definition of dominance, $w_{j,i-1} \cdot \alpha_i \in \Phi^-$ and hence $w_{j,i} \notin G$, which implies that $w_i \notin G$, contrary to the choice of k. So $\beta_k \in \Delta'$ for $j + 1 \le i \le k$ as claimed.

We have seen that w is equal to the shorter word obtained by deleting the generators r_j and r_k from w. So, provided that $K \ge 2$, to show that w K-fellow travels with this shorter word, it suffices to show that $v := w_{j,k}$ K-fellow travels with $u := w_{j+1,k-1}$. We note that the i-th prefixes of the two words are $v(i) := w_{j,j+i-1}$ and $u(i) := w_{j+1,j+i}$. For $i < i'$, we write $w_{i,i'}^R$ to denote the reverse of the word $w_{i,i'}$; of course $w_{i,i'}^R$ represents the inverse of the element represented by $w_{i,i'}$.

For $1 \le i \le |v| = k - j + 1$, using $r_j =_G u r_k u^{-1} =_G w_{j+1,k} w^R_{j+1,k-1}$, we have

$$
\begin{aligned}
v(i)^{-1} u(i) &= w^R_{j,j+i-1} w_{j+1,j+i} = w^R_{j+1,j+i-1} r_j w_{j+1,j+i} \\
&=_G w^R_{j+1,j+i-1} w_{j+1,k} w^R_{j+1,k-1} w_{j+1,j+i} \\
&=_G w_{j+i,k} w^R_{j+i+1,k-1} \\
&=_G w_{j+i,k-1} r_k w^R_{j+i,k-1} r_{j+i} \\
&=_G r_{\beta_{j+i}} r_{j+i},
\end{aligned}
$$

since $\beta_{j+i} = w_{j+1,k-1} \cdot \alpha_k$. But as we saw above, $\beta_{j+i} \in \Delta'$, which is a finite set, so $|v(i)^{-1} u(i)|_G \le L + 1$, where L is the maximum length of a geodesic word for r_β with $\beta \in \Delta'$. Hence G has FFTP.   □

7.4 Garside groups

In this section, we prove that Garside groups have FFTP with respect to a certain generating set X. This proof was published previously by Holt [148]. It is proved by Charney and Meier [66] that these groups have regular geodesics with respect to X, and some of our arguments are based on this proof. See also the paper by Dehornoy and Paris [78] for a general treatment of Garside groups, including the proof of their biautomaticity. [77] is another source for this topic. This section is entirely self-contained; we restrict ourselves to proving those properties of Garside groups that are needed for the proof of FFTP.

For the remainder of this section we let G be a Garside group, as defined in 1.10.6. Let $\widetilde{X}$ be the set of all left divisors of the Garside element Δ, and let $X = \widetilde{X} \setminus \{1\}$. We shall prove that G has FFTP with respect to the generating set X. We recall that $\widetilde{X}$ is also equal to the set of all right divisors of Δ, and is closed under conjugation by Δ. We define $A := X^{\pm}$ and $\widetilde{A} = A \cup \{1\}$. Note that, since no word over $\widetilde{X}$ that contains 1 can be a geodesic, we have $\mathcal{G}(G, \widetilde{X}) = \mathcal{G}(G, X)$.

We use the symbol $\le$ (rather than $\le_L$) for left division; we shall not need a symbol for right division in this section. For $g, h \in G^+$, we write $g \wedge h$ for the greatest common left divisor of g and h.

7.4.1 Lemma *For $g \in X$ and $h \in G^+$, $\Delta \wedge (gh) = \Delta \wedge (g(\Delta \wedge h))$.*

Proof Clearly any common left divisor of Δ and $g(\Delta \wedge h)$ is a common left divisor of Δ and gh. Conversely, let $\Delta \wedge (gh) = k$, so $gh = kl$ for some $l \in G^+$. Since $g \in X$, we have $g \le \Delta$ and hence $g \le k$, so $k = gm$ for some $m \in G^+$. Hence $gh = gml$ and, by the cancellation law, $h = ml$. Now m is a right divisor of k which is itself a left and hence right divisor of Δ, so $m \in \widetilde{X}$, and hence $m \le \Delta \wedge h$. So $k = gm \le g(\Delta \wedge h)$, which completes the proof.   □

Remark: By expressing an arbitrary $g \in G^+$ as a word in X^*, we can use a straightforward induction argument on the length of the word to prove that the above result holds for all $g \in G^+$, but we do not need that result.

Let $g \in G^+$ be represented by the word $w = \mu_1\mu_2 \cdots \mu_n \in X^*$. Then we can define $\eta_i \in X$ for $1 \leq i \leq n$ by $\eta_n = \mu_n$ and $\eta_i = \Delta \wedge (\mu_i\eta_{i+1})$ for $1 \leq i < n$. By repeated application of Lemma 7.4.1, we see that $\eta_i = \Delta \wedge (\mu_i \cdots \mu_n)$ for $1 \leq i \leq n$. In particular $\eta_1 = \Delta \wedge g$. (The generator η_1 is often called the *left front* of g.)

Now for each $i < n$ we have $\mu_i\eta_{i+1} = \eta_i v_{i+1}$ for some $v_{i+1} \in G^+$. But since $\mu_i \leq \Delta$ and $\eta_i = \Delta \wedge (\mu_i\eta_{i+1})$, we have $\mu_i \leq \eta_i$ so v_{i+1} is a right divisor of η_{i+1}, which implies that $v_{i+1} \in \widetilde{X}$.

We have now derived an alternative word $\eta_1 v_2 \cdots v_n \in \widetilde{X}^*$ that represents g, which we denote by $L(w)$, and we note that $\eta_i v_{i+1} \cdots v_n = L(w_i \cdots w_n)$ for $1 \leq i \leq n$. Since $\eta_1 = \Delta \wedge g$, we obtain the following result.

7.4.2 Lemma *If $g \in G^+$ is represented by $w \in X^*$ and $\Delta \leq g$, then $g =_G \Delta v$ for some $v \in X^*$ with $|v| < |w|$.*

Remark: By repeating the rewriting process starting with the element $\eta_1^{-1}g$, represented by $v_2 \cdots v_n$, we can find a word $\eta_1'\eta_2' \cdots \eta_k'$ for g with $k \leq n$, where $\eta_1' = \eta_1$ and each $\eta_i' = \Delta \wedge (\eta_i' \cdots \eta_k')$. This is known as the *left greedy normal form* for g, and plays an important role in the automatic structure for G, but we shall not pursue that further here.

For $\mu \in \widetilde{X}$, we have $\Delta =_G \mu\mu^*$ where, since $\widetilde{X}$ is the set of right divisors of Δ, we have $\mu^* \in \widetilde{X}$. Similarly, there exists $^*\mu \in \widetilde{X}$ with $\Delta =_G {}^*\mu\mu$. (The generators $^*\mu$ and μ^* are often called the *left* and *right complements* of μ.)

It follows from the cancellation laws in G that the maps $\mu \mapsto \mu^*$ and $\mu \mapsto {}^*\mu$ are permutations of $\widetilde{X}$. Note also that $\Delta\mu^* =_G {}^*\mu\mu\mu^* =_G {}^*\mu\Delta$, so there is a permutation σ of $\widetilde{X}$, which restricts to a permutation of X, for which $\Delta\mu =_G \sigma(\mu)\Delta$ for all $\mu \in \widetilde{X}$, and $^*\mu =_G \sigma(\mu^*)$. We have $\Delta^r \in Z(G^+)$ and hence $\Delta^r \in Z(G)$, where r is the order of the permutation σ. In the braid groups, for example, we have $r = 2$.

Throughout the remainder of this section, g denotes an arbitrary element of G, which is represented by the word $w = \mu_1^{\epsilon_1}\mu_2^{\epsilon_2} \cdots \mu_n^{\epsilon_n} \in \widetilde{A}^*$, where each $\mu_i \in \widetilde{X}$ and $\epsilon_i \in \{1, -1\}$. (We assume that $\epsilon_i = 1$ if $\mu_i = 1$.) Let $p := p(w)$ and $m := m(w)$ be the total number of ϵ_i equal to 1 and -1, respectively; so $p + m = n$.

Suppose that $m \geq 1$ and k is maximal with $\epsilon_k = -1$. Then, using the relations $\Delta\mu =_G \sigma(\mu)\Delta$ and $\mu^{-1}\Delta =_G \mu^*$ for $\mu \in \widetilde{X}$, we have

$$w\Delta =_G \mu_1^{\epsilon_1} \cdots \mu_{k-1}^{\epsilon_{k-1}}\mu_k^*\sigma^{-1}(\mu_{k+1}) \cdots \sigma^{-1}(\mu_n),$$

and we denote this word of length n in $\widetilde{A}^*$ by $\rho(w, \Delta)$; note that $m(\rho(w, \Delta)) = m - 1$.

If $m \geq 2$, then we can define $\rho(w, \Delta^2)$ as $\rho((\rho, \Delta), \Delta)$ and hence by recursion we can rewrite $w\Delta^r$ for any $1 \leq r \leq m$, and so define a word $\rho(w, \Delta^r)$ of length n with $m - r$ exponents equal to -1. For $1 \leq s \leq p$, a similar process rewrites $w\Delta^{-s}$ from the right to define a word $\rho(w, \Delta^{-s})$, of length n in $\widetilde{A}^*$, with $p - s$ exponents equal to $+1$. We also define $\rho(w, \varepsilon) := w$, and hence we have defined $\rho(w, \Delta^r)$ for all $-p(w) \leq r \leq m(w)$ as a word of length $|w|$ in $\widetilde{A}^*$ representing $w\Delta^r$.

We can rewrite in much the same way from the left and so define a word $\rho'(\Delta^r, w)$ for $-p(w) \leq r \leq m(w)$ of length $|w|$ in $\widetilde{A}^*$ that represents $\Delta^r w$. It is straightforward to verify that $\rho(w, \Delta^{-r}) = \rho'(\Delta^r, w^{-1})^{-1}$ and $\rho'(\Delta^{-r}, w) = \rho(w^{-1}, \Delta^r)^{-1}$.

For $g \in G^+$, it is conceivable that the length $|g|_{G^+}$ of the shortest word in X^* that represents g could be longer than the length $|g|_G$ of the shortest word in A^* that represents g. Our next aim is to prove that this does not happen, and so $|g|_{G^+} = |g|_G$.

7.4.3 Lemma *Let $g \in G^+$ with $|g|_{G^+} = n$. Then $|\Delta^r g|_{G^+} = n + r$ for any $r \geq 0$.*

Proof This is certainly true with $r = 0$, and it is sufficient to prove it for $r = 1$. Let $v_1 \cdots v_k$ be a shortest word in X^* that represents Δg. Then, by Lemma 7.4.2, we can assume that $v_1 = \Delta$, but then cancelling Δ gives $g =_{G^+} v_2 \cdots v_k$, so $k - 1 \geq n$. But clearly $k \leq n + 1$, so $k = n + 1$ as claimed.   □

7.4.4 Proposition *Let $g \in G^+$ with $|g|_{G^+} = n$, and suppose that $g =_G w$ with $w \in A^*$. Then $|w| - m(w) \geq n$. In particular, $|w| \geq n$ and so $|g|_G = n$.*

Proof Let $m = m(w)$. Then $\rho'(\Delta^m, w)$ is a word in $\widetilde{X}^*$ of length $|w|$ which is equal in G to $\Delta^m g$. So Lemma 7.4.3 implies that $|w| \geq m + n$.   □

The next two results are parts of Lemma 3.2 and Proposition 3.3 of [66].

7.4.5 Lemma *Let $g \in G^+$ with $|g|_G = n$ and let $k \in \mathbb{N}$ with $1 \leq k \leq n$. Then $|g\Delta^{-k}|_G = |\Delta^{-k}g|_G \leq n$ with equality if and only if $\Delta \not\leq g$.*

Proof We have $g\Delta^{-k} =_G \Delta^k(\Delta^{-k}g)\Delta^{-k}$, and X is closed under conjugation by Δ, so $|g\Delta^{-k}|_G = |\Delta^{-k}g|_G$.

Let $w \in X^*$ represent g with $|w| = n$. Since $|\rho'(\Delta^{-k}, w)| = n$, we have $|\Delta^{-k}g|_G \leq n$. It follows from Lemma 7.4.2 that if $\Delta \leq g$ then $|\Delta^{-k}g|_G < n$ (note that $k > 0$). Conversely, suppose that $\Delta^{-k}g =_G v$ with $v \in A^*$ and $|v| < n$. So $g =_G \Delta^k v$. By Proposition 7.4.4, we have $n = |g|_G \leq k + |v| - m$ where $m = m(v)$, so $m < k$. Hence $g =_G \Delta^{k-m}\rho'(\Delta^m, v) \in \widetilde{X}^*$, so $\Delta \leq g$.   □

7.4.6 Lemma *Let $g \in G^+$ and let $w \in \widetilde{X}^*$ with $w =_G g$ and $|w| = n$. Then $w \in \mathcal{G}(G, X)$ (that is, $|g|_G = n$) if and only if $\Delta \not\leq \Delta^n w^{-1}$.*

Proof If $|g|_G < n$ and $g =_G v$ with $v = v_1 \cdots v_k \in X^*$ and $k < n$, then $\Delta^n w^{-1} =_G \Delta^{n-k} \rho'(\Delta^k, v^{-1})$, so $\Delta \leq \Delta^n w^{-1}$.

Conversely, if $\Delta \leq \Delta^n w^{-1}$ then, by Lemma 7.4.2, we can write $\Delta^n w^{-1} =_G \Delta v$, for some $v \in X^*$ with $|v| < |\rho'(\Delta^n, w^{-1})| = n$, so

$$g =_G w =_G v^{-1} \Delta^{n-1} =_G \rho(v^{-1}, \Delta^{|v|}) \Delta^{n-1-|v|},$$

which is a word of length at most $n - 1$. Hence $|g|_G < n$. $\qquad\square$

As before, let $w = \mu_1^{\epsilon_1} \mu_2^{\epsilon_2} \cdots \mu_n^{\epsilon_n} \in \widetilde{A}^*$ represent $g \in G$, where each $\mu_i \in \widetilde{X}$ and $\epsilon_i \in \{1, -1\}$, and let $p := p(w)$ and $m := m(w)$. Define $\overline{w} := \rho(w, \Delta^m)$ and $\underline{w} = \rho(w^{-1}, \Delta^p)$. Then $\overline{w}$ and $\underline{w}$ are words in $\widetilde{X}^*$ of length n that represent $g\Delta^m$ and $g^{-1}\Delta^p$ respectively.

7.4.7 Proposition *For $w \in \widetilde{A}^*$, $m := m(w)$, $p := p(w)$, we have*

(i) $\overline{w} \in \mathcal{G}(G, \widetilde{X}) \iff \Delta \not\leq \Delta^p g^{-1}$;
(ii) $\underline{w} \in \mathcal{G}(G, \widetilde{X}) \iff \Delta \not\leq g\Delta^m \iff \Delta \not\leq \Delta^m g$.

Proof (i) follows from Lemma 7.4.6 and the fact that $\Delta^n \overline{w}^{-1} =_G \Delta^p g^{-1}$. Lemma 7.4.6 tells us that $\underline{w} \in \mathcal{G}(G, \widetilde{X})$ if and only if $\Delta \not\leq \Delta^n \underline{w}^{-1} =_G \Delta^m g$. Since $g\Delta^m =_G \Delta^{-m}(\Delta^m g)\Delta^m$, and X is closed under conjugation by Δ, we have $\Delta \leq \Delta^m g$ if and only if $\Delta \leq g\Delta^m$, and hence (ii) is proved. $\qquad\square$

The following result, which is Lemma 3.6 of [66], characterises the words in $\mathcal{G}(G, X) = \mathcal{G}(G, \widetilde{X})$.

7.4.8 Proposition *Let $w \in A^*$ represent $g \in G$, with $|w| = n$ and $m = m(w)$. Then $w \in \mathcal{G}(G, X)$ if and only if one of the following holds:*

(i) *$m = 0$ and $w = \overline{w} \in \mathcal{G}(G, X)$;*
(ii) *$m = n$ and $w^{-1} = \underline{w} \in \mathcal{G}(G, X)$;*
(iii) *$0 < m < n$ and both $\overline{w}$ and $\underline{w}$ lie in $\mathcal{G}(G, X)$.*

Proof The cases $m = 0$ and $m = n$ are obvious. So suppose that $0 < m < n$ and that $w \in A^*$ represents $g \in G$, with $|w| = n$ and $m = m(w)$. Let $k = |\overline{w}|_G$, so $k \leq |\overline{w}| = n$.

First suppose that $k < m$. Then $g =_G \overline{w}\Delta^{-m} =_G \rho(v, \Delta^{-k})\Delta^{-m+k}$, where v is a word in X^* of length k representing $\overline{w}$. So $|g|_G \leq m$ and, since $m < n$, neither w nor $\overline{w}$ is geodesic.

So now suppose that $m \leq k \leq n$. Then by Lemma 7.4.5 applied with $\overline{w}$ in place of g and k, m in place of n, k, we have $|g|_G = |\overline{w}\Delta^{-m}|_G \leq k$, with equality if and only if $\Delta \not\leq \overline{w}$. Now w is geodesic if and only if $|g|_G = n$ which, by the above, is true if and only if both $k = n$ and $\Delta \not\leq \overline{w}$. Clearly $k = n$ is equivalent to $\overline{w} \in \mathcal{G}(G, A)$. That $\Delta \not\leq \overline{w}$ is equivalent to $\underline{w} \in \mathcal{G}(G, X)$ follows from Proposition 7.4.7 (ii), since $\overline{w} =_G g\Delta^m$.   □

We are now ready to prove our main result.

7.4.9 Theorem *If $w \in A^*$ with $w \notin \mathcal{G}(G, X)$, then w 3-fellow travels with a G-equivalent shorter word. So G has FFTP with respect to X.*

Proof Let $w = \mu_1^{\epsilon_1} \mu_2^{\epsilon_2} \cdots \mu_n^{\epsilon_n}$ with each $\mu_i \in X$, $\epsilon_i \in \{1, -1\}$, and set $p := p(w)$ and $m := m(w)$. Let g be the element of G represented by w. We may assume that the maximal proper prefix $w(n-1)$ of w is geodesic.

By Propositions 7.4.8 and 7.4.7 either $p > 0$ and $\Delta \leq \Delta^p g^{-1}$ or $m > 0$ and $\Delta \leq \Delta^m g$. These two conditions are clearly symmetrical, so we assume that the first one holds.

We need to compute $\rho'(\Delta^p, w^{-1}) = \rho'(\Delta^p, \mu_n^{-\epsilon_n} \cdots \mu_1^{-\epsilon_1})$. For $0 \leq i \leq n$, let $p(i) := p(w(i)) = |\{j \mid 1 \leq j \leq i, \epsilon_j = 1\}|$. We recall that $\Delta\mu =_G \sigma(\mu)\Delta$ and $\Delta\mu^{-1} =_G {}^*\mu =_G \sigma(\mu^*)$. So $\Delta^{p(i)}\mu_i =_G \sigma^{p(i)}(\mu_i)\Delta^{p(i)}$, while

$$\Delta^{p(i)}\mu_i^{-1} =_G \Delta^{p(i)-1}\Delta\mu_i^{-1} =_G \Delta^{p(i)-1\,*}\mu_i =_G \sigma^{p(i)-1}({}^*\mu_i)\Delta^{p(i)-1} =_G \sigma^{p(i)}(\mu_i^*)\Delta^{p(i)-1}.$$

Hence $\rho'(\Delta^p, w^{-1}) = \lambda_n \lambda_{n-1} \cdots \lambda_1$ where each $\lambda_i \in \widetilde{X}$ and, for $1 \leq i \leq n$, $\lambda_i = \sigma^{p(i)}(\mu_i^*)$ when $\epsilon_i = 1$ and $\lambda_i = \sigma^{p(i)}(\mu_i)$ when $\epsilon_i = -1$.

Now we let

$$L(\lambda_n \cdots \lambda_1) = \eta_n \nu_{n-1} \cdots \nu_1$$

where $L()$ is the function defined just before Lemma 7.4.2. Then we have generators η_i for $1 \leq i \leq n$ and ν_i for $1 \leq i < n$ in $\widetilde{X}$ which satisfy $\eta_1 = \lambda_1$ and $\lambda_{i+1}\eta_i = \eta_{i+1}\nu_i$ for $1 \leq i < n$, where $\eta_{i+1} = \Delta \wedge (\lambda_{i+1}\eta_i)$. Then $\lambda_n \cdots \lambda_1 =_G \eta_n \nu_{n-1} \cdots \nu_1$ and $\eta_i = \Delta \wedge (\lambda_i \cdots \lambda_1)$ for $1 \leq i \leq n$. In particular, $\eta_n = \Delta \wedge g = \Delta$, since we are assuming that $\Delta \leq \Delta^p g^{-1}$.

We shall now define a word $u = \kappa_1 \cdots \kappa_{n-1} \in \widetilde{A}^*$, which 2-fellow travels with w and is equal in G to w. Let k be maximal such that $k \leq n$ and $\epsilon_k = 1$. We are assuming $p > 0$, so such a k exists.

Define $q(i) := p(i)$ for $0 \leq i < k$ and $q(i) := p(i) - 1 = p - 1$ for $k \leq i < n$, and then define κ_i to be the element represented by $\Delta^{-q(i-1)}\nu_i^{-1}\Delta^{q(i)}$. Since, for $1 \leq i < n$, we have $q(i) = q(i-1)$ or $q(i-1) + 1$, the definition of X as the set of divisors of Δ together with the fact that it is closed under conjugation by Δ ensures that $\kappa_i \in \widetilde{A}$. Since $\eta_n = \Delta$ and $q(n-1) = p - 1$, we have

$$u =_G \nu_1^{-1}\nu_2^{-1} \cdots \nu_{n-1}^{-1}\Delta^{p-1} =_G \lambda_1^{-1} \cdots \lambda_n^{-1}\Delta^p =_G w.$$

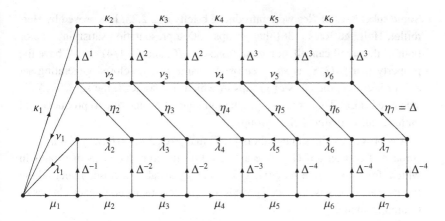

Figure 7.1 w and u fellow travel

The complete configuration is illustrated in Figure 7.1 for a word

$$w = \mu_1\mu_2\mu_3^{-1}\mu_4\mu_5\mu_6^{-1}\mu_7^{-1}$$

in which $p = 4$, $m = 3$ and $k = 5$.

For $1 \le i < n$, we have

$$
\begin{aligned}
w(i)\Delta^{-p(i)} &=_G \lambda_1^{-1}\cdots\lambda_i^{-1} \\
&=_G v_1^{-1}v_2^{-1}\cdots v_{i-1}^{-1}\eta_i^{-1} \\
&=_G v_1^{-1}v_2^{-1}\cdots v_i^{-1}\Delta^{q(i)}\Delta^{-q(i)}v_i\eta_i^{-1} \\
&=_G \kappa_1\cdots\kappa_i\Delta^{-q(i)}v_i\eta_i^{-1} = u(i)\Delta^{-q(i)}v_i\eta_i^{-1}.
\end{aligned}
$$

(Recall that $\eta_i v_{i-1}\cdots v_1 = L(\lambda_i\cdots\lambda_1)$ and so $\eta_i v_{i-1}\cdots v_1 =_G \lambda_i\cdots\lambda_1$.) So $u(i)^{-1}w(i) =_G \Delta^{-q(i)}v_i\eta_j^{-1}\Delta^{p(i)}$. Since $p(i) = q(i)$ or $q(i) + 1$ for all i, $\eta_j^{-1}\Delta \in \widetilde{A}$, and $\widetilde{A}$ is closed under conjugation by Δ, this proves that $|u(i)^{-1}w(i)|_G \le 2$.

From our assumption that $w(n-1) \in \mathcal{G}(G, X)$, we have $|w|_G \ge n - 2$, which means that $\kappa_i = 1$ for at most one value of i. By removing such an element κ_i from v if it exists, we obtain a word that is G-equivalent to w in A^* and which 3-fellow travels with w. This completes the proof. $\square$

7.5 Groups with geodesics lying in some subclass of $\mathcal{R}eg$

Some subclasses of $\mathcal{R}eg$ were introduced briefly in 2.2.23. It is proved by Hermiller, Holt and Rees [136] that groups with a presentation satisfying one or both of the small cancellation conditions $C'(1/6)$ and $C'(1/4) + T(4)$ have the property that $\mathcal{G}(G, X)$ is *star-free* on the same inverse-closed generating set A. Furthermore, the class of groups in which, for some X, the set $\mathcal{G}(G, X)$ is star-free is closed under taking graph (and hence free and direct) products, and includes all virtually abelian groups.

It is proved by the same authors [137] that $\mathcal{G}(G, X)$ is 1-*locally testable* for some A if and only if G is free abelian. Furthermore, the class of groups in which, for some X, the set $\mathcal{G}(G, X)$ is k-locally testable is closed under taking finite direct products, and such groups have finitely many conjugacy classes of torsion elements.

It is proved by Gilman, Hermiller, Holt and Rees [99] that $\mathcal{G}(G, X)$ is *locally excluding* for some A if and only if it is virtually free.

7.6 Conjugacy geodesics

The question of whether there are regular languages corresponding to shortest representatives of the conjugacy classes of the group G is investigated by Ciobanu, Hermiller, Holt and Rees [70].

A natural set to consider with respect to a given ordered inverse-closed generating set A of G is the set of shortlex least representatives of the conjugacy classes. This set, however, appears very rarely to be regular. For example, it is not regular for a non-abelian free group on a free generating set. The only class of groups for which it has been proved regular for some A is the class of virtually cyclic groups, for which it is regular for all A.

However, the set of all geodesic words that represent shortest elements of their conjugacy classes turns out to be regular in many of the situations in which $\mathcal{G}(G, X)$ is regular. This is proved in [70] for hyperbolic groups for any finite generating set X, virtually abelian groups, Artin groups of extra-large type, and homogeneous Garside groups. Furthermore, Ciobanu and Hermiller [69] prove that the class of groups for which this set is regular for some X is closed under taking arbitrary graph products. Furthermore, it is proved by Antolín and Ciobanu [8, Corollary 1.9] that groups hyperbolic relative to a family of virtually abelian subgroups have this property.

7.6.1 Conjugacy growth series Recall from 1.6.5 that the growth function $\gamma_{G,X}$ of G with respect to X is defined by setting $\gamma_{G,X}(n)$ equal to the number of elements of G with X-length at most n, and the corresponding growth series is $\sum_{n=0}^{\infty} \gamma_{G,X}(n)t^n$. We observed in Section 4.1 that a power series $\sum_{n=0}^{\infty} a_n t^n$ defines a rational function of t whenever a_n is the number of words of length n in a regular language. In particular, any geodesic automatic group has a rational growth function.

We can define analogously other growth series related to groups $G = \langle X \rangle$, such as the geodesic and conjugacy geodesic growth series. These will be rational whenever the associated language is regular, but the converse is not true in general: the rationality of a growth series need not imply that the associated language of words over X is regular.

We observed above that the language of shortlex least representatives of the conjugacy classes of G is rarely regular, which raises the question of the rationality of $\sigma(G, X) := \sum_{n=0}^{\infty} c_n t^n$, where c_n is the number of conjugacy classes of G that contain elements of X-length at most n.

In fact Rivin conjectured [220, Conjecture 13.1] that, for a hyperbolic group G and any finite generating set X for G, $\sigma(G, X)$ is rational if and only if G is cyclic. The 'if' part of this conjecture was proved in [70, Theorem 14.2], and the 'only if' part was recently proved by Antolín and Ciobanu [9].

8

Subgroups and coset systems

8.1 Rational and quasiconvex subsets of groups

Our principal source for this section is Gersten and Short's paper [93]. However, Theorem 8.1.3 and Propositions 8.1.6, 8.1.7 and 8.1.8 are also proved for hyperbolic groups in [5, Section 3] and in [203, Section 7].

Let X be a finite set of generators of a group G and $A := X^{\pm}$. If the regular language $L \subseteq A^*$ maps onto G, then we call (X, L) a *rational structure for G*; we call (X, L) a *rational normal form for G* if L maps bijectively onto G. So, for example, any automatic structure L for G is a rational structure.

For a language $L \subseteq A^*$ and a subset B of G, we shall abuse notation and write $B \cap L$ for the subset of L that maps onto B. For a rational structure (X, L) for G and $B \subseteq G$, we say that B is *L-rational* if $B \cap L$ is a regular language.

For an arbitrary subset L of A^* that maps onto G, the subset B of G is called *L-quasiconvex* if there is a constant $k \geq 0$ such that, for any $w \in L$ that evaluates to an element of B, the word (path) w lies in a k-neighbourhood of B in the Cayley graph $\Gamma(G, X)$. The concept of quasiconvex subsets was introduced for hyperbolic groups by Gromov [125].

We recall from 1.6.1 that, for a given subgroup $H \leq G$, the *Schreier graph* $\Gamma(G, H, X)$ is a directed labelled graph in which the vertices are labelled by the right cosets Hg of H in G, with an edge labelled a from Hg to Hga for $a \in A$. As with any graph with edges labelled by elements of A, the Schreier graph can be regarded as an automaton and, by making H the start state and the single accepting state, it accepts the language of all words that represent elements of H. Of course it has infinitely many states in general.

8.1.1 Theorem [93, Theorem 2.2] *For a rational structure (X, L) of the group G, a subgroup $H \leq G$ is L-rational if and only if it is L-quasiconvex.*

Proof Suppose that H is L-rational and let $w \in L \cap H$. Then, since $L \cap H$ is

184

regular, there is a constant k such that, for any prefix v of w, there is a word u with $|u| \leq k$ and $vu \in L \cap H$, and hence w lies within a k-neighbourhood of H.

Conversely, if all words $w \in L \cap H$ lie in a k-neighbourhood of H, then the prefixes v of such words w all lie in a finite set C of right cosets of H in G, and we can define a finite state automaton accepting $L \cap H$ by intersecting the automaton accepting L with the finite subgraph of $\Gamma(G, H, X)$ with vertices in C. □

8.1.2 Proposition [93, Proposition 2.3] *An L-quasiconvex subgroup H of G is finitely generated.*

Proof The prefixes of words $w \in L \cap H$ lie in a finite set C of right cosets of H in G. Let U be a prefix closed set of representatives of the cosets in C and, for $g \in G$ with $Hg \in C$, let $\overline{g}$ be the unique element in $U \cap Hg$. Then it follows from the equation (†) in the proof of Theorem 1.4.1 that the subset $\{ux\overline{ux}^{-1} : Hu, H\overline{ux} \in C, x \in X\}$ of the set of Schreier generators of H in G generates H. □

8.1.3 Theorem [93, Theorem 3.1] *Let (X, L) be a rational structure of the group G and let H be an L-rational (or, equivalently, an L-quasiconvex) subgroup of G. If L is an automatic (respectively asynchronously automatic, biautomatic) structure of G, then H is automatic (respectively asynchronously automatic, biautomatic). If, in addition, G is hyperbolic and L is the language of geodesic words, then H is also hyperbolic.*

Proof We use the finite set $Y = \{ux\overline{ux}^{-1} : Hu, H\overline{ux} \in C, x \in X\}$ of generators for H from the previous proof, and let

$$B := Y^{\pm 1} = \{ua\overline{ua}^{-1} : Hu, H\overline{ua} \in C, a \in A\}.$$

Note that this usually includes the identity of H. Let M be an fsa with alphabet A accepting the language $L \cap H$ and let Σ be the set of states of M. We construct a new automaton M' with alphabet B that accepts the set of words that arise by rewriting words in $L \cap H$ over B, as described in the proof of Theorem 1.4.1, Equation (†). The state set of M' is $\Sigma \times C$ and the start state is (σ_0, H), where σ_0 is the start state of M. The accepting states are those of the form (σ, H), where σ is an accepting state of M. The only transitions from the state (σ, Hu), with $u \in U$, are labelled by letters in B of the form $ua\overline{ua}^{-1}$, and they go to the state $(\sigma^a, H\overline{ua})$.

Now, for a word $w \in L \cap H$, the rewritten word $w' \in B^*$ with $w' \in L(M')$ has the same length as w and synchronously k-fellow travels with w in $\Gamma(G, X \cup Y)$, where $k = \max\{|u| : u \in C\}$. It follows that, for words $v, w \in L \cap H$ that synchronously or asynchronously k-fellow travel in $\Gamma(G, X)$, the rewritten words

$v', w' \in L(M')$ synchronously or asynchronously k'-fellow travel in $\Gamma(G, Y)$ for some constant k'. By using Exercise 5.2.5, it can now be seen that, if $L(M)$ is the language of an automatic, asynchronously automatic or biautomatic structure for G, then $L(M')$ is the same for H. We refer the reader to [93, Theorem 3.1] for further details.

Now suppose that G is hyperbolic and that L consists of all geodesic words. Since H is L-quasiconvex, there exists k such that all words in $L \cap H$ lie in a k-neighbourhood of H in $\Gamma(G, X)$. We enlarge our inverse-closed finite generating set $B = Y \cup Y^{-1}$ of H so that it contains all elements of H that have X-length at most $2k + 1$. Let $w = a_1 \cdots a_n \in L$ with $w \in H$. Then, for $0 \le i \le n$, there exists $u_i \in A^*$ with $|u_i| \le k$ and $a_1 \cdots a_i u_i \in H$, and $u_0 = u_n = \varepsilon$. Then $w =_G b_1 b_2 \cdots b_n$ with $b_i =_G u_{i-1}^{-1} a_i u_i$ and $b_i \in H$, so $b_i \in B$. So, for all $w \in L \cap H$, we have $d_{\Gamma(H,Y)}(1, w) \le d_{\Gamma(G,X)}(1, w)$.

Let τ be a geodesic triangle in $\Gamma(H, Y)$. We need to prove that τ is δ-slim for some constant δ. For each $b \in B$, choose $w_b \in A^*$ with $w_b =_G b$, and let $\lambda = \max\{|w_b| : b \in B\}$. Replace edges labelled b in τ by paths labelled w_b to give a triangle τ' in $\Gamma(G, X)$.

We claim that the three paths connecting the corners of τ' are quasigeodesics in $\Gamma(G, X)$. For let u be one of these paths, and let $p, q \in \tau'$. Then there are vertices $p', q' \in \tau'$ that correspond to vertices in τ with $d_u(p, q) \le d_u(p', q')$ and $d_u(p', p), d_u(q', q) \le \lambda$. Then

$$d_u(p, q) \le d_u(p', q') \le \lambda d_{\Gamma(H,Y)}(p', q') \le \lambda d_{\Gamma(G,X)}(p', q') \le$$
$$\lambda(d_{\Gamma(G,X)}(p, q) + 2\lambda),$$

which proves the claim.

So, by Theorem 6.7.1, there is a constant M and a geodesic triangle τ'' with the same corners as τ' such that the corresponding paths connecting the corners of τ' and τ'' lie within M-neighbourhoods of each other. Since G is hyperbolic, τ'' is δ-slim for some constant δ and hence τ' is $(\delta + 2M)$-slim, and then τ is $(\delta + 2M + \lambda + 1)$-slim, since each point on any of the three edges of τ' is distance at most $\lambda + 1$ from a vertex on the corresponding edge of τ. So H is hyperbolic. $\qquad \square$

We can solve the generalised word problem for L-quasiconvex subgroups of automatic groups. The KBMAG package [146] contains programs for computing the finite state automaton for $L \cap H$ when H is an L-quasiconvex subgroup for a shortlex automatic structure L. In particular, we can compute this language $L \cap H_1 \cap H_2$ for the intersection of two such subgroups H_1 and H_2, just by using standard operations on finite state automata, and we can then compute the associated Schreier generators for $H_1 \cap H_2$.

We call the subgroup H of G *A-quasiconvex* or simply *quasiconvex* in G, if it is L-quasiconvex, where L is the language of all geodesic words over X.

For hyperbolic groups, A-quasiconvexity of H is equivalent to it being L-quasiconvex for any automatic structure L of G and since, for any fixed λ, ϵ, (λ, ϵ)-quasigeodesics representing the same group element fellow travel, it is independent of the generating set X.

In general quasiconvexity of a subgroup can depend on X.

8.1.4 Example $G = \langle a, b, c \mid ac = ca, bc = cb \rangle \cong F_2 \times \mathbb{Z}$ and $H = \langle a, b \rangle$. Then H is quasiconvex in G with respect to $X = \{a, b, c\}$, but not with respect to $X = \{a, b, c, ac\}$, as can be seen by considering the elements $(ac)^n b(ac)^{-n} \in H$.

Rips [219] presents a straightforward construction of a finitely generated subgroup H of a hyperbolic group G that is not quasiconvex in G. In fact G can be chosen to satisfy arbitrarily strict small cancellation conditions. The subgroup H is not finitely presented, and the generalised word problem of H in G is not soluble.

8.1.5 Proposition [93, Proposition 4.1, attributed to Anisimov and Seifert] *All finitely generated subgroups of free groups are quasiconvex.*

Proof Let G be free on X and $A = X^{\pm}$. Then the set L of geodesic words in A^* is equal to the set of freely reduced words. Let $H = \langle Y \rangle$ for a finite subset $Y \subset L$. Any word $w \in L \cap H$ is the free reduction of a word $v \in Y^*$ and, since v lies in a $k/2$-neighbourhood of H, where $k = \max\{|y|_X : y \in Y\}$, so does w. □

8.1.6 Proposition [93, Proposition 4.3] *The centraliser $C_G(S)$ of a finite subset S of a group G with biautomatic structure L is L-rational, and hence biautomatic. If in addition G is hyperbolic, then so is $C_G(S)$.*

Proof By biautomaticity and Exercise 5.2.5, for any $w \in A^*$, the set

$$L(w) := \{(u, v)^p : u, v \in L, wu =_G vw\}$$

is a regular subset of $A^p \times A^p$. For each $g \in S$, choose $w(g) \in A^*$ to be a word representing g. Then $w \in C_G(S)$ if and only if

$$(w, w) \in \left(\cap_{g \in S} L(w(g)) \right) \cap \{(u, u) : u \in L\},$$

which is regular.

If G is hyperbolic then, by Corollary 6.3.3, the language of geodesic words over any finite generating set of G is a biautomatic structure for G, and so $C_G(S)$ is hyperbolic by Theorem 8.1.3. □

An immediate corollary is that centralisers of finitely generated subgroups of biautomatic or hyperbolic groups are biautomatic or hyperbolic, as is the centraliser of any rational subgroup (using Proposition 8.1.2), and hence also the centre of a rational subgroup. In particular, if $g \in G$ with G biautomatic, then the abelian group $Z(C_G(g))$ is L-rational, and hence finitely generated.

These results imply some restrictions on the structure of subgroups of biautomatic and hyperbolic groups. No such restrictions appear to be known on the subgroup structure of automatic groups.

8.1.7 Proposition [93, Proposition 5.1] *A hyperbolic group G cannot contain a free abelian subgroup of rank 2.*

Proof Suppose that $\mathbb{Z}^2 \cong \langle x, y \rangle \leq G$. Then $K := C_G(\{x, y\})$ is a rational subgroup of G and hence is hyperbolic, and so is the abelian group $Z(K)$, which contains $\langle x, y \rangle$. So $Z(K)$ is finitely generated, and hence it has a free abelian subgroup of rank at least 2 of finite index, which is again hyperbolic by Proposition 6.5.4. But we saw in Example 5.8.5 that $\mathbb{Z}^m$ is not strongly geodesically automatic for $m \geq 2$, so this contradicts Theorem 6.3.2.   □

8.1.8 Proposition [93, Theorem 5.2, attributed to Gromov] *The centraliser in a hyperbolic group of an element of infinite order has a cyclic subgroup of finite index.*

We defer the proof until the end of the next section.

8.1.9 Proposition [93, Proposition 5.3] *Let ϕ be an automorphism of a hyperbolic group G with $\phi^n = 1$ for some $n \geq 1$. Then the subgroup*

$$\mathrm{Fix}(\phi) := \{g \in G : \phi(g) = g\}$$

is hyperbolic.

Proof The split extension E of G by $\langle t \mid t^n \rangle$, where t acts as ϕ on G is hyperbolic because $|E : G|$ is finite. Now $C_E(t)$ is a rational subgroup of E and hence hyperbolic, and is equal to the direct product of $\mathrm{Fix}(\phi)$ with $\langle t \rangle$. Since $|C_E(t) : \mathrm{Fix}(\phi)|$ is finite, $\mathrm{Fix}(\phi)$ is hyperbolic.   □

8.1.10 Definition For $g \in G$, the *translation number* $\tau_{G,X}(g)$ is defined by

$$\tau_{G,X}(g) := \liminf_{n>0}(|g^n|_X/n).$$

It is proved in [93, Section 6] that, if G is biautomatic and $g \in G$ has infinite order, then $\tau_{G,X}(g) > 0$. Here is an outline of the argument. Let L be a biautomatic structure for G. Then, as we saw above, the subgroup $C := Z(C_G(g))$ is finitely generated abelian and L-quasiconvex. So a word $w \in L$ representing

g^n must remain within a bounded distance of C. Then, since the length of g^n as an element of C grows linearly with n, it follows that w cannot be shorter than $n|g|_X$ by more than a constant factor. Since, by Theorem 5.2.11, accepted words in an automatic structure can only be longer than equivalent geodesic words by a constant factor, we have $\tau_{G,X}(g) > 0$.

This result is used in [93] to show that certain groups, including finitely generated nilpotent groups (and more generally, but a little less easily, polycyclic groups) that are not virtually abelian, as well as Baumslag–Solitar groups $\mathrm{BS}(m,n)$ with $|m| \neq |n|$, cannot be isomorphic to subgroups of biautomatic groups.

8.2 Automatic coset systems

A natural question to ask is whether we can define automaticity of G relative to a subgroup H of G. Rather than having a regular language mapping surjectively onto G, we would require only that it maps surjectively onto the set of right cosets of H in G.

In Chapter 5, we observed that there are two equivalent definitions of automaticity in groups, the first, in Definition 5.1.1, using the fellow traveller property and the second, in Theorem 5.2.3, using multiplier automata. We can generalise either of these to cosets of subgroups but, although we know of no example that exhibits this fact, the two generalisations in 8.2.1 and 8.2.4 below do not appear to be equivalent.

8.2.1 Definition An *automatic coset system* for a subgroup H of G with respect to a finite generating set X of G consists of fsa W (the *coset word-acceptor*) and M_a (the *coset multiplier automaton*) for each $a \in A \cup \{\varepsilon\}$ (where $A := X^{\pm 1}$), such that the language L of W contains a representative of each right coset Hg of H in G and, for each a and $v, w \in A^*$, the pair $(v, w) \in L(M_a)$ if and only if $v, w \in L$ and $Hva = Hw$.

If an automatic coset system exists, then we say that the pair (G, H) is *coset automatic*, or that H is *coset automatic* in G.

This definition was used by Redfern in his PhD Thesis [217], where he devised and implemented procedures that attempted to construct automatic coset systems. The procedures had the disadvantage that, although they appeared to work correctly and efficiently on a number of interesting examples, they had no verification procedure, so the computed results could not be formally proved correct.

If H is normal in G, then an automatic coset system for H in G clearly defines an automatic structure for G/H, which proves the following result.

8.2.2 Proposition *If (G, H) is coset automatic with $H \trianglelefteq G$, then G/H is automatic.*

Redfern showed in his thesis that the property of H being coset automatic in G is independent of the chosen generating set X of G. The proof is a straightforward adaptation of that for automatic groups (the case $H = 1$) given in 5.2.9 and, in more detail, in [84, Section 2.4].

Essentially the same argument as for Theorem 5.2.11 proves that the generalised word problem for H in G can be solved in quadratic time when (G, H) is coset automatic.

8.2.3 Definition A *shortlex* automatic coset system is one in which W accepts precisely one word w for each coset Hg, and this is the shortlex least word w that represents an element of Hg with respect to some ordering of A. If such a coset system exists, then we say that (G, H) is shortlex coset automatic with respect to the ordered set A.

8.2.4 Strong automatic coset systems Let $G = \langle X \rangle$ and let $\Gamma := \Gamma(G, X)$ be its Cayley graph. A *strong automatic coset system* for a subgroup H of G with respect to a finite generating set X of G consists of an fsa W (the *coset word-acceptor*) and a constant K, such that

(i) the language L of W contains a representative of each right coset Hg of H in G;
(ii) if $v, w \in L$ and $h \in H$ with $d_\Gamma(v, hw) \leq 1$, then $d_\Gamma((v(i), hw(i)) \leq K$ for all $i \geq 0$.

If a strong automatic coset system exists, then we say that the pair (G, H) is *strongly coset automatic*. We say that it is *strongly shortlex coset automatic* with respect to the ordered set A if there exists a strong shortlex automatic coset system.

Strong automatic coset systems were introduced by Holt and Hurt [149], although they were not called that there. Again it is not difficult to show that this property is independent of the chosen generating set X of G. We prove below that every strong automatic coset system is an automatic coset system, but we do not know whether the converse is true.

Apart from those mentioned already, very few theoretical results have been proved about automatic coset systems. We prove in Theorem 8.2.8 below that, if H is a quasiconvex subgroup of a hyperbolic group, then a strong automatic coset system exists for H in G.

8.2.5 Algorithms One advantage that strong automatic coset systems have over general automatic coset systems is that generalisations exist of the algorithms to compute shortlex automatic structures described by Epstein, Holt and Rees [86] and by Holt [145], which can construct and verify the correctness of strong shortlex automatic coset systems. These programs are described in detail in [149]. They can be used in practice to show that there are many interesting examples other than quasiconvex subgroups of hyperbolic groups.

One interesting feature of these algorithms, which arises in the proof of the following theorem, is that they involve the manipulation of fsa that have deterministic transition tables (that is, for any state σ and alphabet symbol a, there is at most one transition from σ on a), but more than one start state. A word w is accepted by such an fsa if and only if there is an accepting path labelled w from at least one of the start states to an accepting state. So the fsa is non-deterministic, but not for the usual reason, and the algorithm described in Proposition 2.5.2 to compute a deterministic fsa with the same language typically runs more quickly than for general non-deterministic fsa; in fact it is polynomial with degree equal to the number of start states.

8.2.6 Theorem *A strong automatic coset system for a subgroup H of G is an automatic coset system.*

Proof Let $L = L(W)$ be the language of a strong automatic coset system. Then, by definition, the set D of all group elements of the form $v(i)^{-1}hw(i)$ with $h \in H$, and $v, w \in L$ with $va =_G hw$ for some $a \in A \cup \{\varepsilon\}$ is finite.

We now construct a 2-variable fsa over A with state-set D, in which the transitions are $f \xrightarrow{(a,b)} g$ for $f, g \in D$, $a, b \in A^p$ and $g =_G a^{-1}fb$. The set S of start states is equal to the set of all elements $h \in H$ that arise in the definition of D. For any subset $C \subseteq D$, we define the fsa $\mathrm{WD}_{G,H}(D, C)$ by making C the set of accept states.

For each $a \in A \cup \{\varepsilon\}$, we define the multiplier fsa M_a to be the 2-variable fsa with language $\{(v, w)^p : v, w \in L(W)\} \cap L(\mathrm{WD}_{G,H}(D, \{a\}))$. It is then straightforward to verify that W together with the M_a form an automatic coset system for H in G. □

8.2.7 Subgroup presentations It is proved by Holt and Hurt [149] that, if (G, H) is strongly shortlex coset automatic, then H has a finite presentation, and an algorithm is described there to compute such a presentation. In fact the proof works in the more general situation that (G, H) has a strong automatic coset system on an inverse-closed generating set A such that the language of the word-acceptor is prefix-closed and contains a unique representative of each coset of H in G.

Here is a brief summary of this algorithm. It is straightforward to see that the set S in the proof of Theorem 8.2.6 of elements $h \in H$ that arise in equations $va =_G hw$ with $v, w \in L$ and $a \in A \cup \{\varepsilon\}$ is equal to the set of Schreier generators (see 1.4.6) of H in G and their inverses, and it is proved that the set of defining relators of H arising from the Reidemeister–Schreier Theorem (1.4.7) is equal to the set of words $h_1 h_2 \cdots h_k$ with $h_i \in S$ that arise as *active* start states in the composite automata $M_{a_{i_1} a_{i_2} \cdots a_{i_k}}$ (see 2.10.8), where $a_{i_1} a_{i_2} \cdots a_{i_k}$ is a defining relator of G. A start state is said to be active if there is a path in the fsa from it to an accept state.

This algorithm has been implemented in KBMAG [146] for the special case when G itself is shortlex automatic and (G, H) is strongly shortlex coset automatic with respect to A. It should in principle be possible also to compute a presentation on the given generators of H, and this can be done in Alun Williams' MAF package [252].

8.2.8 Theorem *Let H be a quasiconvex subgroup of a hyperbolic group G. Then (G, H) is strongly shortlex coset automatic with respect to any finite generating set X of G.*

Proof Let L be the set of words over $A := X^{\pm 1}$ that are the shortlex least representatives of their cosets Hg. Suppose that $va =_G hw$ with $v, w \in L$ and $a \in A \cup \{\varepsilon\}$.

We claim that $|h|$ is uniformly bounded. Let u be a geodesic word representing h. Then we have a hyperbolic triangle with sides labelled by u, va and w if va is geodesic, u, v and wa^{-1} if wa^{-1} is geodesic, or by u and extensions of u and v to the mid-point of the edge labelled a otherwise. This triangle is δ-thin for some fixed $\delta \in \mathbb{N}$. If $|u| > 2\delta + 1$, then there is a vertex on the edge labelled u at distance greater than δ from both endpoints of the edge. But this vertex is at distance at most δ from its corresponding vertex on the edge labelled v or on the edge labelled w. So there is a word in A^* lying in the same coset as v or w but shorter than v or w, contrary to the choice of L (see Figure 8.1). So $|h| \leq 2\delta + 1$.

Now the δ-thinness of the above triangle implies that, for the specified equation $va =_G hw$, the paths labelled v and w fellow travel, and so the set D of all group elements of the form $v(i)^{-1} hw(i)$ with $h \in H$, and $v, w \in L$ with $Hva = Hw$ for some $a \in A \cup \{\varepsilon\}$ is finite, and we can construct the fsa $\mathrm{WD}_{G,H}(D, C)$ for $C \subseteq D$ as in the proof of Theorem 8.2.6.

It remains to prove that L is regular. If $u \notin L$, then it has a minimal prefix

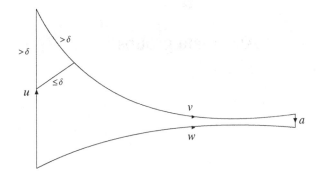

Figure 8.1 $|h|$ is bounded

not in L, which has the form va for some $v \in L$ and $a \in A$. Then $va =_G hw$ for some $w \in L$ with $w < va$ in the shortlex ordering used to define L. So $(v, w) \in L(\mathrm{WD}_{G,H}(D, \{a\}))$. So $A^* \setminus L$ consists of those words with prefixes va with $a \in A$ such that there exists $w \in A^*$ with $(v, w) \in L(\mathrm{WD}_{G,H}(D, \{a\}))$ and $w < va$, and hence L is regular by 2.10.5 and Proposition 2.5.10. □

Proof of Proposition 8.1.8 Let G be hyperbolic and let g be an element of G having infinite order. Then $C := C_G(g)$ is a quasiconvex subgroup of G and hence hyperbolic by Proposition 8.1.6. If $C/\langle g \rangle$ has an element of infinite order, then C has a subgroup isomorphic to $\mathbb{Z}^2$, contradicting Proposition 8.1.7. So $C/\langle g \rangle$ is a torsion group.

Now by Proposition 8.1.6 again, $Z := Z(C)$ is quasiconvex in C and, since $\langle g \rangle \leq Z$, C/Z is a torsion group. Now, by Theorem 8.2.8, (C, Z) is strongly coset automatic, so C/Z is automatic by Proposition 8.2.2, and hence it is finite by Exercise 5.2.8. □

9

Automata groups

In contrast to the previous chapters, where the languages we dealt with were languages over a symmetric generating set for a group, we will now use automata to define group elements. The basic idea, which dates back at least to Alešhin [3], is that certain mappings from A^* to itself can be described by automata with output capabilities, so called Mealy machines.

The real starting point for a systematic study of such groups was the discovery of the first groups of intermediate growth in this class of groups by Grigorchuk [114] and Gupta and Sidki [129]. These groups naturally act on homogeneous rooted trees and many of them reflect the self-similar structure of the tree (see 9.3.13), which often allows for inductive arguments.

The realisation of iterated monodromy groups of partial self-coverings of orbispaces as automata groups by Nekrashevych [201] has given the subject another boost. This strongly links dynamical systems of post-critically finite rational functions and their Julia sets to contracting self-similar automata groups and their limit dynamical systems; see Section 9.3.16. Their study from a more automata theoretic point of view is a more recent development; see Sections 9.4 and 9.6.

9.1 Introducing permutational transducers

9.1.1 Finite state transducers A *finite state transducer* is a finite state automaton with an output function. So besides the input alphabet A there is now also an output alphabet B whose letters will be written to an auxiliary write-only output-tape. Also, each transition must now specify a finite, possibly empty, output string over B that will be written to the output-tape when the transition is made. We shall abbreviate finite state transducer and its plural by fst.

194

An fst simply translates strings over A into strings over B. As such, it does not accept input words, and so there is no need for accepting states. However, it may sometimes be useful to allow accepting states in order to validate the input and/or output; see 9.1.4.

We will usually also not specify a start state, even though this is part of the definition of an fsa, because we are interested in the collection of fst obtained by choosing each of the states as start state in turn; see 9.1.2.

An fst is *deterministic* if the underlying fsa, which is obtained by ignoring the output function, is deterministic. We abbreviate deterministic finite state transducer, and its plural, by dfst.

In the literature, fst are also known as *generalised sequential machines*, and dfst as *Mealy machines*. Since generalised sequential machines were defined to be deterministic in 2.2.25, we prefer to call them transducers in this chapter.

A graphical representation of an fst can be obtained easily by labelling each transition/edge by two strings, the input letter (or ε) to be read and the output string that is to be written to the output-tape. We shall always display the input letter to the left of the corresponding output string and separate them by a vertical bar; see Figure 9.1, for example.

9.1.2 Transducer mappings Each state σ of a dfst, when selected as its start state, can be interpreted as a (possibly partial) map from A^* to B^*: the image of an input word is the output word produced by the transducer after processing the input completely having started in σ. The only requirement is that there is no sequence of ε-moves starting and ending at the same state and that all possible ε-moves are carried out even after the end of the input has been reached. In the same way we can view σ as a (partial) map from A^ω to $B^\omega \cup B^*$. (Here we ignore the problem that an actual physical machine cannot read infinitely many symbols in finite time.) Abusing notation we shall denote the map defined by σ also by σ.

In general, a state of an fst defines a map from A^* to the set of languages over B, but we will not study such maps here.

9.1.3 Binary adding machine As an example consider the dfst in Figure 9.1. Its input and output alphabets are $A = \{0, 1\}$. The state τ is simply the identity map, while σ maps, for example, 11010 to 00110. It is easy to verify that σ maps every input, when viewed as binary representation of an integer with the least significant bit at the left, to the binary representation of the input plus one truncated to the same length (number of bits) as the input. This explains why this dfst is known as the *binary adding machine*.

196

Automata groups

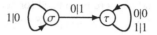

Figure 9.1 The binary adding machine

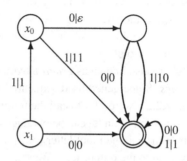

Figure 9.2 The standard generators for Thompson's group F

9.1.4 Thompson's group
The automaton in Figure 9.2 is an example of a dfst whose states x_0 and x_1 define bijections of A^ω but not of A^*, as x_1 maps 1 and 10 both to 1. However, the image of each element $w \in A^\omega$ is determined once the automaton has reached the state indicated by the double circle; this is a scenario where permitting accept states is useful.

It is easy to see that the maps x_0 and x_1 are the standard generators of Thompson's group F [62], and the notation agrees with that in Exercise 1.10.12.

9.1.5 Composing transducers
The output of one fst can be made the input of another fst provided the input alphabet of the latter contains the output alphabet of the first machine.

More precisely, suppose T_i is an fst with state set Σ_i, input alphabet A_i and output alphabet A_{i+1}, for $i = 1, 2$, and $\sigma_i \in \Sigma_i$. Let T be the fst with state set $\Sigma_2 \times \Sigma_1$ and a transition from (τ, σ) to (τ', σ') with input $a \in A_1$ and output $w \in A_3^*$ if and only if σ' can be reached from σ in T_1 while reading a and producing output $u \in A_2^*$ and T_2 can reach state τ' from state τ whilst reading u such that it outputs w.

It is an exercise to show that T is well-defined and that its state (σ_2, σ_1) is the composition σ_1 followed by σ_2 of the maps σ_1 and σ_2. (Here we adjusted the notation to fit our convention that maps are composed from right to left.)

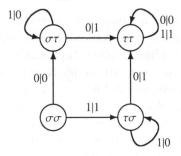

Figure 9.3 Two adding machines composed

As an example, the result of composing the adding machine with itself is given in Figure 9.3. Note that this transducer could be simplified by identifying the states labelled $\sigma\tau$ and $\tau\sigma$.

9.1.6 Inverse transducer Suppose T is an fst with state set Σ, input alphabet A and output alphabet B. We would like to define an fst T^{-1} with input alphabet B and output alphabet A by swapping the inputs and outputs of each transition in T. However, the result is not always an fst, because a transition may now have a word of length greater than one over B as input, which is not allowed. This can be rectified easily by adding new states; for example $\sigma \xrightarrow{bb'|a} \sigma'$ is replaced by $\sigma \xrightarrow{b|\varepsilon} \tau \xrightarrow{b'|a} \sigma'$, where τ is a newly added state. For convenience, we also relabel each state σ of this new automaton as σ^{-1}.

The fst T^{-1} obtained in this fashion is called the *inverse* of T. It is now an exercise to verify that, if the state σ of T defines an injective map, then the state σ^{-1} of T^{-1} defines the left inverse map, in the sense that $\sigma^{-1}(\sigma(w)) = w$ whenever $\sigma(w)$ is defined.

9.1.7 Permutational transducers If we want a state σ of an fst to be a group element, then the input and output alphabets must be the same and σ must define a bijection of some set; this could be A^* or A^ω, or possibly just subsets of these, for which the fst can verify membership with the aid of accept states. We will concentrate on the first two cases.

Possibly the simplest way to ensure invertibility of an fst mapping is to require that the transitions/edges out of any given state define a permutation of the input alphabet A. More precisely, each output string is of length one, each state has exactly $|A|$ edges emanating from it, and each letter in A occurs once as an input and once as an output on one of those edges. We call such a machine a *permutational transducer*. Note that such an automaton is deterministic by definition. We call a state of a permutational transducer *active* if the edges emanating from it define a non-trivial permutation of A, and *inactive* otherwise.

The binary adding machine (Figure 9.1) is an example of a permutational transducer with one active state. (In the literature permutational transducers are also known as *synchronous* and *invertible* dfst; synchronous means that output strings are of length one.)

Composition and inversion are a little easier for permutational transducers than in the general case described in 9.1.5 and 9.1.6; the inverse fst does not require any additional states. In particular, compositions and inverses of permutational transducers are again permutational transducers.

9.1.8 Action on tree Given a finite alphabet A, the 'Cayley graph' of the free monoid A^* is a rooted tree, which we also denote by A^*; this is a slight abuse of notation but keeps it less overloaded. The root is the vertex corresponding to the identity element of A^*, i.e. the empty string. The vertices at distance k from the root correspond to the elements of A^k and form the *k-th level* of the tree; as usual each edge has length one. The set of vertices at level k or less form a finite rooted tree, denoted $A^{\leq k}$.

As an exercise, the reader should verify that permutational transducer mappings preserve this tree structure.

9.1.9 Tree automorphisms The automorphism group of the tree $A^{\leq k}$ is easily seen to be the k-fold iterated permutational wreath product (see 1.5.5) of $\mathrm{Sym}(A)$ acting on A. More precisely, $\mathrm{Aut}(A^{\leq k}) = \mathrm{Aut}(A^{\leq k-1}) \wr_\phi \mathrm{Sym}(A)$ and $\mathrm{Aut}(A^{\leq 1}) = \mathrm{Sym}(A)$, where ϕ is the natural action of $\mathrm{Sym}(A)$ on A.

Restricting the elements of $\mathrm{Aut}(A^{\leq k})$ to $A^{\leq k-1}$ induces a homomorphism onto $\mathrm{Aut}(A^{\leq k-1})$ and the collection of all these homomorphisms (and their compositions) forms an inverse system. The inverse limit of this inverse system is the automorphism group of the tree A^*, which is thus a profinite group.

Observe that $\mathrm{Aut}(A^*) = \mathrm{Aut}(A^*) \wr_\phi \mathrm{Sym}(A) \cong \mathrm{Sym}(A) \ltimes \prod_{a \in A} \mathrm{Aut}(A^*)$, so that the state σ of a permutational transducer defines the element $\pi_\sigma \beta_\sigma \in \mathrm{Aut}(A^*)$, where $\pi_\sigma \in \mathrm{Sym}(A)$ is the permutation of A encoded by the labels of the edges emanating from σ and $\beta_\sigma : A \to \mathrm{Aut}(A^*)$ is defined by $\beta_\sigma(a) = \sigma_a$ for $a \in A$, where σ_a is the state reached from σ after reading the input letter a. For example, the adding machine σ from Figure 9.1 is given by the recursive definition $\sigma = (0, 1)(\sigma, 1)$, as τ defines the trivial map.

9.1.10 Proposition *The set* $\mathrm{Ptm}(A^*)$ *of all permutational transducer mappings over the alphabet A is a countable subgroup of* $\mathrm{Aut}(A^*)$.

Proof We already saw that the set of permutational transducer mappings is closed under composition and inverses, and it is a countable set, because there are only finitely many permutational transducers with a fixed (finite) number of states over A. □

9.1.11 Other types of transducers It is possible to drop the condition that a transducer has only finitely many states. Then every element of Aut(A^*) becomes a transducer mapping. When this is done, as in [237, 201] for example, then our group Ptm(A^*) from Proposition 9.1.10 is called the group of *finite state automorphisms* of the tree. This full generality has some advantages, although most authors quickly restrict to finite state automorphisms. Besides that, the ability to use a computer to help with calculations is lost.

One could also think of transducers modelled on pda, where the additional power of the stack would increase the 'complexity' of the resulting maps. This has, to our knowledge, not been explored yet.

9.2 Automata groups

An *automaton group* is a group that can be generated by the set of all states (viewed as maps) of some permutational transducer.

In particular, automata groups are certain finitely generated subgroups of Ptm(A^*) (see 9.1.10) for some alphabet A, and hence residually finite.

9.2.1 Exercise Show that every finite group is an automaton group, and that the infinite cyclic group is an automaton group, even over a two-letter alphabet.

9.2.2 The Grigorchuk group One of the groups that initiated the study of automata groups is 'the' Grigorchuk group, whose defining automaton is depicted in Figure 9.4. This group is one of a whole family of groups constructed by Grigorchuk [114], of which many are examples of groups of intermediate growth. The name we use was introduced by de la Harpe [74]. The fact that this group has intermediate growth is proved in [116, 74, 12] and we will not reproduce it here. There is a detailed account of the facts and open problems around this group by Grigorchuk [120].

Instead, let us demonstrate how computations are carried out in automata groups. Let T denote the defining automaton (Figure 9.4). Then we read off that T is a permutational transducer over $A = \{0, 1\}$ and that the group is generated by $\{a, b, c, d\}$, because the unlabelled state defines the identity of Aut(A^*). Observe also that the inverse automaton T^{-1} is identical with T, which means that every generator has order 2.

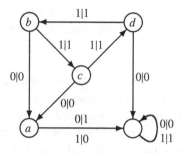

Figure 9.4 'The' Grigorchuk group

We could use transducer composition in order to find information about
words of length two or more over the generators but it is more convenient to
exploit the wreath product structure of Aut(A^*). According to 9.1.9, the transducer T yields $\pi_a = \alpha$ and $\pi_b = \pi_c = \pi_d = 1$, so that

$$a = \alpha(1,1), \quad b = (a,c), \quad c = (a,d) \quad \text{and} \quad d = (1,b),$$

where α is the non-trivial element of Sym(A) and the pairs are $(\beta_\sigma(0), \beta_\sigma(1)) =$
(σ_0, σ_1). Now one can easily multiply the generators, keeping in mind that
$\alpha^{-1}(x,y)\alpha = (y,x)$, according to the definition of the wreath product. For example

$$bcd = (a,c)(a,d)(1,b) = (1,cdb),$$
$$cdb = (a,d)(1,b)(a,c) = (1,dbc) \quad \text{and}$$
$$dbc = (1,b)(a,c)(a,d) = (1,bcd),$$

which together imply that $bcd = cdb = dbc = 1$, so that b, c and d are the
non-trivial elements of a Klein 4-group, that is isomorphic to $C_2 \times C_2$.

9.2.3   Theorem   *The Grigorchuk group is a finitely generated infinite 2-
group, i.e. the order of each element is some power of two.*

Proof   Let G denote the Grigorchuk group. It is clear that G is finitely generated. From

$$aba^{-1} = \alpha(1,1)(a,c)(1,1)^{-1}\alpha = (c,a)$$

it follows that the projection of the subgroup generated by b, c, d and b^a onto
the second component is isomorphic to G. Since b, c, d and b^a are all inactive
(which implies that they fix the first level of the tree point-wise) the subgroup

they generate is different from G (in which a is active). This shows that G has a proper subgroup that surjects onto G, whence G is infinite.

That G is a 2-group is shown by induction on the length of $g \in G$ as a word over the generators. The case $|g| = 1$ was established above. So assume that $|g| > 1$ and observe that g is, up to conjugation by a generator (which does not affect the order), of the form $a \star a \star \cdots a\star$, where each $\star$ is a placeholder for an element of $\{b, c, d\}$. Each block of length 4 is of the form $a \star a\star$ and hence equal to (u, v) where $|u|, |v| \leq 2$, because the components of b, c and d have length one and the components of a are trivial. So if g contains an even number of a's, then $g = (u, v)$ with $|u|, |v| \leq |g|/2$ and the order of g is the least common multiple of the orders of u and v, which is a power of 2 by the induction hypothesis.

If, on the other hand, g contains an odd number of a's, then g^2 contains an even number of a's and $g^2 = (u, v)$ with $|u|, |v| \leq |g^2|/2 = |g|$. Unfortunately, we cannot apply the induction hypothesis yet, but the fact that d has a trivial component saves us. Namely if g involves a d, then g^2 involves at least two d's and, what is more, there exist two occurrences of d in g^2 such that the number of a's preceding them is odd for one of them and even for the other. In this situation, $g^2 = (u, v)$ with $|u|, |v| < |g|$ and we are done. If g does not contain a d, then we continue with the two components u and v, and the induction step can be done as above, unless both have the same length as g, an odd number of a's and still do not involve any d. However, this is only possible if g did not involve any c either, as $c = (a, d)$. So then $g = (ab)^n$ with odd n and we leave it as an exercise to verify that ab has order 16.   □

9.2.4 The Gupta–Sidki p-groups
Shortly after Grigorchuk's work appeared, Gupta and Sidki [129] showed that for every prime $p \geq 3$ the group generated by

$$a = (0, 1, \ldots, p - 1) \in \text{Sym}(A) \quad \text{and} \quad t = (a, a^{-1}, 1, \ldots, 1, t)$$

acting on the tree $\{0, 1, \ldots, p - 1\}^*$ is an infinite p-group. That these groups are also of intermediate growth is more difficult to establish; see [12, 16].

9.2.5 Groups of small automata
A permutational transducer with only one state defines an automorphism of the form $g = \pi_g(g, g, \ldots, g)$, which is easily seen to be the automorphism of A^* induced by the permutation π_g of A; it maps every letter a of the input word to $\pi_g(a)$.

Let us briefly classify automata groups generated by permutational transducers with two states over an alphabet of size two, which can also be found in [123, 122, 259]. Automata groups over a two-letter alphabet with three states

i)

iii)

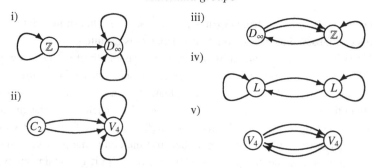

iv)

ii)

v)

Figure 9.5 Automata groups with precisely one active state over $\{0, 1\}$

are classified to some extent in [29, 27, 28] and those with four states are studied by Caponi [63].

If the permutational transducer (also called automaton here) is not connected, then we get a subgroup of $\mathrm{Sym}(A)$ extended to A^*, by the observation above. (With more than two states, this solves the first part of Exercise 9.2.1.)

Recall from 9.1.7 that a state σ is active, if $\pi_\sigma \neq 1$. From now on assume that the automaton is connected, with states a and b, and let $\alpha = (0, 1)$ be the non-trivial permutation of $A = \{0, 1\}$. If a and b are both inactive or both active, then the group is trivial, or cyclic of order two (generated by α extended to A^*), respectively, as one can easily verify.

The remaining five possibilities for the underlying graph are shown in Figure 9.5, where each state is labelled by the isomorphism type of the resulting group if that state is active while the other is not; here C_2 denotes a cyclic group of order two, V_4 denotes a Klein four group (isomorphic to $C_2 \times C_2$), $\mathbb{Z}$ and D_∞ denote the infinite cyclic and the infinite dihedral group (isomorphic to $C_2 * C_2$), respectively, and L denotes the lamplighter group $C_2 \wr_R \mathbb{Z}$. These are all easy to verify except for the following two cases.

Firstly, consider the automaton in Figure 9.5(iii). Let a and b denote the states on the left- and right-hand sides, respectively, and assume that $\pi_a = 1$. Then, up to permuting A, which does not affect a, we have $a = (b, b)$ and $b = \alpha(a, b)$ and

$$[a, b] = aba^{-1}b^{-1} = (b, b)\alpha(a, b)(b^{-1}, b^{-1})(a^{-1}, b^{-1})\alpha = (1, [b, a]),$$

which shows that $[a, b] = 1$ and hence $b^2a = (b^2a, b^2a)$, which in turn means that $b^2a = 1$. By this equality, the group is cyclic and generated by b. Suppose, for a contradiction, that b has finite order m. Then m must be even, because $\pi_b = \alpha$ implies $\pi_{b^m} = \alpha^m$. So $m = 2k$ for some k and $1 = b^{2k} = a^{-k} =$

(b^{-k}, b^{-k}), as $b^2 a = 1$ and $a = (b, b)$. Therefore, $b^{m/2} = 1$, which contradicts the minimality of m, and hence the group is infinite cyclic.

9.2.6 The lamplighter group The second difficulty is posed by the automaton in Figure 9.5(iv). Without loss of generality $a = \alpha(b, a)$ and $b = (b, a)$. (The other option is $b = (a, b)$ but both choices yield isomorphic groups; invert the automaton and then swap 0 and 1, that is, consider the elements $a^{-\hat{\alpha}} = \hat{\alpha} a^{-1} \hat{\alpha}^{-1}$ and $b^{-\hat{\alpha}}$, where $\hat{\alpha} = \alpha(\hat{\alpha}, \hat{\alpha})$.)

Recall from 1.5.5 that $C_2 \wr_R \mathbb{Z}$ is presented by $\langle s, t \mid s^2, [s, t^i s t^{-i}] (i \in \mathbb{N}) \rangle$. Note that $\alpha = ab^{-1} = ba^{-1}$ has order two. For $i \in \mathbb{N}$, put $\alpha_i := \alpha^{b^i} = b^i \alpha b^{-i}$. Then

$$a^i b^{-i} = a^{i-1} b^{-(i-1)} \alpha_{i-1} = \ldots = \alpha \alpha_1 \cdots \alpha_{i-1}.$$

By induction on $i \geq 0$, one can now verify

$$[\alpha, \alpha_i] = [\alpha, \alpha^{b^i}] = [\alpha, b^i]^2 = (a^i b^{-i}, b^i a^{-i})^2 = 1,$$

which implies that the homomorphism $\phi : s \mapsto \alpha$, $t \mapsto b$ is a surjection from $C_2 \wr_R \mathbb{Z}$ onto $\langle \alpha, b \rangle$.

For $i \in \mathbb{N}$, define $E_i := \langle \alpha_j \mid 0 \leq j < i \rangle$. Then $E_i = \langle \alpha, \alpha \alpha_j \mid 1 \leq j < i \rangle$ and $\alpha \alpha_j = [\alpha, b^j] = (a^j b^{-j}, b^j a^{-j})$. Assume, for a contradiction, that i is minimal with $\alpha_i \in E_i$. Then $\alpha \alpha_i \in E_i$, and hence $a^i b^{-i} \in \langle a^j b^{-j} \mid 1 \leq j < i \rangle = E_{i-1}$, which in turn gives $\alpha_{i-1} \in E_{i-1}$, contradicting the minimality of i. So $\alpha_i \notin E_i$ for $i \geq 1$. The first consequence of this is that the α_i ($i \in \mathbb{Z}$) are pairwise distinct and generate an elementary abelian 2-group, and that b has infinite order. Secondly, an arbitrary word over $\{\alpha, b\}$ can, after conjugation by a power of b, be written as $w = \alpha_{i_1} \alpha_{i_2} \cdots \alpha_{i_m} b^k$ with $0 \leq i_1 < i_2 < \cdots < i_m$. So if w represents the trivial element, then $b^{-k} = \alpha_{i_1} \alpha_{i_2} \cdots \alpha_{i_m}$ has order two, and so $k = 0$. But we just proved that $\alpha_{i_1} \alpha_{i_2} \cdots \alpha_{i_m}$ cannot represent the trivial element unless it is the empty word. This means that ϕ is injective.

9.3 Groups of tree automorphisms

In this section we introduce some notions for tree automorphisms that have proven useful in the past. The emphasis is on tree automorphisms rather than automata or permutational transducers.

9.3.1 Sections and automaton of tree automorphism Let $g \in \text{Aut}(A^*)$ and assume that $g = \pi_g \beta_g$; see 9.1.9. For a vertex v of the tree A^* define the *section*

g_v *of g at v* recursively by $g_\varepsilon = g$, $g_a = \beta_g(a)$ for $a \in A$, and $g_v = (g_u)_a$ for $v = ua$ with $u \in A^*$ and $a \in A$. Note that $g_v \in \mathrm{Aut}(A^*)$ for each $v \in A^*$.

Using the set $S_g := \{g_v : v \in A^*\}$ of all sections of g as state set with the transitions $g_v \xrightarrow{a|\pi_{g_v}(a)} g_{va}$ for all $v \in A^*$ and $a \in A$, we get an automaton, possibly with infinitely many states, whose state $g = g_\varepsilon$ defines the same map as g. We call this automaton the *automaton of g*, keeping in mind that it might be infinite. Observe that the automaton of g is finite if and only if $g \in \mathrm{Ptm}(A^*)$.

The idea of studying automorphisms of A^* according to the cycle structure and growth properties of their automata is due to Sidki [237]. A *circuit* in the automaton of $g \in \mathrm{Aut}(A^*)$ (or in a permutational transducer) is a directed cycle or loop. A circuit is *simple* if it does not intersect itself (except at its endpoints of course), and it is trivial if it is an edge from a state σ to itself with $\sigma = 1$.

9.3.2 Bounded and polynomially growing automorphisms
The following hierarchy of subgroups of $\mathrm{Aut}(A^*)$ was defined by Sidki [237] and also plays a role in [201].

The *activity index, $\theta(g,k)$, of g on level k* is the number of active sections of g at vertices of the k-th level, that is

$$\theta(g,k) = |\{v \in A^k : \pi_{g_v} \neq 1\}|.$$

Given $g, h \in \mathrm{Aut}(A^*)$, the identities $(gh)_v = g_{h(v)} h_v$ and $(g^{-1})_v = (g_{g^{-1}(v)})^{-1}$, which are direct consequences of the composition in wreath products, immediately imply that

$$\theta(gh,k) \leq \theta(g,k) + \theta(h,k) \quad \text{and} \quad \theta(g^{-1},k) = \theta(g,k).$$

Thus for each $d \geq 0$ the set $B_d(A^*)$ of all those $g \in \mathrm{Aut}(A^*)$ with $\theta(g,k) \leq p(k)$ for some polynomial $p(k)$ (depending on g) of degree at most d forms a group. In particular, $B_0(A^*)$ is the group of *bounded automorphisms*.

Remark: Nekrashevych [201, Section 3.9.2] uses $\hat{\theta}(g,k) = |\{v \in A^k : g_v \neq 1\}|$ instead of $\theta(g,k)$ and claims that both definitions are equivalent. However, this equivalence holds inside $\mathrm{Ptm}(A^*)$, but there exist $g \in \mathrm{Aut}(A^*)$, with $\theta(g,k) = 1$ and $\hat{\theta}(g,k) = |A|^k$ for all $k \geq 1$. We leave the verification of these facts as an exercise for the reader.

It is also convenient to define the *finitary automorphism group $F(A^*)$* as the group of those $g \in \mathrm{Aut}(A^*)$ for which $\theta(g,k) = 0$ for all sufficiently large k. Note that $F(A^*)$ is the direct limit of the groups $\mathrm{Aut}(A^{\leq k})$ when we identify $\mathrm{Aut}(A^{\leq k-1})$ with the subgroup of $\mathrm{Aut}(A^{\leq k})$ consisting of the elements with trivial sections on level k. So $F(A^*)$ is locally finite, as a direct limit of finite groups.

For example, all generators of the Grigorchuk group, as well as the adding machine generator, are bounded, while the generators of the lamplighter group defined by Figure 9.5(iv) are not bounded, as explained by the following result.

9.3.3 Proposition *An automorphism $g \in \mathrm{Ptm}(A^*)$ is bounded if and only if the automaton of g does not have two distinct non-trivial simple circuits that are connected by a (possibly empty) directed path.*

Proof Suppose that the automaton of g has two distinct non-trivial simple circuits that are connected to each other. Then we can find states σ and τ and words $t, u, v, w \in A^*$ such that u and w are not empty and $g_t = \sigma$, $\sigma_u = \sigma$, $\sigma_v = \tau$ and $\tau_w = \tau$. Moreover, we can assume that the first letters of u and v are different if $v \neq \varepsilon$, or that the first letters of u and w are different if $v = \varepsilon$. Observe also that none of the states along the path labelled by $tuvw$ defines the trivial map. Choose $m, n \geq 1$ with $|u^m| = |w^n|$ and let $r \geq 1$. Put $z_i = tu^{mi}vw^{n(r-i)}$ for $0 \leq i \leq r$. Then $|z_i| = |t| + |v| + r|u^m| = l$, and $g_{z_i} = \tau \neq 1$ for $0 \leq i \leq r$. Moreover, $z_i \neq z_j$ for $i \neq j$, by our assumptions on the first letters of u and v or w, which implies that $\hat{\theta}(g, l) \geq r$. Since r was arbitrary and $g \in \mathrm{Ptm}(A^*)$, the automorphism g is not bounded.

Conversely, assume that the automaton of g does not have two distinct non-trivial simple circuits that are connected. Since $g \in \mathrm{Ptm}(A^*)$, the automaton of g has Q states with $Q \in \mathbb{N}$. Every path u with $|u| > Q$ starting in state g must visit some state at least twice, or in other words wind around a circuit. The point to note is that as soon as a path leaves a non-trivial circuit, it must reach the trivial state after at most Q steps, because it cannot enter a non-trivial circuit again. So for $u \in A^*$, we can have $g_u \neq 1$ only if $u = vw^m z$ with $|v|, |z|, |w| < Q$ and w a label of a non-trivial circuit. The number of words of length k of this form is bounded by Q^3, and hence g is bounded. □

9.3.4 Directed automorphisms An automorphism $g \in \mathrm{Aut}(A^*)$ is *directed* if g is not finitary and there is an infinite word $w \in A^\omega$ and a constant $K > 1$ such that $g(w) = w$ and $g_v = 1$ for all $v \in A^*$ at distance at least K from w. In other words, the section g_v is finitary of *depth* less than K whenever v is not a prefix of w and, if v is a prefix of w, then the section g_v is directed. In particular, every directed automorphism is bounded, but its automaton is not necessarily finite.

Examples of directed automorphisms are the generators b, c, d of the Grigorchuk group and t of the Gupta–Sidki group.

This definition with $K = 2$ appears in [119, 16], where groups generated by directed automorphisms and automorphisms of the form $g = \pi_g$ (called *rooted* automorphisms) are studied; in [119] the focus is on torsion criteria and conditions ensuring that the resulting group is a branch group (see 9.3.13), while [16] also covers presentations, the lower central series, the derived series, growth, torsion growth, width, associated Lie algebras and more.

9.3.5 Circuitous automorphisms Sidki [237] defines an automorphism $g \in$ $\mathrm{Ptm}(A^*)$ to be *circuitous* if it is bounded and g lies on a circuit of its automaton.

It is clear that, if g is a directed permutational transducer mapping, then there exists a unique $v \in A^*$ of minimal length such that g_v is circuitous, but one cannot swap direct and circuitous here. For example, the generator σ of the binary adding machine is circuitous but none of its sections are directed.

The following result is due to Sidki [237].

9.3.6 Proposition *The group* $\mathrm{Ptm}_0(A^*) := B_0(A^*) \cap \mathrm{Ptm}(A^*)$ *of bounded permutational transducer mappings is generated by finitary and circuitous tree automorphisms.*

Proof By definition, finitary and circuitous automorphisms are bounded and elements of $\mathrm{Ptm}(A^*)$, so the group generated by all finitary and circuitous permutational transducer mappings is contained in $\mathrm{Ptm}_0(A^*)$.

A closer inspection of the arguments in the proof of Proposition 9.3.3 shows that, given $g \in \mathrm{Ptm}_0(A^*)$, there is a set S of words of minimal length (bounded by the number of states of the automaton of g) such that g_s is either finitary or circuitous for each $s \in S$ and every element of A^ω has precisely one element of S as prefix.

Notice that sections of finitary automorphisms are finitary and sections of a circuitous element are also finitary except for one of them that is circuitous (by Proposition 9.3.3). So if k is the maximal length of the elements of S, then g factors as $g = \alpha g_k$ with $\alpha \in \mathrm{Aut}(A^{\leq k})$, so α is finitary, and g_k is the element of $\prod_{v \in A^k} \mathrm{Aut}(A^*)$ (the stabiliser of the k-th level) whose pairwise commuting components are the sections of g on level k. It therefore suffices to show that, if c is circuitous, $v \in A^k$, and $^v c$ is the element of $\mathrm{Aut}(A^*)$ that fixes the k-th level and whose only non-trivial section on level k is $(^v c)_v = c$, then $^v c$ is a product of finitary and circuitous elements. Since c is circuitous, there is a state b in the automaton of c and $u \in A^k$ with $b_u = c$. In other words, $b = \beta(^u c)$ with β finitary and $^u c$ defined similarly to $^v c$. So $^u c = \beta^{-1} b$ is a product of finitary and circuitous elements. Finally, there is a finitary element γ with $\gamma(u) = v$, so $\gamma \beta^{-1} b \gamma^{-1} = {}^v c$ and the proof is complete. $\square$

9.3.7 Corollary *A finitely generated group of bounded permutational transducer mappings is a subgroup of a group generated by finitely many finitary and circuitous permutational transducer mappings.*

9.3.8 Amenability A group is called *amenable* if there exists a left (or right, if you prefer) invariant, finitely additive, normalised measure on the group. To be precise, there is function μ from the power set of G to the non-negative real numbers such that, for any disjoint subsets U, V of G, $\mu(gU) = \mu(U)$ (invariance), $\mu(U \cup V) = \mu(U) + \mu(V)$ (finitely additivity) and $\mu(G) = 1$ (normalised).

Since Grigorchuk [114] discovered his groups of intermediate growth, it has been of interest to decide which automata groups are amenable. Note that groups of subexponential growth are amenable (see [74, VII.34]). Until about a decade ago, amenability, as well as subexponential growth, have been established mostly on a case by case basis. One exception is the work of Bartholdi and Šunić [18], which introduced a large class of subgroups of $Ptm_0(A^*)$ of subexponential growth.

Bartholdi, Kaimanovich and Nekrashevych [17] proved that $Ptm_0(A^*)$ is amenable. An alternative proof is given by Brieussel [48]. This result was extended by Amir, Angel and Virág [6] to $Ptm_1(A^*) := Ptm(A^*) \cap B_1(A^*)$, the group of permutational transducer mappings with linear activity growth.

9.3.9 Self-similar groups A subgroup G of $Aut(A^*)$ is *self-similar* if all sections g_v of all $g \in G$ at every vertex $v \in A^*$ are themselves elements of G.

By definition, every automaton group is self-similar, which is one of the reasons to study automaton groups. Self-similarity often allows inductive arguments. Conversely, if G is self-similar, then it can be viewed as the state set of an automaton with possibly infinitely many states. Hence, self-similar groups are also called *state-closed*, for example in [239, 200]. The monograph [201] by Nekrashevych is the standard reference for self-similar groups.

The general linear groups $GL_n(\mathbb{Z})$ have self-similar faithful actions on A^* with $|A| = 2^n$, as shown by Brunner and Sidki [56]; a very similar construction was given by Scott [231].

The main idea behind these embeddings is a general construction based on *virtual endomorphisms*, which are homomorphisms $\phi: H \to G$, where H is a subgroup of G of finite index. Every self-similar subgroup G of $Aut(A^*)$ gives rise to the virtual endomorphism $g \mapsto g_a$ from the vertex stabiliser $Stab_G(a)$ of $a \in A$ in G to G; see [201, Section 2.5].

9.3.10 Contracting groups A subgroup G of $Aut(A^*)$ is *contracting* if it is self-similar and there is a finite subset N of G such that, for every $g \in G$, there is a $k \geq 0$ with $g_v \in N$ for all $v \in A^*$ with $|v| \geq k$. The minimal such N is called the *nucleus of* G. We note that a contracting group is a subgroup of $Ptm(A^*)$.

One of the most useful features of a contracting group is the fact that the sections of elements are eventually shorter (with respect to word length) than the element itself. This was used in the proof of Proposition 9.2.3.

9.3.11 Proposition   *Let G be a finitely generated contracting group. Then there is a constant $M \in \mathbb{N}$ such that for every constant $K > 1$ there exists $k \in \mathbb{N}$ with $|g_v| \leq \frac{|g|}{K} + M$ for all $v \in A^k$.*

Proof   Let M be the maximal length of the elements of the nucleus of G and fix $K > 1$. Since the set of elements of G of length at most KM is finite, there exists k such that the sections at level k of those words are in the nucleus. Now factor an arbitrary element g of G as $g = g_1 g_2 \cdots g_{l+1}$, where $|g_i| = KM$ for $1 \leq i \leq l$ and $|g_{l+1}| \leq KM$. Then $l = \left\lfloor \frac{|g|}{KM} \right\rfloor$ and every section g_v of g at level k satisfies $|g_v| \leq M(l + 1) \leq \frac{|g|}{K} + M$, as required.   □

Holt and Röver [157] proved that finitely generated subgroups of $\mathrm{Ptm}_0(A^*)$ have indexed (and thus context-sensitive) co-word problem (see Sections 12.2 and 14.3). Bondarenko and Nekrashevych [25] proved that every finitely generated self-similar subgroup of $\mathrm{Ptm}_0(A^*)$ is contracting. There one can also find a proof of the following result.

9.3.12 Exercise   Show that the nucleus of a self-similar finitely generated subgroup of $\mathrm{Ptm}_0(A^*)$ consists of all sections of elements belonging to non-trivial circuits in the automaton of G.

9.3.13 Branch and weakly branch groups   Let G be a subgroup of $\mathrm{Aut}(A^*)$. We say that G is *level transitive* if G acts transitively on A^k for every $k \in \mathbb{N}$.

For $v \in A^*$, the *rigid (vertex) stabiliser of v in G* is the subgroup comprising those elements that act trivially outside the subtree vA^*; that is

$$\mathrm{Rist}_G(v) := \{g \in G \colon g_w \neq 1 \Rightarrow w \in vA^*\},$$

and the *rigid level stabiliser* of the k-th level is the direct product of the rigid vertex stabilisers of all vertices on level k, i.e.

$$\mathrm{Rist}_G(k) = \prod_{v \in A^k} \mathrm{Rist}_G(v).$$

The group G is a *weakly branch group* if it is level transitive and $\mathrm{Rist}_G(v) \neq 1$ (or equivalently is infinite) for every $v \in A^*$, and G is a *branch group* if it is level transitive and $\mathrm{Rist}_G(n)$ has finite index in G for every $n \in \mathbb{N}$.

Examples of branch groups are the full automorphism group $\mathrm{Aut}(A^*)$, the finitary automorphism group $F(A^*)$ and the group of permutational transducer mappings $\mathrm{Ptm}(A^*)$. Finitely generated examples of branch groups are provided by the Gupta–Sidki p-groups (9.2.4) and the Grigorchuk group (9.2.2). The adding machine is level transitive but not even weakly branch; an infinite cyclic group has no non-trivial commuting subgroups with trivial intersection. In fact a weakly branch group cannot be virtually nilpotent; this is observed in [16] and makes a nice exercise.

The class of branch groups appears as one of three cases in Wilson's classification [253] (or [254]) of just infinite groups; that is, infinite groups all of whose proper quotients are finite. Grigorchuk [119] introduced the term 'branch groups', and provided an equivalent purely algebraic definition; in this setting the corresponding trees are spherically homogeneous, i.e. all vertices of a given level have the same degree, which may vary from level to level.

A branch group G is called *a regular branch group over* K if K is a subgroup of finite index in G and $K \subseteq \mathrm{Rist}_K(a)$ for each $a \in A$. Again, the groups $\mathrm{Aut}(A^*)$, $F(A^*)$ and $\mathrm{Ptm}(A^*)$ are examples that all branch over themselves. The Grigorchuk group is a regular branch group over the normal closure of $(ab)^2$, and the Gupta–Sidki group is a regular branch group over its commutator subgroup. For these facts, which could be attempted as an exercise, and further information on branch groups, we refer to [119, 254, 16].

Lavreniuk and Nekrashevych [177] proved that the automorphism group of a weakly branch group is isomorphic to the normaliser of the group in the homeomorphism group of the boundary of the tree; see also [201, Section 2.10.2]. Röver [224] generalised this result, showing that the abstract commensurator of a weakly branch group is isomorphic to the relative commensurator of the group in the homeomorphism group of the boundary of the tree. Both works also provide technical criteria that force the automorphism group or the abstract commensurator to be contained in the automorphism group of the tree. Since this is the only place where these notions occur in this book, we restrict ourselves to the vague definition that relative and abstract commensurators are generalisations of normalisers and automorphism groups, respectively, when doing things 'up to finite index', and refer the reader to [224] or [74] for the details.

9.3.14 Recurrent or fractal groups A subgroup G of $\mathrm{Aut}(A^*)$ is called *fractal* (or *recurrent* in [201]), if it is level transitive and $\mathrm{Stab}_G(a)$, the stabiliser of a vertex a on the first level, projects onto G under the section map; that is $G = \{g_a : g \in \mathrm{Stab}_G(a)\}$.

This is a strengthening of self-similarity, in the sense that self-similarity is equivalent to saying that G is a subgroup of $G \wr_A \mathrm{Sym}(A)$ but does not imply that the section map restricted to the stabiliser of a vertex must be surjective.

9.3.15 Exercise Show that the adding machine, and the Gupta–Siki groups are fractal groups; for the Grigorchuk group this is contained in the proof of Theorem 9.2.3.

9.3.16 Limit space In his very interesting and ground breaking work [201], Nekrashevych describes a connection between contracting self-similar groups and certain dynamical systems. The important object in this context is the limit space $\mathcal{J}_G$ of a contracting self-similar group G, which is the quotient of the set $A^{-\omega}$ of left-infinite sequences over A by the asymptotic equivalence. Two left-infinite sequences $\cdots a_3 a_2 a_1$ and $\cdots b_3 b_2 b_1$ are asymptotically equivalent if there is a left-infinite path $\cdots e_3 e_2 e_1$ in (the graph of) the automaton of the nucleus of G such that the edge/transition e_i has label $a_i | b_i$. The space $\mathcal{J}_G$ is connected if the group is finitely generated and level-transitive.

It turns out that the asymptotic equivalence can also be characterised in terms of the Schreier graphs of the action of the group on the levels of the tree. Here, the Schreier graph of a group acting on a set with respect to a fixed generating set is the graph whose vertices are the permuted set and in which two of these vertices are connected if one is the image of the other under one of the fixed generators; for transitive actions this coincides with the definition in 1.6.3. Now two left-infinite sequences $\cdots a_3 a_2 a_1$ and $\cdots b_3 b_2 b_1$ happen to be asymptotically equivalent if and only if there is a global bound K such that $a_n \cdots a_2 a_1$ and $b_n \cdots b_2 b_1$ are at distance at most K in the Schreier graph of the action on level n for every $n \geq 1$. So in some sense, the Schreier graphs approximate the limit space.

The action of an automaton group on the n-th level of the tree A^* is given by the automaton over the alphabet A^n, which one obtains from the original automaton by viewing each path of length n as a single edge. The Schreier graph on level n is then the graph of the dual automaton (see next section) of this automaton over A^n.

Iterated monodromy groups provide the opposite route from certain dynamical systems to self-similar groups but are beyond the scope of this book.

9.4 Dual automata

In the definition of a permutational transducer (or just a synchronous fst) the roles of the states and the input letters can be interchanged, leading to the definition of the dual automaton. More precisely, if M is a synchronous fst with state set Σ over the alphabet A, then the dual automaton $\mathfrak{d}(M)$ of M is the fst with state set A and alphabet Σ, with a transition $a \xrightarrow{\sigma|\tau} b$ if and only if $\sigma \xrightarrow{a|b} \tau$ is a transition of M.

It is easy to verify that the dual of a synchronous dfst is also a synchronous dfst, but the dual of a permutational transducer is not necessarily a permutational transducer; for example, the dual of the adding machine (Figure 9.1) is not a permutational transducer. The permutational transducers in Figure 9.6 are dual to each other. Alešhin [3] used dual automata implicitly, while an explicit definition is in [185].

Observe that the states of a synchronous dfst generate a semigroup in general. The following easy result appears in [227], for example.

9.4.1 Theorem *The automaton (semi)group generated by a dfst M is finite if and only if the automaton (semi)group generated by its dual $\eth(M)$ is finite.*

Proof Let M have state set Σ and alphabet A. Assume that the (semi)group S generated by Σ is finite. Now let w be any element of the (semi)group $\eth(S)$ generated by A. So $w \in A^*$ and, for every $\mu \in \Sigma^*$, we have $w_\mu = \mu(w)$, by the definition of the dual. This means that the number of different sections of w is bounded by the size of S, which implies that $\eth(S)$ is finite, as there are only finitely many synchronous dfst with a fixed number of states over a given alphabet. Since $\eth(\eth(M)) = M$, this proves the theorem. $\quad\square$

9.4.2 Bireversible automata A permutational transducer is called *reversible* if its dual is also a permutational transducer. This notion was introduced by Macedońska, Nekrashevych and Sushchansky [185].

The permutational transducer M is *bireversible* if M and its inverse M^{-1} are both reversible.

We invite the reader to check that the automaton in Figure 9.5(v) is bireversible, and that one of the two automata generating the lamplighter group (Figure 9.5(iv)) is reversible but not bireversible.

9.4.3 VH-square complex Glasner and Mozes [103] exploit the fact that each transition $\sigma \xrightarrow{a|b} \tau$ can be encoded in a square with oriented edges; see the next section. Thus every permutational transducer gives rise to a VH-square complex, that is a square complex in which every vertex link is a bipartite graph. From this point of view, a permutational transducer is bireversible if and only if the vertex links in the square complex are in fact complete bipartite graphs.

9.5 Free automata groups

Bhattacharjee [23] showed that most (in a technical sense) n-tuples of elements of Aut(A^*) generate a non-abelian free group of rank n. However, it is not quite so easy to find non-abelian free groups in Ptm(A^*), let alone amongst automata groups. (Alešhin [3] claimed that the group generated by the automaton on the left in Figure 9.6 is a free group of rank three, but his proof was incomplete.)

As mentioned before, Brunner and Sidki [56] showed that the general linear groups $GL_n(\mathbb{Z})$ are subgroups of Ptm(A^*) when $|A| = 2^n$, which implies that Ptm(A^*) with $|A| \geq 4$ has non-abelian free subgroups. However, these examples are not automata groups yet; they are not generated by all states of an automaton.

Since non-abelian free groups are not amenable, the results of 9.3.8 imply that $\mathrm{Ptm}_1(A^*)$ has no non-abelian free subgroups. In fact, the following theorem is a special case of a result by Sidki [238] in which, under an additional hypothesis, the alphabet A is allowed to be infinite.

9.5.1 Theorem *For every n, the group $\mathrm{Ptm}_n(A^*) := B_n(A^*) \cap \mathrm{Ptm}(A^*)$ of permutational transducer mappings with polynomial activity growth of degree n (see 9.3.2) has no non-abelian free subgroups.*

The first non-abelian free automata groups were constructed by Glasner and Mozes [103] using VH-square complexes that are related to bireversible automata; see 9.4.3. However, since they use symmetric generating sets, the ranks of the groups are half the numbers of states of the defining transducers.

Vorobets and Vorobets [248] finally lifted the fog around Alešhin's automaton group from [3], by proving that the group defined by the Alešhin automaton on the left in Figure 9.6 is a free group of rank 3. The same authors [249] exhibited a countable family of non-abelian free automata groups whose ranks are odd and equal to the number of states. For even rank at least four, such a family is given by Steinberg, Vorobets and Vorobets [242]. All of these groups act on the binary tree, i.e. $|A| = 2$, and there it is impossible that the rank is two and equal to the number of states, by 9.2.5.

It is interesting to note that Alešhin's automaton is the only transducer with three states over a two-letter alphabet that generates a free group (see [28]).

9.5.2 Free products Savchuk and Vorobets [227] constructed a family of bireversible permutational transducers with n ($n \geq 4$) states whose automata groups are free products of n cyclic groups of order two. These groups have the recursive definition

$$a = (c, b), \ b = (b, c), \ c = \alpha(d_0, d_0), \ d_i = \alpha_i(d_{i+1}, d_{i+1}), \ d_k = \alpha(a, a),$$

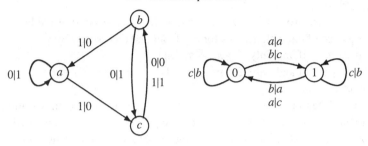

Figure 9.6 Alešhin's free group (left) and its dual

where $k \geq 0$, $0 \leq i < k$, and $\alpha_i \in \{1, \alpha\}$; the number of states is $k + 4$. These transducers generalise the case $n = 3$, which is known as the Bellaterra group or automaton and treated in [201, Section 1.10.3] and [28].

Free products of finitely many finite groups are realised as subgroups of $\mathrm{Ptm}(A^*)$ (for suitable A), but not as automata groups, by Gupta, Gupta and Oliynyk [128].

9.6 Decision problems

In this section we collect the current knowledge on decision problems in automata groups and semigroups. An automaton semigroup is a semigroup generated by the states of a synchronous dfst. Since many techniques rely on dual automata, dealing with groups and semigroups simultaneously is par for the course.

Here we should also mention the two GAP packages, FR by Bartholdi [15] and automgrp by Muntyan and Savchuk [199], which are geared towards calculations with automata groups and semigroups.

9.6.1 The word problem It is clear that a state σ of a permutational transducer defines the trivial map if and only if, for each vertex $v \in A^*$, the section σ_v is inactive. Since the composition of two transducers can be computed (see 9.1.5), it follows that the word problem is soluble for every subgroup of $\mathrm{Ptm}(A^*)$. However, this naive approach can be improved dramatically using the following minimisation process.

9.6.2 Minimisation via the Nerode equivalence Generalising the minimisation process for dfsa described in Theorem 2.5.4, we define the *Nerode equiv-*

alence of a synchronous fst M with state set Σ and alphabet A as the limit $\equiv$ of the sequence of equivalences $\equiv_k$ for $k \geq 0$ defined recursively by

$\sigma \equiv_0 \tau$ if and only if $\pi_\sigma = \pi_\tau$, and

$\sigma \equiv_{k+1} \tau$ if and only if $\sigma \equiv_k \tau$ and $\sigma_a \equiv_k \tau_a$ for all $a \in A$,

where $\sigma, \tau \in \Sigma$.

Since M has only finitely many states, this sequence becomes stationary from some point on, whence the Nerode equivalence can be computed. The *minimisation* $\mathfrak{m}(M)$ of M is the transducer obtained from M by identifying Nerode equivalent states with the obvious transitions. It is an exercise to verify that Nerode equivalent states define the same permutational transducer mapping, so that M and $\mathfrak{m}(M)$ define the same automaton (semi)group. Clearly $\mathfrak{m}(\mathfrak{m}(M)) = \mathfrak{m}(M)$ and we call M *minimal* if $\mathfrak{m}(M) = M$.

9.6.3 The conjugacy problem It is proved by Šunić and Ventura [244] that the conjugacy problem for automata groups is undecidable in general. Their construction involves a free group and hence the resulting example of an automaton group with insoluble conjugacy problem does not have polynomial activity growth (see 9.3.2), by Theorem 9.5.1. In fact, the conjugacy problem is soluble in the class of bounded automorphisms, as is proved by Bondarenko, Bondarenko, Sidki and Zapata [26].

Previously, Rozhkov [225] and Leonov [179] independently proved that the Grigorchuk group (9.2.2) has soluble conjugacy problem; Leonov's result covers a whole family of groups. For the Gupta–Sidki group (9.2.4) the solubility of the conjugacy problem follows from the stronger result by Wilson and Za-lesskii [257] that these groups, among others, are conjugacy separable.

9.6.4 Open problem Is the conjugacy problem soluble for automata groups with polynomial activity growth?

9.6.5 The finiteness problem The finiteness problem for a class of (semi) groups C is soluble, if there is a procedure which, given a (semi)group G in C, decides whether G is finite or not.

It is proved by Gillibert [95] that the finiteness problem for automaton semi-groups is insoluble, in general. For automaton groups the general case is still an open problem.

Nevertheless, there are a number of partial results. For example, it is proved by Klimann [168] that a semigroup generated by a reversible synchronous dfst with two states is either finite or free, which leads to a decision procedure if the automaton is also invertible, i.e. when it generates a group. The same author showed in joined work with Picantin [169] that reversible automata without cycles with exit generate finite groups. The multi-author paper [2] introduced mᔰ-reduction of an automaton and proved that an automaton with trivial mᔰ-reduction generates a finite (semi)group. This result is a straightforward consequence of Theorem 9.4.1 and the fact that an automaton and its minimisation generate the same (semi)group. For this purpose, the pair $(M, ᔰ(M))$ of an automaton M and its dual is called mᔰ-reduced if both automata are minimal. If the pair is not reduced, then replace it by $(m(M), ᔰ(m(M)))$ or by $(m(ᔰ(M)), ᔰ(m(ᔰ(M))))$; this process is confluent, so it does not matter which option one chooses, if there is an option. An automaton has trivial mᔰ-reduction if repeating this process leads to an automaton with one state and one letter. Another result of that paper is that the automaton group of a reversible but not bireversible permutational transducer is always infinite.

As an exercise one can verify that all the transducers from Figure 9.5(ii) and (v) have trivial mᔰ-reductions. It is also easy to check that an automaton with trivial mᔰ-reduction generates a finite (semi)group, using Theorem 9.4.1 and the fact that M and $m(M)$ generate the same (semi)group.

9.6.6 The order problem In contrast to the previous problems, the order problem is only concerned with a single permutational or deterministic and synchronous transducer mapping σ. It asks whether there exists an algorithm that decides if such a mapping has finite or infinite order.

Klimann, Picantin, Savchuk [171] recently introduced orbit automata and used these to decide the order problem in more situations than could be dealt with previously. In general, the order problem is still open.

The same authors proved in [170] that the automaton group of a connected 3-state reversible permutational transducer cannot be a Burnside group. In other words, if the group is infinite then it has an element of infinite order.

9.7 Resolving famous problems

Subgroups of $\mathrm{Ptm}(A^*)$ for finite A (see 9.1.10) have provided answers to many long standing problems in group theory. In this final section of the chapter we give a very brief summary of some of these and references for further reading.

9.7.1 Burnside groups In 1902, Burnside [59] asked whether a finitely generated torsion group must be finite. That this is not the case was first proved by Golod and Shafarevich [107, 106] (see also [140, Theorem 8.1.4]), but much simpler examples are groups like the Grigorchuk or Gupta–Sidki groups (see 9.2.2, 9.2.4 and Theorem 9.2.3). Other examples, which are not all p-groups, were constructed earlier by Sushchansky [245]. Additional information and references can be found in [74], for example.

9.7.2 Intermediate growth In 1968, Milnor [194] asked whether a finitely generated group has either an exponential or a polynomial growth function (see 1.6.5). As mentioned before, the Grigorchuk group is one member of an uncountable family of groups of intermediate growth, as shown by Grigorchuk [115, 116], and Bartholdi [12] proved that the Gupta–Sidki group also has intermediate growth. Grigorchuk [117] also found the first torsion-free groups of intermediate growth, while Erschler [88] constructed the first examples that are not residually finite. For the historic development of all important results on growth functions of (finitely generated) groups we refer to the monograph by Mann [187].

9.7.3 Non-elementary amenability We gave the definition of an *amenable* group in 9.3.8. All finite groups and all abelian groups are amenable and the class of amenable groups is closed under the following operations: passing to subgroups, taking quotients, group extensions (see 1.5.4) and direct limits. The smallest class of groups that contains all finite and all abelian groups and is closed under these four operations is known as the class of *elementary amenable groups*. In 1957, Day [73] asked whether every amenable group is elementary amenable.

In fact, in the definition of the class of elementary amenable groups it suffices to require closure under extensions and direct limits. It then follows that every elementary amenable torsion group is locally finite. The Grigorchuk group is amenable (see 9.3.8), but it is not elementary amenable, as it is not locally finite, by Theorem 9.2.3, and hence answers Day's question.

Lysënok [184] proved that the Grigorchuk group G has the presentation

$$\langle a, b, c, d \mid a^2, b^2, c^2, d^2, bcd, \phi^i(ad)^4, \phi^i(adacac)^4 \text{ for } (i \geq 0) \rangle$$

where ϕ is the injective endomorphism of G given by

$$\phi(a) = aca, \quad \phi(b) = d, \quad \phi(c) = b, \quad \phi(d) = c.$$

Grigorchuk [118] observed that this means that the HNN-extension

$$\langle G, t \mid a^2, b^2, c^2, d^2, bcd, (ad)^4, (adacac)^4, a^t = aca, b^t = d, c^t = b, d^t = c \rangle$$

of G is a finitely presented amenable but not elementary amenable group.

Here we have to point out that Bartholdi [13] proved that contracting regular branch groups (see 9.3.10 and 9.3.13) are not finitely presentable but have finite endomorphic presentations. In particular, the Grigorchuk group is not finitely presentable. A finite *endomorphic presentation* of a group consists of a finite number of finite sets of relators $R_0, \ldots, R_m$ and endomorphisms $\phi_1, \ldots, \phi_m$ such that $R_0 \cup \bigcup_{1 \leq i \leq m, j \geq 0} R_i^{\phi^j}$ is a set of defining relators of that group.

9.7.4 Subgroup growth A finitely generated group has only finitely many subgroups of given index $n \in \mathbb{N}$. This leads to the *subgroup growth function* $\eta_G(n) = |\{U \subseteq G : |G : U| \leq n\}|$. Lubotzky, Mann and Segal [180] proved that a group has polynomial subgroup growth if and only if it is virtually soluble of finite rank. Lubotzky, Pyber and Shalev [181] conjectured that, for a group that does not have polynomial subgroup growth, there is a constant $c > 0$ such that $\eta_G(n) \geq n^{(c \log n)/(\log \log n)^2}$ for infinitely many n. This conjecture was refuted by Segal [233], using specially crafted just-infinite branch groups in the automorphism group of a spherically homogeneous tree (see 9.3.13). His construction also produces groups whose finite quotients have a prescribed set of finite simple groups as composition factors.

9.7.5 Non-uniformly exponential growth Recall from 1.6.5 the definition of the growth function $\gamma_{G,X}(n)$ of a finitely generated group G. The *growth rate* of G is defined as $\gamma(G, X) = \lim_{n \to \infty} \sqrt[n]{(\gamma_{G,X}(n)}$. The group G has exponential growth if and only if $\gamma(G, X) > 1$ for some (or equivalently every) finite generating set X of G. The group is of *uniformly exponential growth* if there exists an $\epsilon > 0$ such that $\gamma(G, X) \geq 1 + \epsilon$ for every finite generating set X of G. This definition is due to Gromov [127] where he asked whether exponential growth implies uniformly exponential growth. Wilson [256] constructed the first group of non-uniformly exponential growth as a subgroup of $\mathrm{Aut}(A^*)$, which he then generalised [255] to an uncountable family of such groups. A similar construction was given by Bartholdi [14].

9.7.6 Strong Atiyah Conjecture Defining the notions needed in this paragraph is beyond the scope of this book, but we still like to mention the results. Grigorchuk and Żuk [123] calculated the spectrum of the random walk operator of the lamplighter group with respect to (conjugates of) the generators given in 9.2.6, which subsequently (despite the publication dates) led them, in joint work with Linnell and Schick [121], to a counterexample to the Strong Atiyah Conjecture about Betti numbers.

PART THREE

THE WORD PROBLEM

10
Solubility of the word problem

10.1 The Novikov–Boone theorem

We devote this section to an outline of the proof of the following theorem.

10.1.1 Theorem (Novikov–Boone, 1955) *There exists a finitely presented group with insoluble word problem.*

The account that we give here is based on the more detailed treatment in Rotman's book [223, Chapter 12], although we have changed some of the notation. An explicit example of a presentation of a group with insoluble word problem, with 10 generators and 27 relators of total length 427, can be found in Collins' article [71].

We let

$$S = S(M) = \mathrm{Sgp}\langle C \cup \Sigma \cup \{\hat{\sigma}\} \mid \phi_i \sigma_i \psi_i = \theta_i \sigma'_i \kappa_i, \ i \in I \rangle,$$

with $\phi_i, \psi_i, \theta_i, \kappa_i \in C^*$, $\sigma_i, \sigma'_i \in \Sigma \cup \{\hat{\sigma}\}$, be a finitely presented semigroup constructed as described in Corollary 2.9.8, for which there is no decision process to decide, given arbitrary strings $u, v \in C^*$ and $\sigma \in \Sigma$, whether $u\sigma v =_S \hat{\sigma}$. Let $\Sigma' := \Sigma \cup \{\hat{\sigma}\}$.

Given a word $w = c_1^{\epsilon_1} \cdots c_k^{\epsilon_k}$ with each $c_i \in C$ and $\epsilon_i = \pm 1$, we define the word $\bar{w}$ to be $c_1^{-\epsilon_1} \cdots c_k^{-\epsilon_k}$ and, given two such words u, v and $\sigma \in \Sigma'$, we define the word $(u\sigma v)^\dagger$ to be $\bar{u}\sigma v$. The reason for using $(u\sigma v)^\dagger$ rather than $u\sigma v$ in the group relations that we are about to introduce will become clear later.

The plan now is to adapt the semigroup presentation of S to construct a finitely presented group with insoluble word problem. First, for each $i \in I$, we introduce a new generator r_i corresponding to the relation $\phi_i \sigma_i \psi_i = \theta_i \sigma'_i \kappa_i$, and then replace that relation in S by a relation $r_i^{-1} (\phi_i \sigma_i \psi)^\dagger r_i = (\theta_i \sigma'_i \kappa_i)^\dagger$ in the group. Then we introduce a further generator x, together with a set of new

221

group relations

$$\{c^{-1}xc = x^2, \; r_i^{-1}cxr_i = cx^{-1} : c \in C, \; i \in I\}.$$

Let $R = \{x\} \cup \{r_i : i \in I\}$ be the new generators. The group

$$H := \langle C \cup \Sigma' \cup R \mid r_i^{-1}\bar{\phi}_i\sigma_i\psi_ir_i = \theta_i\sigma'_i\kappa_i, \forall i \in I,$$
$$r_i^{-1}cxr_i = cx^{-1}, \forall c \in C, \; \forall i \in I,$$
$$c^{-1}xc = x^2, \forall c \in C \rangle$$

is called G_2 in [223].

Suppose that u, v are positive words in C^* and $\sigma \in \Sigma$, let $\gamma = u\sigma v$, and suppose that $\gamma =_S \hat{\sigma}$. We claim that $\gamma^\dagger$ is equal in H to a word of the form $\lambda\hat{\sigma}\rho$, where λ and ρ are words over R.

For within S, the word $\gamma = u\sigma v$ can be transformed to $\hat{\sigma}$ by a sequence of applications of relations to subwords each containing a single element of Σ. The first such application replaces a subword of $u\sigma v$ equal to one side of a relation $\phi_i\sigma_i\psi_i = \theta_i\sigma'_i\kappa_i$ by the other. Suppose first that the left-hand side is replaced by the right-hand side; that is, that $u\sigma v = u_1\phi_i\sigma_i\psi_iv_1$ is transformed to $u_1\theta_i\sigma'_i\kappa_iv_1$. Then, within H, we apply the analogue of that substitution to $\gamma^\dagger = \bar{u}_1\bar{\phi}_i\sigma_i\psi_iv_1$ to derive $\bar{u}_1r_i\theta_i\sigma'_i\kappa_ir_i^{-1}v_1$. Using the relations of H that involve x, we now transform $\bar{u}_1r_i$ to $\lambda_1\bar{u}_1$, and $r_i^{-1}v_1$ to $v_1\rho_1$, where λ_1 and ρ_1 are words over R, so that $(u\sigma v)^\dagger$ has been transformed to $\lambda_1\bar{u}_1\theta_i\sigma'_i\kappa_iv_1\rho_1$.

(Note that, for $c \in C$ and $i \in I$, we have $c^{-1}r_i =_H xr_ixc^{-1}$, but there is no such transformation for cr_i, which is the reason why we have used $\bar{u}\sigma v$ rather than $u\sigma v$ in our group relations.)

Now suppose, alternatively, that the first substitution in S replaces the right-hand side of the relation by the left-hand side, so that $u\sigma v = u_1\theta_i\sigma'_i\kappa_iv_1$ is transformed to $u_1\phi_i\sigma_i\psi_iv_1$. Then, within H, the word $(u\sigma v)^\dagger$ can be transformed first to $\bar{u}_1r_i^{-1}\bar{\phi}_i\sigma_i\psi_ir_iv_1$ and then to $\lambda_1\bar{u}_1\phi_i\sigma_i\psi_iv_1\rho_1$, where λ_1 and ρ_1 are again words over R.

Proceeding analogously through the sequence of applications of relations of S, we find corresponding relations of H that ultimately can transform $\gamma^\dagger$ to a word of the form $\lambda\hat{\sigma}\rho$, as claimed.

We now define a group G, by adding two further generators t, k and some commutativity relations involving them to the presentation for H. Using the notation $X(K)$ to denote our chosen generating set of an arbitrary group K, we then have $X(G) = C \cup \Sigma' \cup \{x, t, k, r_i \ (i \in I)\}$, and we define

$$G = \langle X(G) \mid c^{-1}xc = x^2, \ \forall c \in C, \ r_i^{-1}cxr_i = cx^{-1}, \forall c \in C, \ \forall i \in I,$$
$$r_i^{-1}\phi_i\sigma_i\psi_i r_i = \bar{\theta}_i\sigma_i'\kappa_i, \ \forall i \in I,$$
$$tr_i = r_i t, \ \forall i \in I, \ tx = xt,$$
$$kr_i = r_i k, \ \forall i \in I, \ kx = xk, \ k(\hat{\sigma}^{-1}t\hat{\sigma}) = (\hat{\sigma}^{-1}t\hat{\sigma})k\rangle.$$

The ordering of the relations in the presentation has been chosen for convenience in 10.1.3.

It remains to show that G has insoluble word problem. This follows from the following result known as Boone's Lemma.

10.1.2 Lemma *Suppose that the semigroup S and the group G are as defined above. Let $\gamma = u\sigma v$, where $u, v \in C^*$. Then $\gamma =_S \hat{\sigma}$ if and only if $w = \gamma^\dagger$ satisfies the equation $[w^{-1}tw, k] =_G 1$.*

For if there were a decision process to determine whether a word over $X(G)$ was the identity in G, then we could use this process on words $\gamma = u\sigma v$ with $u, v \in C^*$ to determine whether $\gamma =_S \hat{\sigma}$, thereby contradicting Corollary 2.9.8. Hence the proof of Theorem 10.1.1 is completed by the proof of Boone's Lemma.

The proof of one of the two directions of the equivalence of the lemma is straightforward. For suppose that $\gamma = u\sigma v$ satisfies $\gamma =_S \hat{\sigma}$. We established above that there are words λ and ρ over R for which $\gamma^\dagger =_H \lambda\hat{\sigma}\rho$ and hence $\gamma^\dagger =_G \lambda\hat{\sigma}\rho$. It follows from the relations of G that k commutes with

$$\rho^{-1}\hat{\sigma}^{-1}t\hat{\sigma}\rho =_G (\lambda\hat{\sigma}\rho)^{-1}t(\lambda\hat{\sigma}\rho) =_G (\gamma^\dagger)^{-1}t(\gamma^\dagger).$$

In other words, the word $w = \gamma^\dagger$ is a solution of the equation $[w^{-1}tw, k] =_G 1$.

The proof of the opposite implication of this lemma occupies several pages in [223], and is based on the decomposition of G as a sequence of HNN-extensions. We now present a summary of that proof.

10.1.3 Summary of the proof of the 'if' part of Boone's Lemma. Let S, G, γ, u, v, and $w = \gamma^\dagger = (u\sigma v)^\dagger$ be as in Boone's Lemma, and suppose that w is a solution of the equation $[w^{-1}tw, k] =_G 1$. We split the proof that $\gamma =_S \hat{\sigma}$ into four steps.

Step 1. We express G as a chain of HNN-extensions. (We find it convenient here to have one more subgroup in this chain than in the proof in [223], which means that our notation diverges once again from that of [223].)

We define $G_0, G_1, G_2, G_3, G_4, G_5 = G$ to be groups defined by presentations, where the generating sets are

$$X(G_0) = \{x\}, \quad X(G_1) = X(G_0) \cup C, \quad X(G_2) = X(G_1) \cup \Sigma',$$

$$X(G_3) = X(G_2) \cup \{r_i \mid i \in I\}, \quad X(G_4) = X(G_3) \cup \{t\},$$

$$X(G_5) = X(G_4) \cup \{k\},$$

and the defining relations of each G_i are those defining relations of G that involve only the generators of G_i.

We claim that each G_i, for $1 \leq i \leq 5$, is a sometimes multiple HNN-extension with base group G_{i-1}. In particular, this allows us to identify each G_i with the subgroup of G that is generated by its generators, giving the chain of subgroups $1 < G_0 < G_1 < G_2 < G_3 = H < G_4 < G_5 = G$.

The relations of $G = G_5$ involving k all specify that k commutes with certain elements of G_4, so G is an HNN-extension of G_4 with stable letter k. Similarly, G_4 is an HNN-extension of $H = G_3$ with stable letter t. It is clear from the relations $c^{-1}xc = cx^2$ that G_1 is an HNN-extension with base G_0 and stable letters C, and $G_2 = G_1 * F(\Sigma')$, where $F(\Sigma')$ is the free group on Σ', is an HNN-extension of G_1 with stable letters Σ'.

Finally, since C generates a free subgroup of G_1 modulo $\langle x^{G_1} \rangle$ and, for $\epsilon = \pm 1$, the sets $C_\epsilon := \{cx^\epsilon \mid c \in C\}$ project onto C in $G_1/\langle x^{G_1} \rangle$, we see that for each $i \in I$ the two sets $C_1 \cup \{\bar{\phi}_i\sigma_i\psi_i\}$ and $C_{-1} \cup \{\bar{\theta}_i\sigma_i'\kappa_i\}$ both freely generate subgroups of G_2, and hence G_3 is an HNN-extension of G_2 with stable letters r_i, as claimed. Note that G_3 is the subgroup H defined in the previous section.

Step 2. Our next aim is to use our assumption that $[w^{-1}tw, k] =_G 1$ to derive an equation of the form $\hat{\sigma} =_{G_3} \lambda w \rho$, where λ, μ are words over R. (Note that the derivation of this equation in G was also an intermediate step in the proof of the other implication of Boone's Lemma, above.)

Application of Britton's Lemma (Corollary 1.5.18) to G as an HNN-extension of G_4 implies immediately that $w^{-1}tw$ is in the subgroup of G_4 generated by $R \cup \{\hat{\sigma}^{-1}t\hat{\sigma}\}$. So we have

$$w^{-1}tw\rho_0\hat{\sigma}^{-1}t^{\epsilon_1}\hat{\sigma}\rho_1\hat{\sigma}^{-1}t^{\epsilon_2}\hat{\sigma}\rho_2 \cdots \rho_{n-1}\hat{\sigma}^{-1}t^{\epsilon_n}\hat{\sigma}\rho_n =_G 1,$$

where each ρ_i is a word over R and $\epsilon_i = \pm 1$, and we choose such an expression with n minimal. By Britton's Lemma applied to G_4 as an HNN-extension of G_3 with stable letter t, this word must contain a pinch $t^{-\epsilon}gt^\epsilon$ with $g \in \langle R \rangle$ and, since t commutes with all elements of R, this pinch evaluates to g in G_4.

Suppose first that the pinch does not involve the first occurrence of t in the word. Then it must be a subword $t^{\epsilon_i}\hat{\sigma}\rho_i\hat{\sigma}^{-1}t^{\epsilon_{i+1}}$ for some i with $0 < i < n$ and $\epsilon_{i+1} = -\epsilon_i$, and for which $\hat{\sigma}\rho_i\hat{\sigma}^{-1} =_{G_3} g \in \langle R \rangle$. Then, since t commutes with g, this subword is equal in G_4 to g. But the subword is part of the longer subword $\rho_{i-1}\hat{\sigma}^{-1}t^{\epsilon_i}\hat{\sigma}\rho_i\hat{\sigma}^{-1}t^{\epsilon_{i+1}}\hat{\sigma}\rho_{i+1}$, which is equal in G_4 to $\rho_{i-1}\hat{\sigma}^{-1}g\hat{\sigma}\rho_{i+1} =_{G_3}$ $\rho_{i-1}\rho_i\rho_{i+1}$. So the whole word is equal in G_4 to a word of the same form with a smaller value of n, contradicting the minimality of n.

So the pinch must involve the first t in the word, and hence $w\rho_0\hat{\sigma}^{-1} \in \langle R \rangle$, from which our claim follows immediately.

Step 3. So now we have $\hat{\sigma} =_{G_3} \lambda w\rho = \lambda\bar{u}\sigma v\mu$ with λ, μ words over R, and we may clearly assume that λ and μ are freely reduced. The remainder of the proof takes place within the group $H = G_3$.

By Britton's Lemma in the HNN-extension G_3 of G_2 with stable letters r_i, the word $\lambda\bar{u}\sigma v\mu\hat{\sigma}^{-1}$ must either be a word over $X(G_2)$, or it must contain a pinch of the form $r_j^{-\epsilon}gr_j^{\epsilon}$ for some $j \in I$, where g is a word over $X(G_2)$. In this step, we show that a pinch of this form cannot be contained entirely within either of the words λ or μ. If it were then, since x is the only generator in $X(G_2) \cap R$, g would have to be a power of x. But we saw in Step 1 that the subgroup of G_2 associated with the stable letter r_j is freely generated by $C_1 \cup \{\bar{\phi}_j\sigma_j\psi_j\}$ and $C_{-1} \cup \{\bar{\theta}_j\sigma'_j\kappa_j\}$, so they contain no non-trivial powers of x, and the claim follows.

Step 4. We now complete the proof, and show that $u\sigma v =_S \hat{\sigma}$. The proof is by induction on the number m of letters in λ from the set $\{r_i^{\pm 1} : i \in I\}$. Suppose first that $m = 0$. Since, by Step 3, there is no pinch contained entirely within μ, there can be no occurrence of any $r_i^{\pm 1}$ in μ. So $\lambda = x^k$ and $\mu = x^l$ for some $k, l \in \mathbb{Z}$, and $x^k\bar{u}\sigma v x^l =_{G_2} \hat{\sigma}$. But we saw in Step 1 that $G_2 = G_1 * F(\Sigma')$, and so this is only possible if $\sigma = \hat{\sigma}$ and $x^k\bar{u} =_{G_1} 1 =_{G_1} vx^l$ which, from the structure of G_1 implies $k = l = 0$ and $u = v = \varepsilon$. So $u\sigma v$ and $\hat{\sigma}$ are equal as words, and are thus certainly equal in S.

So assume that $m > 0$. Then, by Britton's Lemma applied to G_3, the word $\lambda\bar{u}\sigma v\mu\hat{\sigma}^{-1}$ contains a pinch $r_j^{-\epsilon}gr_j^{\epsilon}$ for some $j \in I$ which, by Step 3, does not lie entirely within λ or μ. So, since λ and μ are the only subwords that could contain $r_j^{\pm 1}$, the pinch must involve the final letter of λ from the set $\{r_i^{\pm 1} : i \in I\}$, and the first such letter of μ. So the pinch has the form $r_j^{-\epsilon}x^k\bar{u}\sigma v x^l r_j^{\epsilon}$ with $k, l \in \mathbb{Z}$. So

$$\lambda\bar{u}\sigma v\mu = \lambda' r_j^{-\epsilon}x^k\bar{u}\sigma v x^l r_j^{\epsilon}\mu' \qquad (*)$$

where λ', μ' are words over R and λ' contains only $m - 1$ letters from the set $\{r_i^{\pm 1} : i \in I\}$.

We assume that $\epsilon = 1$; the case $\epsilon = -1$ is similar. So $x^k\bar{u}\sigma v x^l$ is equal in G_2 to an element in the subgroup freely generated by $C_1 \cup \{\bar{\phi}_j\sigma_j\psi_j\}$. Since $G_2 = G_1 * F(\Sigma')$, it is not hard to see that $x^k\bar{u}\sigma v x^l =_{G_2} \alpha\bar{\phi}_j\sigma_j\psi_j\beta$, where $\sigma = \sigma_j$ and α, β are (freely reduced) words over C_1, from which it follows that $x^k\bar{u} =_{G_1} \alpha\bar{\phi}_j$ and $vx^l =_{G_1} \psi_j\beta$.

From the second of these equations, we have $x^l\beta^{-1}\psi_j^{-1}v =_{G_1} 1$, where ψ_j and v are both positive words over C. We claim that the whole of ψ_j^{-1} cancels when freely reducing $\psi_j^{-1}v$. If not, then the free reduction of $\psi_j^{-1}v$ would start with c^{-1} for some $c \in C$. Considering G_1 is an HNN-extension of G_0, the word $x^l\beta^{-1}\psi_j^{-1}v$ would contain a pinch. Since α and β can contain no pinches, this pinch would have the form cx^jc^{-1}, with cx^j a suffix of $x^l\beta^{-1}$. But β^{-1} is a word over C_1, so this is only possible for $j = 1$ which is a contradiction, because cxc^{-1} is not a pinch. So (as words) $v = \psi_j v'$ for some $v' \in C^*$, where $v'x^l =_{G_1} \beta$, and similarly $u = u'\phi_j$ with $u' \in C^*$, where $x^k\bar{u}' =_{G_1} \alpha$.

The assignment $x \mapsto x^{-1}$, $c \mapsto c, \forall c \in C$ induces an automorphism of G_1, and conjugation by r_j has the same effect on the elements of C_1 as this automorphism. Since $v'x^l =_{G_1} \beta \in \langle C_1 \rangle$, it follows that $r_j^{-1}v'x^lr_j =_{G_3} v'x^{-l}$ and similarly $r_j^{-1}x^k\bar{u}'r_j =_{G_3} x^{-k}\bar{u}'$.

Returning to the equation (∗) above, and recalling that we have proved that $\sigma = \sigma_j$ and assumed that $\epsilon = 1$, we have

$$\hat{\sigma} =_{G_3} \lambda\bar{u}\sigma_j v\mu = \lambda'r_j^{-\epsilon}x^k\bar{u}\sigma_j vx^l r_j^\epsilon\mu'$$
$$= \lambda'r_j^{-\epsilon}x^k\bar{u}'\phi_j\sigma_j\psi_j v'x^l r_j^\epsilon\mu'$$
$$=_{G_3} \lambda'x^{-k}\bar{u}'r_j^{-\epsilon}\phi_j\sigma_j\psi_j r_j^\epsilon v'x^{-l}\mu'$$
$$=_{G_3} \lambda'x^{-k}\bar{u}'\theta_j\sigma_j'\kappa_j v'x^{-l}\mu'.$$

We can now apply our inductive hypothesis with $\lambda'x^{-k}$ and $x^{-l}\mu'$ in place of λ and μ, with $u'\theta_j$ and $\kappa_j v'$ in place of u and v, and with σ_j' in place of $\sigma = \sigma_j$. From this, we conclude that $u'\theta_j\sigma_j'\kappa_j v' =_S \hat{\sigma}$. Now, since $\phi_j\sigma_j\psi_j =_S \theta_j\sigma_j'\kappa_j$ is a defining relation of S, and $u'\phi_j = u$, $\psi_j v' = v$, we have $u\sigma v =_S \hat{\sigma}$, which completes the proof.

10.2 Related results

The *Grzegorczyk hierarchy* is a subdivision $E_0 \subset E_1 \subset E_2 \subset \cdots$ of the primitive recursive functions, which ranks them according to their growth rates. We shall not give a precise definition here, but simply note that E_1 includes all linearly bounded functions, E_2 includes all polynomially bounded functions, and E_3 (the *elementary functions*) includes all functions that are bounded by an iterated exponential function, with a fixed number of iterations.

By adapting the arguments used in the proof of the Novikov–Boone Theorem, Avenhaus and Madlener [10] prove that, for any $n \geq 3$, there are finitely presented groups for which the word problem is soluble, and for which the time complexity of the solution is bounded above by a function in the class

E_n but not by a function in the class E_{n-1}. In [186, Theorem 4.2], Madlener and Otto proved the stronger result that, for $3 \leq n < m$, the same is true, and furthermore the Dehn function of G is bounded above by a function in E_m but by no function in E_{m-1}.

There is also a hierarchy amongst the recursively enumerable functions that measures their degree of insolubility, and there exist finitely presented groups in which the word problem has any prescribed degree of insolubility. See Miller's survey article [192, Section 2] for references and further results of this type.

The methods involving HNN-extensions that were employed in the construction of groups with insoluble word problem were developed further by Higman and Boone. We recommend Rotman's book [223] for accessible proofs.

We recall from 3.1.2 that a group is said to be *recursively presentable* if it is defined by a presentation with finitely many generators and a recursively enumerable set of relators.

10.2.1 Theorem (Higman 1961, [223, Theorem 12.18]) *Every recursively presentable group can be embedded in a finitely presented group.*

We saw in Theorem 1.5.19 that every countable group G can be embedded in a 2-generator group L. From the proof we see that, if the elements of G can be enumerated recursively, then L is recursively presentable, and hence G can be embedded in a finitely presented group. This result can be used [223, Theorem 12.28] to construct a finitely presented group that contains an isomorphic copy of every countable abelian group. Another application [223, Theorem 12.29] is the construction of a finitely presented group that contains isomorphic copies of all finitely presented groups.

The following result of Boone and Higman provides a fascinating connection between the algebraic properties of a group and the solubility of its word problem.

10.2.2 Theorem (Boone, Higman 1974, [223, Theorem 12.30]) *A finitely generated group has soluble word problem if and only if it can be embedded in a simple subgroup of some finitely presented group.*

Further embedding results are discussed in Section 5 of Sapir's survey article [226].

11

Context-free and one-counter word problems

11.1 Groups with context-free word problem

The following theorem, which says that a group has context-free word problem if and only if it is virtually free, is certainly one of the most fundamental results that inter-relates formal language theory and group theory. The main argument is due to Muller and Schupp, and was published in 1983 [198], but it required a deep result of Dunwoody on the accessibility of finitely presented groups, published in 1985 [79], to complete the proof. We observe also that Dunwoody's result is not needed for torsion-free groups.

We recall from 2.2.7 that the classes of context-free and deterministic context-free languages are denoted by $\mathcal{CF}$ and $\mathcal{DCF}$, respectively.

11.1.1 Theorem (Muller, Schupp 1983) *If G is a finitely generated group, then the following statements are equivalent.*

(i) WP(G) $\in \mathcal{CF}$.

(ii) WP(G) $\in \mathcal{DCF}$.

(iii) G *is virtually free.*

Proof (ii) $\implies$ (i) is true by definition. We described an informal proof of (iii) $\implies$ (ii) in 2.2.8. The formal proof that WP($G, X^{\pm}$) $\in \mathcal{DCF}$ when G is free on X is straightforward modulo a few technicalities: we use the stack to record the reduced form of the word read so far, but we need a bottom of stack symbol b_0, and the pda needs to be able to look at the top two stack symbols, in order to know when the second symbol is b_0. In that case, if the result of the move is to remove the top symbol, then the new top symbol is b_0, and the pda moves into an accept state. This is possible by Proposition 2.6.9. The extension of the result to virtually free groups then follows from Propositions 2.6.34 and 3.4.7.

It remains to prove (i) $\implies$ (iii): if G has context-free word problem then it

228

is virtually free. We give a summary of the proof here and refer the reader to [198] or Chiswell's book [68] for further details. Since the trivial group is free of rank 0, all finite groups are virtually free, and so we may assume that G is infinite. The proof splits into four parts.

Step 1. Let $G = \langle X \rangle$ and $A := X^{\pm}$. Since WP(G, A) is context-free, it is the language generated by a context-free grammar in which the set of terminals is A. Let S be the start symbol of the grammar and V its set of variables. By 2.6.14 we may assume that it is in Chomsky normal form with no useless variables; i.e. each rule has the form $S \to \varepsilon$, $v \to v'v''$ or $v \to a$, where v, v', v'' are variables with $v', v'' \neq S$, and a is a terminal. Also, S does not occur on the right-hand side of any rule.

11.1.2 Lemma *If $v \in V$ and $v \Rightarrow^* w_1$, $v \Rightarrow^* w_2$ with $w_1, w_2 \in A^*$, then $w_1 =_G w_2$.*

Proof Since all variables are useful, there are derivations $S \Rightarrow^* \alpha v \beta \Rightarrow^* \sigma w_1 \tau$ and $S \Rightarrow^* \alpha v \beta \Rightarrow^* \sigma w_2 \tau$ with $\alpha, \beta \in V^*$ and $\sigma, \tau \in A^*$. Hence $\sigma w_1 \tau =_G 1 =_G \sigma w_2 \tau$, and the result follows. □

Suppose that $w = a_1 \cdots a_n \in$ WP(G, A) with $n \geq 3$. We shall construct a planar diagram for w in which the boundary is a simple closed n-gon labelled w. By examining the derivation of w in the grammar, we shall show that there is a constant $K \in \mathbb{N}$ with the property that this boundary loop can be triangulated by triangles all of whose vertices are on the boundary, and in which the edges of each triangle are either edges of the boundary, or are internal chords labelled by words of length at most K.

Now w is derived from S through application of a sequence of rules involving a finite set of variables $v_1, \ldots, v_m$ other than S and, for each v_i, we have $v_i \Rightarrow^* w_i$ for some subword w_i of w. For each such v_i, choose a word $u_i \in A^*$ of minimal length such that $v_i \Rightarrow^* u_i$, and let K be the maximum length of the u_i. Then by Lemma 11.1.2 we have $u_i =_G w_i$ for all i, and so we can draw an internal chord in our diagram labelled u_i from the start to the end vertex of w_i.

By doing this for all v_i for which $1 < |w_i| < n - 1$, we obtain the required triangulation. (But note that, for the variables v_1, v_2 that arise in the first derivation $S \Rightarrow v_1 v_2$, if $|w_1| > 1$ and $|w_2| > 1$ then, in order to avoid getting an internal bigon, we omit the chord labelled u_2.) Figure 11.1 illustrates such a triangulation, where $S \Rightarrow v_1 v_2$ and $v_1 \Rightarrow v_3 v_4$.

Let $L := \max(K, 1)$. Then we have shown that, for each word $w \in$ WP(G, A), there is a van Kampen diagram for w whose internal regions have boundary length at most $3L$. It follows that G has the finite presentation $\langle X \mid R \rangle$, where $R = \{w \in$ WP$(G, A) : |w| \leq 3L\}$.

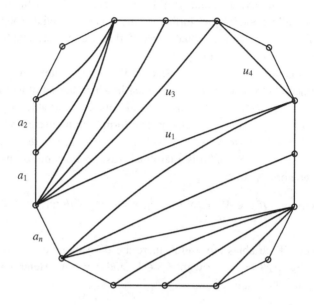

Figure 11.1 Triangulation of the word w

We need a further lemma about triangulations of the form described above.

11.1.3 Lemma   *Suppose that $w \in \mathrm{WP}(G, A)$ with $|w| \geq 3$, and that the boundary edges in the triangulated diagram for w are divided into three consecutive nonempty arcs. Then there exists a triangle in the triangulation with a vertex in each of the three arcs. (A vertex where two arcs meet is in both arcs.)*

Proof   This is clear if there is only one triangle (i.e. if $|w| = 3$), so assume not. The first step is to show that there must be a triangle with two edges on the boundary of the loop. In fact it is easier to prove the stronger assertion that there must be at least two such triangles, which we do by induction on the number of edges on the boundary. Choose a triangle with a single boundary edge, and remove that triangle and boundary edge. The new diagram has two components connected at a single vertex, and we apply induction to both of these. This yields at least one triangle with two edges on the original boundary in each of the two components, including when one or both of these components consists of a single triangle.

So now, choose a triangle with two boundary edges. If it has vertices in all three arcs then we are done. Otherwise, we delete this triangle and its two boundary edges, and the result follows by applying the inductive hypothesis to

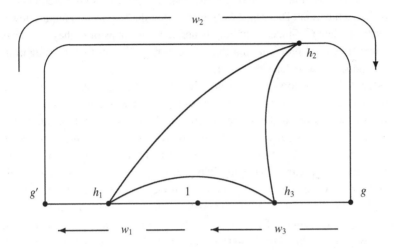

Figure 11.2 Contradiction for step 2

the resulting triangulation of the diagram in which the two boundary edges of the deleted triangle are replaced by its third edge. □

Step 2. We show that $\Gamma(G, X)$ is disconnected by the deletion of a ball of radius $3K/2 + 1$, and hence that G has more than one end. Suppose not, and assume for convenience that K is even.

Since G is infinite, there are geodesic words of arbitrary length. Choose such a word $w_3 w_1$ with $|w_1| = |w_3| = 3K/2 + 1$, and consider the path that it traces in $\Gamma(G, X)$ in which the end of the prefix w_3 is at the base point 1 of $\Gamma(G, X)$. Let g and g' be the start and end points of this path. Then, since we are assuming that removing the (open) ball of radius $3K/2 + 1$ with centre at 1 from $\Gamma(G, X)$ does not disconnect $\Gamma(G, X)$, there is a word w_2 that traces a path joining g' to g, and which lies entirely outside of this ball. By choosing w_2 of minimal length, we ensure that $w := w_1 w_2 w_3$ traces a simple closed loop in $\Gamma(G, X)$.

By Step 1, we can triangulate this loop using internal chords of length at most K, and by Lemma 11.1.3 there is a triangle with vertices h_1, h_2 and h_3 in the paths traced by w_1, w_2 and w_3, respectively. Since $d(1, h_2) \geq 3K/2 + 1$ and $d(h_1, h_2) \leq K$, we have $d(1, h_1) \geq K/2 + 1$ and hence $d(g', h_1) \leq K$. Similarly, $d(g, h_3) \leq K$. But then $d(h_1, h_3) \leq K$ implies $d(g, g') \leq 3K$, a contradiction, since $d(g, g') = |w_3 w_1| = 3K + 2$. Figure 11.2 illustrates this argument.

Step 3. Now we apply Stalling's classification of groups with more than one end, which we stated as Theorem 1.8.1. If G is torsion-free, then this theorem tells us that G is either cyclic or a non-trivial free product. In the second case, Grushko's theorem (Theorem 1.5.2) tells us that the free factors require fewer generators than G. Since, as finitely generated subgroups of G they also have context-free word problem, and they are certainly torsion-free, a straightforward induction completes the proof.

Step 4. If G has torsion and is not virtually cyclic, then Stalling's theorem tells us that it is an HNN-extension or free product with amalgamation, with finite associated or amalgamated subgroups. This case is more difficult and we need to apply a deep result of Dunwoody [79] that all finitely presented groups are *accessible*. (We observed in Step 1 that G can be finitely presented.) This means that G cannot have infinitely many repeated decompositions of this kind, and so we can apply induction to the factors as in Step 3 to conclude that the base group of the HNN-extension or each factor in the amalgamated free product is virtually free, from which it follows using results of Karass, Pietrowski and Solitar [165] and Gregorac [113] that G itself must be virtually free. □

11.2 Groups with one-counter word problem

We recall from 2.2.9 that a language is called (deterministic) one-counter if it is the language of a (deterministic) pda in which the stack alphabet consists of a single element, together with a bottom of stack marker that can only be used as the bottom symbol on the stack. The classes of one-counter and deterministic one-counter languages are denoted by OC and DOC respectively. The following classification of groups G with WP$(G) \in OC$ is due to Herbst [134].

11.2.1 Theorem *The following are equivalent for a finitely generated group* G.

(i) WP$(G) \in OC$.

(ii) WP$(G) \in DOC$.

(iii) G *is either finite or has a subgroup of finite index isomorphic to* $\mathbb{Z}$.

Again (ii) $\implies$ (i) is true by definition. To prove (iii) $\implies$ (ii), note that, for an infinite cyclic group $G = \langle X \rangle$ with $X = \{x\}$, and $A := \{x, x^{-1}\}$, we can accept WP(G, A) by a 1-counter pda that records the reduced form $x^{\pm n}$ of the input word read so far as a stack of height n, and uses a state to record whether this is x^n or x^{-n}. The proof for virtually cyclic groups then follows from the closure operations for OC, which were discussed in 2.6.35.

The original proof of (i) $\implies$ (iii) in [134] uses Theorem 11.1.1 as a first step to show that a group with one-counter word problem must be context-free, and hence depends on the results of Stallings and Dunwoody on ends of groups and accessibility.

An alternative and more elementary proof, with a generalisation to semi-groups with one-counter word problem, is given by Holt, Owens and Thomas [150]. This starts by proving that groups with one-counter word problem have a linear growth function, as defined in 1.6.5, and we include this proof below. There are a number of short and elementary proofs that such groups are virtually cyclic in the literature; see, for example, the papers by Wilkie and van den Dries [251] or Incitti [164].

11.2.2 Proposition *If* WP(G) $\in$ *OC, then G has a linear growth function.*

Proof Let $G = \langle X \rangle$ and $A := X^{\pm}$ with X finite. Let q be the number of states of a one-counter automaton M accepting WP(G, A). We assume that M accepts by empty stack.

For each word $w \in A^*$, choose a shortest computation path $p(w)$ in M that accepts ww^{-1}. We shall show that there is a constant K, which does not depend on w, such that, immediately after reading the prefix w in the path $p(w)$, the stack height $h(w)$ is at most $K|w|$. If we can show this then, for words w of length at most n, there are only $(Kn + 1)q$ possibilities for the pair $(h(w), t(w))$, where $t(w)$ is the state of the machine immediately after reading the prefix w in the path $p(w)$. If $(h(w_1), t(w_1)) = (h(w_2), t(w_2))$ for two words w_1, w_2, then there is an accepting path for $w_1 w_2^{-1}$ so, if w_1 and w_2 represent different group elements, then this cannot happen. Therefore, the growth function $\gamma_{G,X}$ satisfies $\gamma_{G,X}(n) \leqslant (Kn + 1)q$, and the result follows.

We need to prove the claim that $h(w) \leqslant K|w|$ for some $K \geqslant 0$. We can assume, without loss of generality, that all moves in M change the stack height by at most one. Moves are either reading moves, when one input symbol is read, or else non-reading moves. We assume also that the stack height is not changed by reading moves. (We can achieve this by breaking up a reading move that alters the stack into two moves.)

Let w be a word with $|w| = n$. If $h(w) > q(n + 1)$, then, when reading w in $p(w)$, there must be at least one occasion where, between reading two input symbols, or before reading the first symbol, the stack height increases by at least q. Whilst this is happening, M must repeat states. In fact, we can find a subpath of $p(w)$, which we refer to as a *circuit*, linking the repeated states in which the stack height is increased by r for some r with $0 < r \leqslant q$. Similarly, when reading w^{-1} in $p(w)$, there must be a gap between reading input symbols, or before reading the first symbol, or after reading the final symbol, and a circuit linking repeated states in this gap in which the stack height is decreased by u for some u with $0 < u \leqslant q$. If $h(w) > q^3(n + 1)$, then we can find gaps between reading input symbols containing q^2 disjoint circuits of this kind in $p(w)$, in which the stack height is increased by at most q when reading w, and decreased by at most q when reading w^{-1}. Amongst the q^2 circuits in which the height is increased, at least q of them must increase it by the same number $r \leqslant q$. Similarly, amongst the q^2 circuits in which the height is decreased, at least q of them must decrease it by the same number $u \leqslant q$.

The idea is to remove u of the circuits that increase the stack height by r, and r of the circuits that decrease the stack height by u, to produce a shorter path in M accepting ww^{-1}, thereby contradicting the minimality of $p(w)$. For this to work, we have to be sure that the stack cannot become empty at any stage between the removed increasing circuits and the removed decreasing circuits, since this would alter the computation.

To do this, we assume that $h(w) > q^3(n + 2)$. We choose the gap in which we remove the circuits while reading w to be the latest one in which the stack height increases by at least q^3 at some stage during the gap, and we remove the circuits as late as possible during that gap. Similarly, we choose the gap in which we remove the circuits while reading w^{-1} to be the earliest one in which the stack height decreases by at least q^3 at some stage during the gap, and we remove the circuits as early as possible during that gap.

Between the first of these gaps and the end of w, the stack cannot become empty, since otherwise there would certainly be an increase of q^3 in a later gap. Between the beginning of w^{-1} and the place where the decreasing circuits were removed, the stack height decreases by less than q^3 during each gap, and hence by less than $q^3(n + 1)$ altogether. Since we are only removing some of the circuits, it could also decrease by some number less than q^3 during the part of $p(w)$ in which the circuits are removed. But, since $h(w) > q^3(n + 2)$, the stack can never empty.

This contradiction proves that $h(w) \leqslant q^3(n + 2)$. Thus we have proved the claim, and also the theorem. $\qquad\qquad\qquad\qquad\qquad\qquad\qquad\qquad\qquad\qquad\qquad\square$

The next result, which also comes from [150], is a generalisation to groups G for which $\mathrm{WP}(G)$ is the intersection of finitely many one-counter languages.

11.2.3 Proposition *Suppose that $G = \langle X \rangle$ and that, for some $k > 0$, we have* $\mathrm{WP}(G) = \cap_{i=1}^{k} L_i$ *with each $L_i \in OC$. Then the growth function of G is bounded by a polynomial of degree k.*

Proof The argument in the proof of Proposition 11.2.2 shows that the stack height of each of the k one-counter automata after reading the prefix w of a minimal length accepting path for ww^{-1} is bounded by a linear function of $|w|$.

As in Proposition 11.2.2, words w_1 and w_2 representing distinct elements of G must lead to distinct configurations in at least one of the automata, since otherwise we could construct accepting paths for $w_1 w_2^{-1}$ in every automaton. It follows that the growth function of G is polynomial of degree at most k. □

This enables us to classify groups G for which WP(G) is an intersection of finitely many one-counter languages, although we shall need to use the deep theorem of Gromov [124] that groups with polynomial growth are virtually nilpotent.

11.2.4 Theorem [150, Theorem 5.2] *The following are equivalent for a finitely generated group G.*

(i) WP(G) *is the intersection of k one-counter languages for some $k \geqslant 1$.*
(ii) WP(G) *is the intersection of k deterministic one-counter languages for some $k \geqslant 1$.*
(iii) G *is virtually abelian of free abelian rank at most k.*

Proof It is clear that (ii) implies (i). We show first that (iii) implies (ii). Since the class $\mathcal{DOC}$ is closed under inverse GSMs and under intersection with regular sets (see 2.6.35), it follows from Proposition 3.4.7 that it is enough to prove that (ii) holds for the free abelian group $G = \mathbb{Z}^k$ of rank k. Let $X = \{x_1, \ldots, x_k\}$ be a free basis of G and $A := X^{\pm}$. For $1 \leq i \leq k$, let G_i be the subgroup of G generated by $\{x_j : 1 \leq j \leq k, j \neq i\}$. To test whether $w \in A^*$ represents an element of G_i, we can ignore the generators x_j with $j \neq i$ and check that the exponent sum of x_i in w is 0. So, by the same argument as in the proof of Theorem 11.2.1, we see that the languages $L_i := \{w \in A^* : w \in G_i\}$ lie in $\mathcal{DOC}$. Since WP(G) = $\cap_{i=1}^{k} L_i$, this proves (ii).

It remains to prove that (i) implies (iii). Assuming (i), we know from Proposition 11.2.3 that G has growth function bounded by a polynomial of degree k. Now, by Gromov's theorem [124], G is virtually nilpotent. It is proved in [150, Theorem 5.3] that a virtually nilpotent group G for which WP(G) is the intersection of finitely many context-free languages is virtually abelian. (The proof of this requires an application of Parikh's Theorem (Theorem 2.6.23), which is very similar to an application to the co-word problem of these groups that we shall present in Chapter 14, so we shall prove it later as Theorem 14.2.15.) Now it is easy to check that the growth function of the free abelian group $\mathbb{Z}^k$ is polynomial of degree k, and so G is virtually abelian of free abelian rank at most k, which proves (iii). □

12

Context-sensitive word problems

The classes CS and $\mathcal{DCS}$ of context-sensitive and deterministic context-sensitive languages were defined in 2.2.18 as the languages accepted by (deterministic) linearly bounded Turing machines. We saw in Theorem 2.8.2 that CS is also the class of languages generated by a context-sensitive grammar.

It follows from the closure properties of CS and $\mathcal{DCS}$ (2.8.3) and the results in Section 3.4 that the classes of finitely generated groups G with (deterministic) context-sensitive word problem are closed under finite extension and finitely generated subgroups. These classes have been systematically studied by Lakin [174] (see also [175, 176]), who proved in particular that they are closed under free and direct products; we leave these proofs as exercises for the reader.

Lakin observed that all finitely generated linear groups are deterministic context-sensitive. In fact a stronger result is true: their word problems are soluble in logspace. The class of linear groups includes the classes of finitely generated free groups, virtually polycyclic groups and their holomorphs, and torsion-free metabelian groups.

Shapiro had previously proved [235] that automatic groups are deterministic context-sensitive, and we included this result as Corollary 5.2.12. Hence all finitely generated subgroups of automatic groups are deterministic context-sensitive. An example of Lakin that is not of this form is described briefly in the next section.

It is unknown whether there exist groups that are context-sensitive but not deterministic context-sensitive. Indeed it is unknown whether there exists a language of any kind in $CS \setminus \mathcal{DCS}$, although there is no reason at all to believe that these language classes should be equal.

It is not easy to produce examples of finitely presented groups G with soluble word problem for which it can be proved that $\mathrm{WP}(G) \notin CS$. It can be shown that, if $\mathrm{WP}(G)$ is soluble in linearly bounded space, then it is soluble in time

236

c^n for any constant $c > 1$, and so one way of producing examples of the type we are looking for is to use the method of Avenhaus and Madlener mentioned in Section 10.2 to construct groups in which the time complexity of WP(G) is a primitive recursive function not bounded by any exponential function.

12.1 Lakin's example

In [174, Theorem 6.3.16], Lakin constructs an example of a group with word problem in $\mathcal{DCS}$ that does not occur as a finitely generated subgroup of an automatic group. Since this construction involves an interesting application of small cancellation theory, we shall summarise it here.

The basic idea is to start with a language with specified time and space complexity and to use it to construct a finitely generated (but usually not finitely presented) group whose word problem has related complexity.

For an arbitrary finite set Y and language $L \subseteq Y^*$, we introduce 12 new symbols $a_1, \ldots, a_{12}$ assumed not to lie in Y, and let $X := Y \cup \{a_1, \ldots, a_{12}\}$. We then define the group

$$G = G(L) := \langle X \mid R \rangle \quad \text{where} \quad R = \{a_1 \alpha a_2 \alpha \cdots a_{12} \alpha : \alpha \in L\}.$$

Let v be a piece in the symmetrised presentation $\langle X \mid \hat{R} \rangle$ (see Section 1.9). So v is a subword of some $r \in \hat{R}$. Let α be the element of L associated with r. Then v or v^{-1} is either a subword of α, or it has the form $\beta a_i \gamma$ for some i, where β and γ are both subwords of α. In either case $|v| < |r|/6$, so the presentation satisfies the small cancellation condition $C'(1/6)$. Hence, by Greendlinger's Lemma (see 1.9.3), the group $G = \langle X \mid \hat{R} \rangle$ has a Dehn algorithm; that is, any nonempty reduced word in WP($G, X^\pm$) must contain more than one half of a word in $\hat{R}$ as a subword. It follows easily that, for $\alpha \in Y^*$, we have $a_1 \alpha a_2 \alpha \cdots a_{12} \alpha =_G 1$ if and only if $\alpha \in L$.

Now let $\mathcal{F}$ be a class of non-decreasing functions $\mathbb{N}_0 \to \mathbb{N}_0$ with the property that all $f \in \mathcal{F}$ satisfy $f(n) \geq cn$ for some constant $c > 0$. For example, the class of all non-constant polynomial functions with non-negative coefficients, or the classes of such functions of degree at most d for some fixed $d \geq 1$, all satisfy these conditions. Using the notation introduced in 2.2.17, we define

$$\text{DSPACE}(\mathcal{F}) := \bigcup_{f \in \mathcal{F}} \text{DSPACE}(f(n)),$$

and similarly for NSPACE, DTIME and NTIME.

12.1.1 Lemma *Let L, X, R, G and $\mathcal{F}$ be as above and put $A := X^{\pm}$. Then*

$$L \in \text{DSPACE}(\mathcal{F}) \Rightarrow \text{WP}(G, A) \in \text{DSPACE}(\mathcal{F}) \quad and$$
$$L \in \text{NSPACE}(\mathcal{F}) \Rightarrow \text{WP}(G, A) \in \text{NSPACE}(\mathcal{F}).$$

Proof Let $w \in A^*$ with $|w| = n$. As we saw above, if $w \in \text{WP}(G, A)$, then it has a subword equal to more than half of a word in $\hat{R}$. So it has a subword v such that either v or v^{-1} lies in the set L' that consists of words of the form $a_i \alpha a_{i+1} \alpha \cdots \alpha a_{i+6}$ or $\gamma a_i \alpha a_{i+1} \alpha \cdots \alpha a_{i+5} \beta$, where $\alpha \in L$ and subscripts are taken modulo 12, and in the second case β and γ are respectively a suffix and a prefix of α with $|\beta| + |\gamma| > |\alpha|$.

It is not hard to see that, for any $f \in \mathcal{F}$, we have

$$L \in \text{DSPACE}(f(n)) \Rightarrow L' \in \text{DSPACE}(f(n))$$
$$L \in \text{NSPACE}(f(n)) \Rightarrow L' \in \text{NSPACE}(f(n)).$$

So, with an additional space usage of at most n, we can examine every subword v of w in turn and test whether v or v^{-1} has this form. If not, then $w \notin \text{WP}(G, A)$. If so then, again using extra space at most n (since we are not allowed to change the word on the input-tape), we can replace the subword v of w by a shorter word that is equal to v in G. We now repeat the process on the new word w. If $w \in \text{WP}(G, A)$ then we will eventually reduce w to the empty word. If not, then we will at some stage fail to find a subword v of the required form. The result follows. □

In particular, by taking $\mathcal{F}$ to be the set of all linear functions $f(n) = cn + d$ with $c, d \geq 0$, we have

12.1.2 Corollary

$$L \in \mathcal{CS} \Rightarrow \text{WP}(G) \in \mathcal{CS} \quad and \quad L \in \mathcal{DCS} \Rightarrow \text{WP}(G) \in \mathcal{DCS}.$$

Now we saw in Corollary 5.2.12 that the word problem of an automatic group is soluble deterministically in linear space and quadratic time, and the same applies to all finitely generated subgroups of automatic groups. In fact, we can perform this computation on a deterministic Turing machine with an input-tape and two further work-tapes. One of these work-tapes contains the reduction v of the prefix of the input word that had been reduced so far. After computing this reduction, the next generator a is read from the input word, and the other work-tape is used to store the data required for the computation of the reduction of va. On completion of this computation, the reduction of va can be computed (in reversed order) and written onto the first work-tape. The reader will need to refer back to the proofs of Theorem 5.2.11 and 2.10.7 to verify these assertions.

Now [174, Lemma 6.3.14] proves the existence of a language $L \in \mathcal{DCS}$ that cannot be recognised by a deterministic Turing machine with three tapes that operates in space $O(n)$ and time $O(n^2)$. We shall not describe the construction of L here; it involves a diagonalisation argument with Turing machines.

Let $G = G(L)$ be as defined above. Then, by Corollary 12.1.2, we have $\mathrm{WP}(G, A) \in \mathcal{DCS}$. Now if G were a subgroup of an automatic group, then $\mathrm{WP}(G, A)$ would be recognisable by a deterministic Turing machine with three tapes that operates in space $O(n)$ and time $O(n^2)$. But, as we saw above, for $\alpha \in Y^*$, we have $\alpha \in L \Leftrightarrow a_1 \alpha a_2 \alpha \cdots a_{12} \alpha =_G 1$. So, since

$$|a_1 \alpha a_2 \alpha \cdots a_{12} \alpha| = 12|\alpha| + 12,$$

we could use the Turing machine for $\mathrm{WP}(G, A)$ to test membership of L in space $O(n)$ and time $O(n^2)$, a contradiction. So G cannot be a subgroup of an automatic group.

12.2 Some further examples

In Section 14.3 below, we summarise results proved in the paper by Holt and Röver [157], in which certain groups are proved to have indexed co-word problem. These examples include the Higman–Thompson groups (see 1.10.11) and finitely generated groups of bounded permutational transducer mappings, including the Grigorchuk groups, that were discussed in Chapter 9. They also include free products of all known examples of groups with indexed co-word problem.

The groups studied in [157] are all *semi-deterministic co-indexed*, which means that each has co-word problem that is the language of a nested stack automaton (see 2.2.21) that writes a word non-deterministically to the main stack before starting to read the input word, and operates deterministically thereafter. By examining the operation of these nested stack automata, we find that in all cases the original word on the stack need be no longer than the input word and, given that, the total space used by the nested stacks is bounded by a linear function of the input word.

For a nested stack automaton M with these properties, we can construct a deterministic linearly bounded Turing machine T that simulates the operation of M on an input word w with each possible initial stack word v of length at most $|w|$ in turn, and accepts w if and only if M accepts w for some such *v*. So $L(T) = L(M)$. It follows that all of the examples studied in [157] are deterministic context-sensitive.

It is clear that the class $\mathcal{RT}$ of real-time languages is a subset of $\mathcal{DCS}$ and so all real-time groups are deterministic context-sensitive. We shall study real-time groups in Section 13.1. This class includes, for example, hyperbolic groups, geometrically finite hyperbolic groups, and finitely generated nilpotent groups. But note that groups in the first two of these classes are automatic and groups in the third class are linear, so we have alternative reasons why $\mathrm{WP}(G) \in \mathcal{DCS}$ in each of these cases.

12.3 Filling length

In the remainder of this chapter, we first show that groups that satisfy the general geometric property of having linear filling length have (possibly non-deterministic) context-sensitive word problem. We then go on to prove results of Riley [218] that groups with asynchronous combings and almost convex groups have linear filling length.

Let $G = \langle X \mid R \rangle$ be a finite group presentation and as usual put $A := X^{\pm}$. As we observed in 1.2.3, G is isomorphic as a monoid to $A^*/\sim$, where $\sim$ is the congruence generated by $\{(aa^{-1}, \varepsilon) : a \in A\} \cup \{(r, \varepsilon) : r \in R\}$. The same congruence is generated by $\{(aa^{-1}, \varepsilon) : a \in A\} \cup \{(u, v) : uv^{-1} \in \hat{R}\}$, where $\hat{R}$ is the symmetric closure of R, as defined in Section 1.9.

So for any $w \in \mathrm{WP}(G, A)$ there is a sequence of words

$$w = w_0, w_1, \ldots, w_n = \varepsilon,$$

with the property that each w_{i+1} is derived from w_i by applying one of the following three operations:

(i) Remove a subword aa^{-1} from w_i for some $a \in A$;
(ii) Insert a subword aa^{-1} into w_i for some $a \in A$;
(iii) Replace a subword u of w_i by v, where $uv^{-1} \in \hat{R}$.

We call such a sequence (w_i) a *reduction sequence* for w, we define

$$\mathrm{FL}((w_i)) = \max\{|w_j| : 0 \le j \le n\},$$

and let $\mathrm{FL}(w)$ be the minimum value of $\mathrm{FL}((w_i))$ as (w_i) ranges over all reduction sequences for w. Then, following Gersten and Riley [90], we define the *filling length function* $\mathrm{FL} = \mathrm{FL}_{X,R} : \mathbb{N}_0 \to \mathbb{N}_0$ of the presentation by

$$\mathrm{FL}(n) = \max\{\mathrm{FL}(w) : w \in \mathrm{WP}(G, A), |w| \le n\}.$$

An equivalent definition of filling length using van Kampen diagrams is described by Gersten and Riley [90], but we do not need that here. Along with various other filling functions, this concept was introduced by Gromov [126].

In the next section we shall investigate some conditions on G under which FL is linear, which implies that WP(G) is context-sensitive, but let us first consider the invariance properties of FL.

We recall from 3.3.2 that, for two functions $f, g : \mathbb{N}_0 \to \mathbb{N}_0$, we define $f \preceq g$ if there is a constant $C > 0$ such that $f(n) \leq Cg(Cn + C) + Cn + C$ for all $n \in \mathbb{N}_0$, and we write $f \simeq g$ if $f \preceq g$ and $g \preceq f$.

Let $\langle X \mid R_1 \rangle = \langle X \mid R_2 \rangle$ be two presentations of G on the same generating set, and let $w \in$ WP(G, A) with $A := X^{\pm}$. Then we can replace each substitution $u \to v$ of Type (iii) in a reduction sequence using R_1 by a fixed sequence of moves of Types (i), (ii) or (iii) using R_2 to give a reduction sequence for w using R_2 in which the length of the longest word exceeds that of the original by at most a constant. So $\mathrm{FL}_{X,R_1} \simeq \mathrm{FL}_{X,R_2}$.

We can reformulate the definition of filling length in terms of monoid relations that are used for substitutions. If $\mathcal{R} \subseteq A^* \times A^*$ is any set of defining relations for G as a monoid then, for any $w \in$ WP(G, A), there is a reduction sequence for w in which we use substitutions $u \to v$ with $(u, v) \in \mathcal{R}$ or $(v, u) \in \mathcal{R}$, and we define the associated filling length function $\mathrm{FL}_{X,\mathcal{R}}$ as before. By the same argument as in the last paragraph, we have $\mathrm{FL}_{X,\mathcal{R}_1} \simeq \mathrm{FL}_{X,\mathcal{R}_2}$ for any two sets $\mathcal{R}_1$ and $\mathcal{R}_2$ of monoid defining relations for G. So, in particular, with $G = \langle X \mid R \rangle$ as above and

$$\mathcal{R} = \{(aa^{-1}, \varepsilon) : a \in A\} \cup \{(u, v) : uv^{-1} \in \hat{R}\},$$

we have $\mathrm{FL}_{X,\mathcal{R}} = \mathrm{FL}_{X,R}$.

12.3.1 Proposition *Let $\langle X \mid R \rangle$ and $\langle Y \mid S \rangle$ be two finite presentations of the group G. Then $\mathrm{FL}_{X,R} \simeq \mathrm{FL}_{Y,S}$.*

Proof Let $A := X^{\pm}$, $B := Y^{\pm}$, for each $x \in X$ let $\tau(x)$ be a word over Y with $\tau(x) =_G x$, for each $y \in Y$ let $\sigma(y)$ be a word over X with $\sigma(y) =_G y$, and extend τ and σ to monoid homomorphisms $A^* \to B^*$ and $B^* \to A^*$, respectively.

Let $\mathcal{R}$ be the relations of a monoid presentation for G over A; so $\mathrm{FL}_{X,\mathcal{R}} = \mathrm{FL}_{X,R}$, since we saw above that the $\simeq$-class of FL does not depend on the finite set of relators or monoid relations. Let

$$S := \{(\tau(u), \tau(v)) : (u, v) \in \mathcal{R}\} \cup \{(b, \tau(\sigma(b))) : b \in B\}.$$

We claim that S is the set of relations of a monoid presentation of G over B (so $\mathrm{FL}_{Y,S} = \mathrm{FL}_{Y,S}$), and that $\mathrm{FL}_{Y,S} \preceq \mathrm{FL}_{X,\mathcal{R}}$. The reversed inequality will then follow by symmetry, so this is sufficient to prove the result.

Let $M = \text{Mon}\langle B \mid S \rangle$ be the monoid defined by the relations S. Then the left- and right-hand sides of the relations in S are certainly equal in G so, for $w_1, w_2 \in B^*$, we have $w_1 =_M w_2 \Rightarrow w_1 =_G w_2$. To show the converse implication, it suffices to show that $w \in \text{WP}(G, B) \Rightarrow w =_M 1$, because then $w_1 =_G w_2 \Rightarrow w_1 w_2^{-1} =_G 1 \Rightarrow w_1 w_2^{-1} =_M 1$, and, since $w_2^{-1} w_2 =_G 1$ and hence $w_2^{-1} w_2 =_M 1$, this implies $w_1 =_M w_2$.

Now let $w \in \text{WP}(G, B)$. Then $\sigma(w) \in \text{WP}(G, A)$, and so there is a reduction sequence $\sigma(w) = w_0', w_1', \ldots, w_n' = \varepsilon$ for $\sigma(w)$ with substitutions from $\mathcal{R}$ in which each $|w_i'| \leq \text{FL}_{X,\mathcal{R}}(|\sigma(w)|)$. We derive a reduction sequence for w with substitutions from S as follows. First use the relations $(b, \tau(\sigma(b))$ to replace w by $\tau(\sigma(w))$. Now we can replace each substitution $u \to v$ in the reduction sequence for $\sigma(w)$, by the substitution $\tau(u) \to \tau(v)$ to obtain a reduction sequence $\tau(\sigma(w_0)) = \tau(w_0'), \tau(w_1'), \ldots, \tau(w_n') = \varepsilon$ for $\tau(\sigma(w_0))$. So we have shown that $w =_M 1$.

Since there are constants $C = \max\{\sigma(b) : b \in B\}$ and $D = \max\{\tau(a) : a \in A\}$ with $|\sigma(w)| \leq C|w|$ and

$$\tau(w_i') \leq D|w_i'| \leq D\text{FL}_{X,\mathcal{R}}(|\sigma(w)|) \leq D\text{FL}_{X,\mathcal{R}}(C|w|)$$

for all i, we have $\text{FL}_{Y,S} \preceq \text{FL}_{X,\mathcal{R}}$ as claimed, which completes the proof. □

So for a finitely generated group G we can unambiguously talk about the filling length $\text{FL}(G)$ of G. Using similar arguments to the proof of Proposition 3.3.3 it can be shown that, for groups G and H with quasi-isometric Cayley graphs $\Gamma(G, X)$ and $\Gamma(H, Y)$, we have $\text{FL}(G) \simeq \text{FL}(H)$.

12.4 Groups with linear filling length

Our interest in groups with linear filling length comes from the following easy result.

12.4.1 Proposition *If $\text{FL}(G)$ is a linear function of n for some finite monoid presentation of the group G, then $\text{WP}(G)$ is context-sensitive.*

Proof For $w \in \text{WP}(G)$, there is a reduction sequence (w_i) for w in which the maximum value of $|w_i|$ is bounded by a linear function of $|w|$. This reduction sequence constitutes a verification that $w \in \text{WP}(G)$ using space bounded by a linear function of $|w|$. □

In general, a linear filling function for G implies only that $\text{WP}(G)$ is non-deterministic context-sensitive. However, in specific cases we may be able to provide a deterministic algorithm for reducing a word to the identity using the monoid relations. Provided that this algorithm can be executed in linear space, this would imply that WP(G) was deterministic context-sensitive.

We do not know of any specific examples of groups with context-sensitive word problem that do not have linear filling length but it is very likely that such examples exist. There is no reason to believe, for instance, that the class of groups with linear filling length is closed under finitely generated subgroups.

Linear filling length implies that the words occurring in a reduction sequence for $w \in$ WP(G) have length at most $k|w|$ for some constant k. Since we can clearly assume that the words w_i occurring in the chain of reductions are all distinct, the chain has length at most $(1 + 2|X|)^{k|w|}$. Now corresponding to this chain of reductions we can construct a van Kampen diagram Δ for w, whose 2-cells are labelled by those relators $uv^{-1} \in \hat{R}$ for which $u \to v$ is applied at some stage in the chain; this diagram has area at most $(1 + 2|X|)^{k|w|}$, and hence G satisfies an exponential isoperimetric inequality.

We have not had occasion in this book to discuss the *isodiametric function* of a finitely generated group G. The definition is analogous to that of the Dehn function but, in place of the area of a van Kampen diagram Δ, we use its *diameter*, which is defined to be the maximum distance of a vertex in Δ to the base point of Δ.

Now if G has linear filling length, and $w \in$ WP(G) with a reduction sequence $w = w_0, w_1, \ldots, w_n$, each vertex of the van Kampen diagram Δ we described in the previous paragraph is on the boundary of a van Kampen diagram for at least one of the words w_i and so is at distance at most $|w_i|/2$ from the base point of Δ. It follows that a group G with linear filling length has a linear isodiametric function.

Riley [218] studied the effect of almost convexity conditions and the existence of various types of combings on the filling length and on the isodiametric and Dehn functions of G. Here we prove that almost convex groups in the standard sense and groups with an asynchronous combing with a constant fellow traveller condition have linear filling length. These are essentially special cases of Riley's results.

We remark also that the following earlier results of Gersten and Bridson follow immediately from the results about filling length that we prove below. It was proved by Gersten [89, Proposition 3.4 and Corollary 4.3] that any synchronously combable group satisfies a linear isodiametric inequality, and that any almost convex group satisfies a linear isodiametric inequality. Bridson proved in Proposition 8.1 (ii) and Theorem 4.3 (iii) of [40] that an asynchronously combable group satisfies a linear isodiametric inequality and an exponential isoperimetric inequality.

12.4.2 Asynchronously combable groups We recall from 5.1.3 the definition of a (asynchronous) combing for a group G as a language for G for which
the representatives of g and gx (asynchronously) k-fellow travel for some constant k; we note that there are a number of slightly different definitions of
(asynchronously) combable groups in the literature. Our definition is equivalent to what is called a synchronous or asynchronous combing with bounded
width in [40]. The combings defined by Gersten in [89] are synchronous.

It is convenient in this section to reformulate the definition of the fellow
traveller property (see 5.1.2) in a discrete context, as follows. Note that the
values of k and K in the two definitions differ by at most 1.

Let G be a group generated by X and let $\Gamma = \Gamma(G, X)$ be the Cayley graph
of G with respect to X, endowed with the usual metric d in which all edges
have length 1. For non-negative integers t, let $I_t = \{0, 1, \ldots, t\}$. Let $p_0 p_1 \ldots p_m$
and $q_0 q_1 \ldots q_n$ be paths in Γ, where p_i, p_{i+1} and q_i, q_{i+1} are pairs of adjacent
vertices in Γ. These paths asynchronously fellow travel with constant $k \geq 0$
if and only if there exists a subset J of $I_m \times I_n$ that projects surjectively onto
both I_m and I_n, such that $d(p_i, q_j) \leq k$ for all $(i, j) \in J$, and such that, for
$(i, j), (i', j') \in J$, we have $i \leq i' \iff j \leq j'$. Notice that these conditions imply
that $(0, 0), (m, n) \in J$. The paths synchronously k-fellow travel if and only if,
assuming $m \leq n$, we can choose

$$J = \{(i, i) \mid 1 \leq i \leq m\} \cup \{(m, j) \mid m < j \leq n\}.$$

12.4.3 Theorem *Asynchronously combable groups have linear filling length.*

Proof Let L be an asynchronous k-combing for G over a generating set X. By
modifying the fellow traveller constant k if necessary, we can assume that the
combing L contains the empty word ε.

Now suppose that $w = a_1 a_2 \cdots a_n$ is a word of length n with $w =_G 1$. For
$1 \leq i \leq n - 1$, define v_i to be a word in L that represents the prefix of w of
length i, and let $v_0 := v_n := \varepsilon$.

We now construct a planar diagram D for w as follows. The edges are labelled by words over X. The free reduction of the label on the exterior region
is equal to that of w, and the free reductions of the labels on the interior regions
are equal in G to the identity.

For convenenience in drawing D (see Figure 12.1), we place n edges labelled
ε along the bottom of D, with the edges of the word w placed above them, so the
external boundary label of D is actually $w\varepsilon^n$. We denote the bottom boundary
of D, with label ε^n, by e. For $1 \leq i \leq n - 1$, we draw a path labelled v_i from the
appropriate vertex on e to the vertex of w following the edge labelled a_j.

By definition of a combing, the words v_{i-1} and v_i asynchronously k-fellow

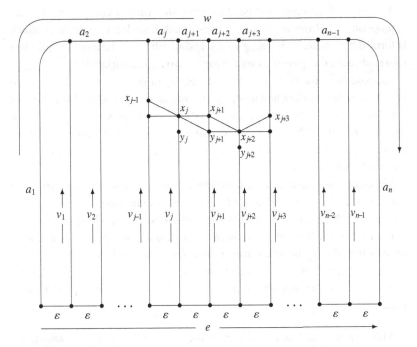

Figure 12.1 Using a combing to find a reduction sequence

travel for some constant k. So the internal region with label $v_{i-1}a_iv_i^{-1}\varepsilon$ can be triangulated in such a way that each triangle contains a side of length 1 joining two adjacent vertices of one of the two lines v_{i-1} and v_i, and two sides of length at most k that join a vertex of v_{i-1} to a vertex of v_i. We call the sides of length 1 *short edges* and the sides of length k *long edges*.

We can now define a set of monoid relations for G such that $\mathrm{FL}_{X,\mathcal{R}}$ is linear. We take $\mathcal{R}$ to be the set of all pairs (u, v) for which $u, v \in A^*$, $u =_G v$, and $|u|, |v| \leq 2k$. We reduce the word w to the identity by applying relations from $\mathcal{R}$ formed by considering pairs of adjacent triangles in D, that is, they share a short edge. We reduce w through a chain of reductions

$$w_0 = w \to w_1 \to \ldots \to w_t = \varepsilon^n = \varepsilon,$$

where each $|w_i| \leq kn$, and each w_i corresponds to a path from left to right formed by concatenating consecutive long edges. A word w_{i+1} is formed from w_i by replacing a subpath of length at most $2k$ along the top of a pair of adjacent triangles by the path in D along the bottom of that pair of triangles. This can be achieved by application of one of the relations in $\mathcal{R}$.

To complete the proof, we need to verify that we can always construct an appropriate w_{i+1} from w_i; that is, if w_i is a path from left to right in D which is formed by concatenating long edges, and which is not the empty word, then some subpath of w_i passes along the top of two adjacent triangles.

Suppose that, for $0 \leq j \leq n$, w_i meets v_j in the vertex x_j. Then, since $v_0 = v_n = \varepsilon$, the path labelled w_i starts and ends at the first and final vertices of e but, since $w_i \neq \varepsilon$, not every x_j is on e. So there exist j_1, j_2 with $0 \leq j_1 < j_2 \leq n$ and $j_2 - j_1 > 1$, such that x_{j_1} and x_{j_2} are on e but, for all j with $j_1 < j < j_2$, x_j is not on e. For each such j with $j_1 < j < j_2$, let y_j be the vertex on v_j immediately below x_j.

We know that x_j and x_{j+1} are joined by a long edge, and it follows from the construction of the triangulation of the region of D between v_j and v_{j+1} that, for each j with $j_1 < j < j_2 - 1$, either x_j is joined to y_{j+1} by a long edge or y_j is joined to x_{j+1} by a long edge. Furthermore, x_{j_1} must be joined to y_{j_1+1} and y_{j_2-1} must be joined to x_{j_2} by long edges. Hence, for some j with $j_1 \leq j < j_2 - 1$, we have long edges from x_j to y_{j+1} and from y_{j+1} to x_{j+2}. So we define w_{i+1} from w_i by replacing the subpath from x_j to x_{j+2} through x_{j+1} by a path through y_{j+1}. $\qquad \square$

Although we shall make no further use of the following observation, we observe that we can compute the length t of the reduction sequence in the above proof in terms of the lengths of the combing words v_j. Where v_i has length l_i, there are $l_{i-1} + l_i$ triangles between v_{i-1} and v_i, hence $2(l_1 + l_2 + \ldots l_{n-1})$ in total in D. Since each reduction traverses two triangles, we have $t = l_1 + l_2 + \cdots + l_{n-1}$.

We observe also that, by Theorem 5.2.11, when G is automatic there is a constant c such that, in the notation of the proof of Theorem 12.4.3, we have $v_i \leq cw(i)$ for all i, and we can compute the v_i in linear space, so WP(G) is deterministic context-sensitive. This was proved originally by Shapiro [235].

However, when G is not automatic the combing words v_i can be exponentially longer than w, and so we cannot store them in a linearly bounded Turing machine. This prevents us from making the process deterministic in any natural way, and so we cannot use this result to show, for example, that the Baumslag–Solitar groups BS(m, n) (see 1.10.10) have deterministic context-sensitive word problem.

We can, however, prove that result directly, using the normal form for HNN-extensions.

12.4.4 Proposition *The Baumslag–Solitar groups* BS(m, n) *with presentation* $\langle x, y \mid y^{-1} x^m y = x^n \rangle$ *have deterministic context-sensitive word problem for all* $m, n \in \mathbb{Z}$.

Proof We can assume that m and n are nonzero. By Proposition 1.5.17, each $g \in \mathrm{BS}(m, n)$ has a unique expression as $y^{j_0} x^{t_1} y^{j_1} x^{t_2} \cdots x^{t_k} y^{j_k} x^{t_{k+1}}$ for some $k \geq 0$ where, for $1 \leq i \leq k$, we have $j_i \neq 0$, $0 < t_i < |m|$ if $j_i > 0$, and $0 < t_i < |n|$ if $j_i < 0$. If w is a word in the generators representing g, then $|j_0| + |j_1| + \cdots + |j_k|$ is at most $|w|$, and hence the length of the normal form, up to but not including the final syllable $x^{t_{k+1}}$, is bounded by a linear function of $|w|$. When we multiply this normal form word on the right by a generator, the exponent t_{k+1} grows by a factor of at most $\max(|m|, |n|)/\min(|m|, |n|)$, so $\log |t_{k+1}|$ is bounded linearly in $|w|$. Hence, by storing $x^{t_{k+1}}$ as the binary representation of t_{k+1}, we can store the nomal form word in space bounded linearly in $|w|$. It is not difficult to see that we can compute this representation of the normal form word of ga for any $a \in \{x^{\pm 1}, y^{\pm 1}\}$ in space $O(\log |t_{k+1}|)$, so the result follows. $\qquad\square$

12.4.5 Almost convex groups As in the previous section, let $\Gamma = \Gamma(G, X)$ be the Cayley graph of a group G over the generating set X. For $n \geq 0$, the sphere $S(1, n)$ and the ball $B(1, n)$ in Γ with centre the identity element 1 correspond to group elements g in which the shortest words over X representing g have length exactly n and at most n, respectively.

Following Cannon [61], a group G is called *almost convex* with respect to X if for any $t > 0$ there is an integer $k(t) > 0$ such that, for all $n \geq 0$, any two vertices in $S(1, n)$ that are at distance at most t apart in Γ are joined by a path in Γ of length at most $k(t)$ that lies entirely within $B(1, n)$; that is, all vertices on the path are within $B(1, n)$ and at most one vertex of any edge on the path is on $S(1, n)$. However, it can be shown (exercise, or see [61]) that, if this condition is satisfied for $t = 2$, then it is satisfied for all t, so to establish almost convexity we just need to prove the existence of the constant $k := k(2)$. But note that Riley [218] considers more general almost convexity conditions in which k may depend on n.

Unlike the other properties discussed in this chapter, it seems likely that the almost convex property is dependent on the generating set X of G. Nonetheless, we have the following result.

12.4.6 Theorem *Suppose that G is almost convex with respect to some generating set X. Then G has linear filling length.*

Proof Let $k = k(2)$ be the constant defined above for G with respect to X. For our monoid relation set $\mathcal{R}$, we take the set of all pairs (u, v) with $u, v \in A^*$ and $u =_G v$, for which $|u| \leq 2$ and $|v| \leq k$. We aim to prove by induction on n that the following statement $P(n)$ holds for all $n \geq 0$.

$P(n)$: If w is a geodesic word over X, $g \in A$, and the path labelled wg starting at the base point 1 of Γ lies within the ball $B(1, n)$ of Γ, then wg can be reduced using the relations in $\mathcal{R}$ to a geodesic representative w' in such a way that the words occurring in the reduction chain have length at most kn.

Since a word w over X with $w =_G 1$ lies entirely within the ball $B(1, m)$ with $m = [(|w| + 1)/2)]$, it follows from $P(m)$ that, by reducing each prefix of w in turn, we can reduce w to ε using the relations in $\mathcal{R}$ such that the words occurring in the reduction chain have length at most $km + |w|$. (The term km bounds the length of any word in the reduction chain of a prefix of w, and $|w|$ bounds the length of the remaining suffix.) This is therefore enough to prove the theorem.

Now $P(0)$ is vacuously true. So suppose that $P(n-1)$ holds for some $n \geq 1$, and let w be a geodesic word and g a generator with wg in the ball $B(1, n)$. We only need to consider three cases, namely where w has length $n-1$ and wg length n or $n-1$, and where w has length n and wg length $n-1$; all other possibilities keep us within $B(1, n-1)$ and so $P(n-1)$ implies that wg can be reduced to a geodesic representative using words of length at most $k(n-1)$.

In the first of our three cases wg itself is a geodesic representative, so there is nothing to prove. In the second case, there exists a word v of length at most k such that $wg =_G wv$ and the path labelled wv from 1 to wg stays within $B(1, n-1)$. We start by applying the substitution $g \to v$ of $\mathcal{R}$ to change wg to wv (of length at most $n-1+k \leq kn$). Let $v = x_1 \ldots x_r x_{r+1}$ with $r+1 \leq k$. Then, by $P(n-1)$, there exists a sequence of reductions

$$wx_1 \to w_1, \ w_1 x_2 \to w_2, \ \ldots, \ w_{r-1} x_r \to w_r,$$

where $w_1, w_2, \ldots w_r$ are geodesics to the vertices on the path from w to wg labelled by v, and each of these reductions involves intermediate words of length at most $k(n-1)$. Since the final edge of v must be within $B(1, n-1)$, the geodesic w_r must have length $n-2$, and hence $w_r x_{r+1}$ is already reduced. So in this second case the reduction of wg to a geodesic representative can be done using words of length at most $k(n-1)+k = kn$. (The $+k$ term bounds the length of each suffix $x_{i+1} \cdots x_{r+1}$ of v.)

In the third case, let $w = w'h$, where w' is the prefix of w of length $n-1$. Now there exists a word v of length at most k such that $wg =_G w'v$ and the path from w' to wg along v stays within $B(1, n-1)$. We start by applying the substitution $hg \to v$ of $\mathcal{R}$ to change wg to $w'v$ (of length $n-1+k \leq kn$). Let $v = y_1 \ldots y_{s+1}$ with $s+1 \leq k$. Then by $P(n-1)$ there is a sequence of reductions

$$w'y_1 \to w'_1, \ w'_1 y_2 \to w'_2, \ \ldots \ w'_{s-1} y_s \to w'_s,$$

each involving intermediate words of length at most $k(n-1)$, just as above, and $w'_s y_{s+1}$ gives us a geodesic representative for wg. This completes the proof that $P(n-1) \Rightarrow P(n)$, and hence of the theorem. $\square$

13
Word problems in other language classes

Word problems in other language classes We have now considered groups with word problems in the most familiar formal language classes in the Chomsky hierarchy, namely the classes $\mathcal{R}eg$, $\mathcal{CF}$ and $\mathcal{CS}$ of regular, context-free and context-sensitive languages. In this chapter, we consider some of the less familiar language classes for which this problem has been investigated, namely the real-time languages $\mathcal{RT}$, the indexed languages $\mathcal{I}$, and the poly-context-free languages $\mathcal{PCF}$.

13.1 Real-time word problems

We recall from 2.2.19 that a *real-time* Turing machine is a deterministic machine with an input-tape that contains the input word, and a finite number of bi-infinite work-tapes that are initially empty. The head on the input-tape reads the word once, scanning it from left to right. After each symbol has been read, each of the work-tapes may independently execute a single operation of reading, writing and possibly moving one place to the left or right. These operations depend on the input letter just read, the state of the machine, and on the set of letters currently being scanned by the work-tapes, as does the ensuing state change. This combination of reading, writing and moving of heads constitutes a single transition of the machine. After reading the final letter of the input word and completing the associated transition, the machine must halt and either accept the input word or not.

A real-time Turing machine clearly operates in linear time and hence, in particular, the word problem of any real-time group (that is, a group G with $\mathrm{WP}(G) \in \mathcal{RT}$) is soluble in linear time by a machine that reads its input at constant speed.

Let us call a Turing machine K-*step real-time* for $K \geq 1$, if it satisfies the

249

above conditions but, in each transition, it is allowed to execute up to K operations on the work-tapes after reading a single input letter. The following result shows that, by increasing the tape-alphabet, we can recognise the same language with a standard real-time Turing machine having the same number of work-tapes. The proof is essentially from Hartmanis and Stearn's paper [132, Theorem 2].

13.1.1 Proposition *Let M be a K-step real-time Turing machine for some $K \geq 0$. Then there is a real-time Turing machine M' having the same number of tapes as M with $L(M') = L(M)$.*

Proof It is sufficient to prove this with $K = 2$, because then we can repeat the construction to handle the case $K = 2^i$ for any $i \geq 0$.

In order for each work-tape operation of M' to simulate the execution of two such operations of M, we compress the contents of the work-tapes and increase the tape-alphabet, in such a way that each cell in a work-tape T' of M' stores the contents of two cells of the corresponding tape T of M. In addition, since M' needs to be able to see the contents of the two neighbouring cells of the cell of T currently being scanned, each cell of T' stores some information that is also stored in the cell to the right of it, and some information that is also stored in the cell to the left of it, so in fact each cell of T is replicated twice within T'.

More precisely, numbering the cells of T by $i \in \mathbb{Z}$, cell number i of T' stores the contents of cells $2i - 2$, $2i - 1$, $2i$ and $2i + 1$ of T. This cell is scanned by M' when either cell $2i - 1$ or $2i$ is being scanned by M, and M' knows which, as part of the information stored in its current state. See Figure 13.1.

Cell of T':	−2	−1	0	1	2
	−6	−4	−2	0	2
Contains contents	−5	−3	−1	1	3
of these cells of T:	−4	−2	0	2	4
	−3	−1	1	3	5

Figure 13.1 Work-tape T' of M' simulating work-tape T of M

Now M' has enough information almost to carry out in a single operation of T' the two (or less) operations of T executed by M on reading an input symbol. At the end of two operations that start with the tape-head of M scanning cell $2i$ of T (the other case, starting at cell $2i - 1$, is similar), that tape-head could be pointing at any one of the cells $2i - 2$, $2i - 1$, $2i$, $2i + 1$, or $2i + 2$. In the first of these cases, the tape-head of T' moves one cell to the left, in the second and third it stays where it is, and in the fourth and fifth cases it moves one cell to the right.

Unfortunately, in order to complete the simulation of the two operations of M on T, M' may also need to update one or both of the adjoining cells $i - 1$ and $i + 1$ of T', which it cannot do as part of the current transition. If the tape-head of T' is still scanning cell i after the move, then it may need to update those parts of the contents of cells $i - 1$ and/or $i + 1$ that are shared with cell i. On the other hand, if the tape-head of T' moves one cell to the left or right, then it will need to update the contents of the cell that it is scanning after the move and, after a move left to cell $i - 1$, it may also need to update part of cell $i + 1$ that is shared with cell i. For example, if T executes a left move followed by a right move, then cells $2i - 1$ and $2i$ of T may be altered, in which case cells $i - 1$ and $i + 1$ of T' may need adjustment after the move. But note that T' will never need to change those parts of any cell that are not shared with cell i.

All of these problems can be handled as follows. Notice that, in each of these situations, after the transition of M', the contents of cell i of T', which was being scanned by M' before the transition, are correct, and these contents can be stored as part of the new state of M'. So, if the tape-head of T' moves to cell $i - 1$, say, then M' knows what the contents of the two components of cell $i - 1$ that are shared with cell i should be, so it can ignore their actual contents, and update them correctly on its next move. The other two components of cell $i - 1$ were not affected by the move, and are guaranteed to be correct. It is possible that parts of cell $i + 1$ of T' could remain uncorrected while the tape-head makes lengthy excursions to the left of cell i but, since it will be forced to pass through cell i again before moving to cell $i+1$, it will see and memorise the correct version of the incorrect parts of cell $i+1$ before encountering them. □

13.1.2 Closure properties of real-time languages The closure properties of real-time languages are studied by Rosenberg [222, Sections 3,4], and we refer the reader there for further details and proofs of the results stated here. It is easily seen that, if L_1 and L_2 are real-time Turing machines over A with k_1 and k_2 work-tapes, respectively, then there are real-time Turing machines with $k_1 + k_2$ tapes that accept each of $L_1 \cup L_2$ and $L_1 \cap L_2$, and a k_1-tape machine that accepts the complement of L_1 in A^*. However, examples are given in [222, Section 4] that show that the class of real-time languages is not closed under reversal, concatenation or Kleene closure. Further, if L_1 is real-time and L_2 is regular, then $L_1 L_2$ is real-time, but $L_2 L_1$ need not be, even when $L_2 = A^*$.

The class $\mathcal{RT}$ is not closed under homomorphism, but it is closed under inverse generalised sequential mappings. To see that, let M be a real-time Turing machine over A with state set Σ, and $\phi : B^* \to A^*$ a GSM defined by the deterministic transducer T with state set Υ. Let K be the length of a longest word in A^* output by a transition of T. We can construct a K-step real-time machine M' accepting $\phi^{-1}(L(M))$ as follows, and then apply Proposition 13.1.1. The state set of M' is $\Sigma \times \Upsilon$. Then M' can use the second component of its states to pass w through T. For each letter of w read, the word over A^* output by T has length at most K, so it can be processed using at most K moves of M, which can be simulated in M' using the first component of its states.

So by Proposition 3.4.3 we can unambiguously talk about a finitely generated group having real-time word problem, and we call such groups *real-time groups*. Furthermore, by Proposition 3.4.5, the class of real-time groups is closed under passing to finitely generated subgroups and, by Proposition 3.4.7, to overgroups of finite index. Closure under direct products and under taking quotients by finite normal subgroups are proved by Holt and Rees [151].

Closure under free products turns out to be less straightforward, and this question remains open. However, a subclass of the class of real-time languages was introduced and called the class of *tidy real-time languages* in [151], and it was shown that the class of groups having tidy real-time word problem is closed under free products. All known examples of real-time groups have tidy real-time word problem.

A language is defined to be *tidy real-time* if it is accepted by a real-time machine in which the work-tapes are right semi-infinite rather than bi-infinite, the tape-heads are always at the right-hand ends of the contents of the work-tapes, and the work-tapes are all empty after accepting a word. (They are called tidy because they clean up after accepting a word, although they may remain in a mess in the case of non-acceptance.)

13.1.3 Groups with generalised Dehn algorithms It is proved by Holt [147] that all groups G with a Dehn algorithm have tidy real-time word problem using four work-tapes. We saw in Sections 6.4 and 6.5 that a group has a Dehn algorithm if and only if it is word-hyperbolic. Holt's result is extended by Holt and Rees [151] to many of the groups that are proved by Goodman and Shapiro [108] to have generalised Dehn algorithms, as defined in 3.5.6. In particular, the result is valid for finitely generated nilpotent groups and for geometrically finite hyperbolic groups.

We confine ourselves here to summarising the arguments involved. We use a real-time Turing machine with four work-tapes to carry out the generalised Dehn algorithm. In Proposition 3.5.5, we described how this could be done in linear time using two stacks, and the reader is advised to read that proof again at this point. Recall, in particular, that the application of the Dehn algorithm is equivalent to reducing the input word w using the rules of a strictly length reducing rewriting system S for G. For a standard Dehn algorithm arising from a Dehn presentation $\langle X \mid R \rangle$ of G, these rules are $\{(u, v) \mid uv^{-1} \in \hat{R}, |u| > |v|\}$, and for a generalised Dehn algorithm the rules are over a larger alphabet than $X^{\pm}$.

Two of the four work-tapes perform the same function as the two stacks used in the proof of Proposition 3.5.5. So Tape 1 is used to store the S-reduced word equal in G to a prefix of the word w read so far, and Tape 2 is used while carrying out S-reductions.

The problem with a machine based on just the two tapes, is that while Tape 2 is being used in the reduction process of Proposition 3.5.5, no letters are being read from the input-tape, and there is no fixed bound on the number of steps that are executed between reading two successive input letters. For our machine to operate in real-time, we need to continue reading input symbols, while we are processing on Tape 2. The third and fourth of our work-tapes are the *queueing-tapes*.

Whenever Tape 2 is being used, one of these two tapes, which we call Tape q, is used to queue letters that we are required to read from the input-tape, but which we are not yet ready to process. When the main algorithm is ready to start reading input again, Tape 2 is empty, but unfortunately the queued symbols on Tape q are in reverse order. So we start by transferring the contents of Tape q to Tape 2, which has the effect of putting the symbols back into the correct order. The main algorithm can then start processing the letters from Tape 2 again, just as it does while processing reductions. While all this is going on, we still need to keep reading input letters, and the other queueing-tape is used for this. From what we said earlier, we can choose a constant K, which can be as large as we like, and only read one input letter for queueing for every K operations that take place during reduction processing and transferring queued symbols from Tape q to Tape 2.

A problem may arise if we reach the end of the input. We are then allowed to perform only K further tape operations. If at that stage Tapes 2, 3 and 4 are all empty (that is, we are not still in the middle of processing reductions) then, since we are executing a Dehn algorithm, we know that the input word $w =_G 1$ if and only if Tape 1 is empty. Otherwise, we need to return an answer. The following result, which we shall not prove here, provides an approach to resolving this problem. The corollary follows immediately from the above discussion.

13.1.4 Proposition [151, Theorem 4.1] *Let $G = \langle X \rangle$ be a group that has a generalised Dehn algorithm with associated rewriting system over an alphabet B containing $X^{\pm}$. Suppose that there are constants α and β with $\alpha > 0$ such* *that, if v ∈ B* is an S-reduced word that results from applying the rules of S to a word u over X, then |u|_G ≥ α|v| + β. Then there exists a constant K such that, if the real-time Turing machine described above reaches the end of an input word w, and Tapes 2,3 and 4 are not all empty after K further tape operations, then w ≠_G 1.*

13.1.5 Corollary   *If G has a generalised Dehn algorithm and constants α and β exist as in the proposition, then G has (tidy) real-time word problem.*

13.1.6 Hyperbolic, nilpotent, and geometrically finite hyperbolic groups
We proved in Theorem 6.4.1 that word-hyperbolic groups $G = \langle X \rangle$ have a Dehn algorithm with S consisting of all rules (u, v) with $u =_G v$ and $k \geq |u| > |v|$, where $k = 4\delta$ and all geodesic triangles in the Cayley graph $\Gamma(G, X)$ are δ-thin.

So the S-reduced words are k-local geodesics as defined before Theorem 6.4.1 and, by Proposition 6.4.4, the hypotheses of Proposition 13.1.4 hold with $\alpha = 1/2, \beta = 1$. Hence G has tidy real-time word problem.

It is proved in Sections 5 and 6 of [151] that the hypotheses of Proposition 13.1.4 also hold for the generalised Dehn algorithms that exist for finitely generated nilpotent groups and for geometrically finite hyperbolic groups, so these groups also have tidy real-time word problems.

13.1.7 Other real-time groups It is proved by Holt and Röver [156] that the Baumslag–Solitar groups $BS(1, n)$ have tidy 5-tape real-time word problem for all $n \in \mathbb{Z}$. It is also proved in that paper that $F_2 \times F_2$ (where F_2 is the free group of rank 2) has 2-tape but not 1-tape real-time word problem.

13.2 Indexed word problems

There have been various attempts to decide whether there exist groups G for which WP(G) is indexed (see 2.2.21) but not context-free, but this issue remains unresolved. In unpublished work, Gilman and Shapiro have proved that, if G is accessible (see comments following Theorem 1.8.1) and WP(G) is accepted by a deterministic nested stack automaton with *limited erasing*, then G is virtually free, and hence is a context-free group.

13.3 Poly-context-free word problems

Recall from 2.2.22 that a language L is (deterministic) *k-context-free* if it is the intersection of k (deterministic) context-free languages L_i, and it is *(deter-*

ministic) *poly-context-free* if it is (deterministic) *k-context-free* for some $k \in \mathbb{N}$. The classes of poly-context-free and deterministic poly-context-free languages are denoted by $\mathcal{PCF}$ and $\mathcal{DPCF}$.

Groups in which WP(G) $\in \mathcal{PCF}$ (that is, poly-context-free groups) are investigated by Brough [54, 55]. It is proved there that, for any $k \in \mathbb{N}$, the class of k-context-free languages is closed under inverse GSMs, union with context-free languages, and intersection with regular languages. So $\mathcal{PCF}$ is closed under all of these operations, and also under intersection and union.

It follows then from the closure results established in Section 3.4 that the class of poly-context-free groups is closed under change of generators, finitely generated subgroups, and finite index overgroups. It is also straightforward to show that the class is closed under direct products. It is proved in [55, Lemma 4.6] that, for $k \in \mathbb{N}$, WP($\mathbb{Z}^k$) is deterministic k-context-free, but not l-context-free for any $l < k$.

Since the class of poly-context-free groups includes the finitely generated free groups, it contains all groups that are virtually direct products of finitely many free groups of finite rank. In fact WP(G) $\in \mathcal{DPCF}$ for each of these examples, and it is conjectured that they are the only poly-context-free groups.

It is proved in [55] that WP(G) $\notin \mathcal{PCF}$ for various types of groups, including torsion-free soluble groups that are not virtually abelian. These proofs involve the application of Parikh's Theorem 2.6.23 in a similar way in which it is used to prove that the co-word problem of various groups is not context-free, which is discussed in more detail below in 14.2.10. We shall prove some of the results from [55] in Theorems 14.2.15 and 14.2.16.

As in the case of context-free co-word probems, it is conjectured but not proved that WP($\mathbb{Z}^2 * \mathbb{Z}$) $\notin \mathcal{PCF}$.

14

The co-word problem and the conjugacy problem

14.1 The co-word problem

The main theme of this final chapter is the *co-word problem* of a finitely generated group, which is defined to be the complement of its word problem. We shall see that for several of the major language classes, including $C\mathcal{F}$ and $\mathcal{I}$, there are more groups with co-word problem in the class than there are with word problem in the class.

We conclude the chapter by considering briefly the formal language properties of the conjugacy problem of a group. This is necessarily short, because there appear to be very few (if any) significant results that have been proved in this area!

14.1.1 General properties of the co-word problem Let A be a monoid generating set for a group G. We define the *co-word problem* for G over A to be the set of words $\mathrm{coWP}(G, A)$ that *do not* represent the identity, that is,

$$\mathrm{coWP}(G, A) := \{w \in A^* : w \neq_G 1\}.$$

So $\mathrm{coWP}(G, A)$ is the complement $A^* \setminus \mathrm{WP}(G, A)$ of the word problem. If $A := X^{\pm}$ with X a group generating set, we may abuse notation and call $\mathrm{coWP}(G, A)$ the co-word problem for G over X.

For those classes C of languages that are closed under complementation, such as $\mathcal{R}eg$, $\mathcal{DCF}$, $\mathcal{DOC}$, $\mathcal{CS}$, $\mathcal{DCS}$, $\mathcal{RT}$, and recursive languages, we have $\mathrm{coWP}(G, A) \in C$ if and only if $\mathrm{WP}(G, A) \in C$. For other classes, such as $C\mathcal{F}$, this need not be the case. Indeed, Corollaries 14.2.1 and 14.2.2 below show that the class of context-free groups is strictly contained in the class of co-context-free groups (i.e. groups with context-free co-word problem).

In each of the following propositions, C denotes a class of formal languages.

The proofs of the first two are omitted because they are essentially the same as for the corresponding results for word problems (Propositions 3.4.3 and 3.4.5).

14.1.2 Proposition *If C is closed under inverse homomorphisms, then the property* $\text{coWP}(G, A) \in C$ *is independent of the chosen finite monoid generating set A of G.*

Since most classes C that we consider are closed under inverse homomorphisms, in general we are able to write $\text{coWP}(G)$ rather than $\text{coWP}(G, A)$, just as we write $\text{WP}(G)$ rather than $\text{WP}(G, A)$. Groups with $\text{coWP}(G)$ in a language class C are often called *co-C-groups*.

14.1.3 Proposition *If C is closed under inverse homomorphisms and intersection with regular languages and* $\text{coWP}(G) \in C$*, then* $\text{coWP}(H) \in C$ *for any finitely generated subgroup* $H \leq G$.

14.1.4 Proposition *Assume that C is closed under inverse deterministic GSMs and under union with regular sets. Let* $G \leq H$ *be groups with* $|H : G|$ *finite and* $\text{coWP}(G) \in C$*. Then* $\text{coWP}(H) \in C$.

Proof This is similar to the proof of Proposition 3.4.7, but now $\text{coWP}(H, B)$ is the union of $\phi^{-1}(\text{coWP}(G, A))$ with the set of words w for which the GSM M is not in state 1 after reading w. □

14.1.5 Proposition *If C is closed under inverse homomorphisms and under union, then the class of co-C-groups is closed under direct products.*

Proof Let $G_i = \langle X_i \rangle$ and $A_i := X_i^{\pm}$ for $i = 1, 2$, where we assume that A_1 and A_2 are disjoint, and suppose that $\text{coWP}(G_i, A_i) \in C$. We identify G_1 and G_2 with the corresponding component subgroups of $G_1 \times G_2$ and use the generating set $X := X_1 \cup X_2$ of G. The projection maps $\phi_i \colon A^* \to A_i^*$ for $i = 1, 2$ are monoid homomorphisms, and

$$\text{coWP}(G, A) = \phi_1^{-1}(\text{coWP}(G_1, A_1)) \cup \phi_2^{-1}(\text{coWP}(G_2, A_2)).$$

□

14.2 Co-context-free groups

14.2.1 Proposition *For a finitely generated group G, the following are equivalent:*

(i) *G is virtually free;*
(ii) *WP(G) is context-free;*

(iii) coWP(G) *is deterministic context-free.*

Proof The equivalence of (i) and (ii) is the Muller–Schupp Theorem (Theorem 11.1.1), and the equivalence of (ii) and (iii) follows from the closure of the class of deterministic context-free languages under complementation (Proposition 2.6.30). We could also prove that (i) implies (iii) more directly, using the deterministic pda for free groups described in 2.2.5 and Proposition 14.1.4. □

The following immediate corollary of Proposition 14.1.5 shows that there are more co-context-free groups than context-free groups.

14.2.2 Corollary *The class of co-context-free groups is closed under direct products.*

It is not known whether this class is closed under free products. We make the following conjecture, which would imply that it is not, but it seems to be difficult to prove and, assuming that it is correct, its proof might require the development of new techniques for proving that languages are not context-free.

14.2.3 Conjecture The group $G = \mathbb{Z}^2 * \mathbb{Z}$ is not co-context-free.

The following results, proved by Holt, Rees, Röver and Thomas [154], give rise to more examples of co-context-free groups, including some that are not finitely presentable, such as $\mathbb{Z} \wr_R \mathbb{Z}$. See 1.5.5 for the definition of wreath products.

14.2.4 Proposition [154, Theorem 10] *Let G and H be finitely generated groups with* coWP(G) *and* WP(H) *both context-free. Then the restricted standard wreath product $G \wr_R H$ is a co-context-free group.*

Proof Suppose that $G = \langle X \rangle$ and $H = \langle Y \rangle$, and let $A := X^{\pm}$, $B := Y^{\pm}$. The restricted standard wreath product $W := G \wr_R H$ is defined in 1.5.5. Recall that the base group N of W is the restricted direct product of copies G_h of G, with one copy for each $h \in H$, where we view N as the set of all functions $\gamma : H \to G$ for which $\gamma(h) = 1$ for all but finitely many $h \in H$. Elements of N are multiplied component-wise. We have a right action $\psi : H \to \mathrm{Aut}(N)$ defined by $\gamma^{\psi(h)}(h') = \gamma(h'h^{-1})$ for $\gamma \in N$ and $h, h' \in H$, and W is the resulting semidirect product $N \rtimes_{\psi} H$. We identify G with the subgroup of the base group of W comprising those elements γ_g with $\gamma_g(1) = g$, but $\gamma_g(h) = 1$ for all $1 \neq h \in H$. Then W is generated by $X \cup Y$. Below, we refer to elements of A and B as A-letters and B-letters, respectively.

Now an input word w represents an element (γ, h), with $\gamma \in N$, $h \in H$; that element is non-trivial if and only if either $h \neq 1$, or $\gamma(h') \neq 1$ for some $h' \in H$. The pda M with language coWP($W, A \cup B$) that we shall now describe decides non-deterministically which of these possibilities it will try to confirm. If it chooses to try and confirm the second possibility, then it also selects a possible h'.

Since WP(H, B) $\in \mathcal{CF}$, we have WP(H, B) $\in \mathcal{DCF}$ by Theorem 11.1.1 and, as we saw in the proof of (iii) $\implies$ (ii) of that theorem, there is a deterministic pda M_H, accepting WP(H, B), that uses a symbol b_0 to mark the bottom of the stack, and the stack is empty apart from b_0 whenever M_H is in an accept state. Let M_G be a possibly non-deterministic pda with $L(M_G) = $ coWP(G, A). The pda M has stack alphabet $A \cup B \cup \{b_0\}$, where b_0 is again used only as a bottom of stack marker, and is never removed.

If M decides to check whether $h \neq 1$, then it ignores all A-letters and simulates M_H, except that M accepts if and only if M_H rejects. In this case M accepts a word representing (γ, h) if and only if $h \neq 1$.

Suppose instead that M decides to check whether $\gamma(h') \neq_G 1$ for some $h' \in H$. This case is more complicated. The pda M needs to simulate each of M_G and M_H during its reading of the input, so its state set is $\Sigma_G \times \Sigma_H$. Before reading its input word w, the pda M chooses the element $h' \in H$. It does this by non-deterministically choosing a word $v \in B^*$ and passing v to M_H for processing. Then $h' \in H$ is the element represented by v and, after v has been processed by M_H, the pda M is put into the state $(\sigma_0, \sigma_v) \in \Sigma_G \times \Sigma_H$, where σ_v is the state that M_H is in after reading v and σ_0 is the start state of M_G.

Now M starts to read w, and operates as follows. Whenever M reads a B-letter, it simulates M_H, but regards an A-letter at the top of the stack as being equivalent to b_0. When M reads an A-letter then, provided the Σ_H-component of its state is accepting for M_H, it simulates M_G; otherwise, it ignores that A-letter and passes to the next symbol of w. Then M accepts w if the Σ_G-component of its state is accepting after reading w, but not otherwise. We claim that, when M is operating in this mode, $w \in L(M)$ if and only if it represents some (γ, h) with $\gamma(h') \neq_G 1$.

For suppose that $w = z_1 z_2 \cdots z_l$, with each $z_i \in A \cup B$, and let $\overline{w}$ denote the element of H represented by the word obtained from w by deleting all A-letters. Write $w(i)$ for the prefix of w of length i. Then w represents (γ, h), where $h = \overline{w}$, and it follows from the expression for multiplication in the wreath product given in 1.5.5 that

$$\gamma = \prod_{z_i \in A} \gamma_{z_i}^{\psi((\overline{w(i)})^{-1})}, \quad \gamma(h') =_G \prod_{z_i \in A} \gamma_{z_i}^{\psi((\overline{w(i)})^{-1})}(h') =_G \prod_{z_i \in A} \gamma_{z_i}(h'\overline{w(i)}).$$

Now let I be the subset of all elements i of $\{1, 2, \ldots, l\}$ such that $h'\overline{w(i)} =_H 1$

and $z_i \in A$; it is precisely for these elements i that $\gamma_{z_i}^{\psi((\overline{w(i)})^{-1})}(h') =_G z_i$, and so $\gamma(h') =_G z_{i_1} z_{i_2} \cdots z_{i_k}$, where $I = \{i_1, i_2, \ldots, i_k\}$ and $i_j < i_{j+1}$ for $1 \le j < k$. In other words, $\gamma(h')$ is equal to the subsequence of w consisting of all A-letters z_i for which $h'\overline{w(i)} =_H 1$.

The construction ensures that the first A-letter z_i that is not ignored by M is the first A-letter z_i for which $h'\overline{w(i)} =_H 1$; that is when $i = i_1$. Then, since for $i > i_j \in I$, we have $\overline{w(i)} =_H \overline{w(i_j)}$ if and only if $i \in I$, the pda M takes action after reading a subsequent A-letter z_i if and only if $i \in I$. So the Σ_G-component of M after reading w is equal to the state of M_G after reading the subsequence of w identified at the end of the last paragraph, which represents $\gamma(h')$. So w is accepted by M operating in this mode precisely when this representative of $\gamma(h')$ is accepted by M_G. $\qquad\qquad\qquad\qquad\qquad\qquad\qquad\qquad\qquad\qquad\qquad\qquad\square$

Note that, for all of the examples of co-context-free groups that we have seen so far, either we have $\mathrm{WP}(G), \mathrm{coWP}(G) \in \mathcal{DCF}$ or else $\mathrm{coWP}(G)$ is the language of a pda that operates non-deterministically before the first input symbol is read and deterministically thereafter. We shall call such groups *semi-deterministic co-context-free*.

14.2.5 Further examples of co-context-free groups New methods are introduced by Lehnert and Schweitzer [178] to prove that certain types of infinite permutation groups, including the *Houghton groups* H_n $(n \ge 2)$ and the *Higman–Thompson groups* $G_{n,r}$ $(n \ge 2, r \ge 1)$ (see 1.10.11) have context-free co-word problem.

In the following definitions and proofs, we shall write the image of a point α under a permutation σ as α^σ and compose permutations from left to right. The groups H_n can defined as follows. For $n \ge 2$, let

$$\Omega_n = \{0\} \cup \{(i, j) : i \in \mathbb{N}, \, j \in Z_n\},$$

where $Z_n = \{1, 2, \ldots, n\}$ with addition modulo n. For each $j \in Z_n$, we define $\sigma_j \in \mathrm{Sym}(\Omega_n)$ by

$$(i, j)^{\sigma_j} = (i - 1, j) \text{ for } i > 1$$
$$(1, j)^{\sigma_j} = 0$$
$$0^{\sigma_j} = (1, j + 1)$$
$$(i, j + 1)^{\sigma_j} = (i + 1, j + 1) \text{ for } i > 0$$
$$(i, k)^{\sigma_k} = (i, k) \text{ for } k \ne j, j + 1.$$

Then H_n is defined to be the subgroup of $\mathrm{Sym}(\Omega_n)$ generated by the σ_j together with all elements of $\mathrm{Sym}(\Omega_n)$ with finite support. But, for $n \ge 3$, it can be checked that the commutator $\sigma_j \sigma_{j+1} \sigma_j^{-1} \sigma_{j+1}^{-1}$ is equal to the transposition $(0, (1, j))$ for $j \in Z_n$ and hence that H_n is generated by the σ_j alone. For $n = 2$, we need an extra generator, for example $(0, (1, 1))$.

The proof that H_2 is co-context-free is easier than for H_n with $n \geq 3$ and is not completely dissimilar to the proof of Proposition 14.2.4, so we shall give that first.

Using a change of notation, we regard $G = H_2$ as the subgroup of Sym($\mathbb{Z}$) generated by the permutations $\sigma: n \mapsto n + 1$ ($n \in \mathbb{Z}$) and the transposition $\tau = (0, 1)$. The normal closure N of τ in G is equal to the group of all permutations of $\mathbb{Z}$ with finite support, and $G = N \rtimes \langle \sigma \rangle$.

14.2.6 Proposition [178, Theorem 6] *The Houghton group $G = H_2$ is semi-deterministic co-context-free.*

Proof As we have just seen, the elements of G are of the form $g = (v, \sigma^k)$ with $v \in N = \langle \tau^G \rangle$ and $k \in \mathbb{Z}$. So $g \neq 1$ if and only if $v \neq 1$ or $k \neq 0$.

The pda M with language coWP($G, \{\sigma^{\pm 1}, \tau\}$) (note that $\tau^{-1} = \tau$) that we shall now describe decides non-deterministically which of these possibilities it will try to confirm. The stack alphabet is $B = \{b_0, A, B, C, D\}$ where as usual b_0 marks the bottom of the stack. Let w be the input word.

The pda M confirms that $k \neq 0$ deterministically by using A and B to keep track of the exponent k of σ in the prefix of w read so far, where the stack contains b_0 together with k symbols A or $-k$ symbols B when $k \geq 0$ or $k \leq 0$, respectively. The generator τ is ignored.

We now describe how M verifies that $v \neq 1$. When doing this, it may assume that the input element g is of the form $(v, 1)$, since otherwise $g \neq 1$ anyway. In other words, we may assume that w contains equal numbers of the symbols σ and σ^{-1}. Since $v \neq 1$ if and only if $k^v \neq k$ for some $k \in \mathbb{Z}$, the check can be carried out by choosing some $k \in \mathbb{Z}$ and verifying that $k^v \neq k$. Now M starts by non-deterministically putting b_0 followed by either an arbitrary number $k \geq 0$ of As or an arbitrary number $-k \geq 0$ of Bs on the stack, where $k \in \mathbb{Z}$ is the chosen integer. Thereafter, M operates deterministically.

Let l be the image of the chosen integer k under the prefix of w read so far. Then the contents of the stack will be of the form $(A^* \cup B^*)(C^* \cup D^*)b_0$, where there are l As if $l \geq 0$ and $-l$ Bs if $l < 0$. When $\sigma^{\pm 1}$ is read from the input, l is increased or decreased by 1, and M simply makes the appropriate adjustment to the top of the stack.

When τ is read, the value of l is changed only if $l = 0$ or 1, and such changes are recorded by using the stack symbols C and D. More precisely, suppose that under the prefix of w read so far there have been i occurrences of τ that mapped

0 to 1, and j that mapped 1 to 0. Then there will be $i - j$ Cs or $j - i$ Ds on the stack when $i - j$ is positive or negative, respectively.

Since we are assuming that w contains equal numbers of the symbols σ and σ^{-1}, we have $l = k$ if and only if there are no Cs or Ds on the stack after reading w. So, for the verification that $k^\nu \neq k$, M needs just two states, an accepting state, meaning that there are some Cs or Ds on the stack, and a fail state, meaning that there are not.

It remains to describe how M maintains the required number of Cs or Ds on the stack. To simplify the description, we shall assume that M can see the top two stack symbols, which is justified by Proposition 2.6.9.

Suppose that the symbol τ is read from w.

- If the top of stack symbol is B or the top two symbols are A, then the current value of l is not 0 or 1 and so nothing is done.
- If the top symbol is A but the second symbol down is not A, then $l = 1$, and l and $i - j$ both need to be decreased by 1. If the second symbol is not C, then we replace the top symbol A by D, and otherwise we just delete AC from the top of the stack.
- If the top of stack symbol is b_0, C or D, then $l = 0$ and l and $i - j$ both need to be increased by 1. If the top of stack symbol is not D, then we put AC on top of the stack. Otherwise, we replace the D by A.

□

The proof in [178] that the groups H_n with $n \geq 3$ and the Higman–Thompson groups $G_{n,r}$ are co-context-free uses the following difficult closure result for context-free languages.

14.2.7 Proposition *The class $\mathcal{CF}$ is closed under cyclic permutations; that is, if $L \subset A^*$ with $L \in \mathcal{CF}$, then*

$$L^{\text{cyc}} := \{vu : u, v \in A^*, \, uv \in L\} \in \mathcal{CF}.$$

This was proved independently by Maslov [188] and Oshiba [208]. It also appears as [159, Exercise 6.4 c)], which is marked as ∗∗, meaning very challenging, but a solution is provided at the end of [159, Chapter 6].

We shall give the proof for H_n ($n \geq 3$) here. The proof for $G_{n,r}$ (defined in 1.10.11) is similar, but we refer the reader to [178] for that. It is not known whether any of these groups are semi-deterministic co-context-free.

14.2.8 Proposition [178, Corollary 2] *The group $G = H_n$ is co-context-free for $n \geq 3$.*

Proof Let $G = H_n$ for some $n \geq 3$, let $A := X^{\pm}$ with $X = \{\sigma_j : j \in Z_n\}$, and let $L = \{w \in A^* : 0^w \neq 0\}$. First we prove that $L \in \mathcal{CF}$ and then that $L^{\text{cyc}} = \text{coWP}(G, A)$. Then Proposition 14.2.7 implies that $\text{coWP}(G, A) \in \mathcal{CF}$.

In fact it is easily seen that $L \in \mathcal{DOC}$. Let (i, j) or 0 be the image of 0 under the prefix of w read so far. We can use the states of a 1-counter pda to keep track of j or 0 and the single stack symbol to keep track of i. So certainly $L \in \mathcal{CF}$.

It remains to show that $L^{\text{cyc}} = \text{coWP}(G, A)$. Let $w = vu \in L^{\text{cyc}}$ with $uv \in L$. Then $0^{v^{-1}w} = 0^u \neq 0^{v^{-1}}$, so $w \in \text{coWP}(G, A)$.

Conversely, let $w \in \text{coWP}(G, A)$. If $0^w \neq 0$ then $w \in L$ and we are done. Also, if $(i, j) \in \Omega_n$ and $(i, j)^w = (i', j')$ with $j \neq j'$, then we must have $(i, j)^v = 0$ for some prefix v of w and then $w = vu$ with $uv \in L$, so $w \in L^{\text{cyc}}$. So we may assume that $(i, j)^w = (i', j)$ for all $j \in Z_n$.

It is easily proved by induction on $|w|$ that, for any $w \in A^*$, there exist integers m_j for $j \in Z_n$ with $\sum_{j \in Z_n} m_j = 0$ such that, for all $j \in Z_n$ and all sufficiently large i, we have $(i, j)^w = (i + m_j, j)$. (In fact multiplying w by σ_j increases m_{j+1} by 1 and decreases m_j by 1.) Now, if the m_j are not all 0, then the fact that w is a permutation implies that there must exist $(i, j) \in \Omega_n$ and $(i, j)^w = (i', j')$ with $j \neq j'$, and we are done. Otherwise, let $(i, j) \in \Omega_n$ with $(i, j)^w = (i', j)$ and $i \neq i'$. Again, if $w = vu$ with $(i, j)^v = 0$, then $uv \in L$ and we are done. Otherwise, the image of (i, j) under prefixes of w is changed only by the generators $\sigma_j^{\pm 1}$ and $\sigma_{j-1}^{\pm 1}$. But then the images of (i'', j) for all $i'' > i$ are changed in the same way, and so $m_j = i' - i \neq 0$, contrary to our assumption. $\square$

14.2.9 Could there be a universal co-context-free group? The interesting possibility of there being a *universal co-context-free group* that contained all co-context-free groups as subgroups was originally suggested by Lehnert. The Higman–Thompson group $V = G_{2,1}$ has been proposed by Bleak, Matucci and Neunhöffer [24] as a candidate; however some negative evidence for the proposal has been presented by Berns-Zieve et al. [22], who describe some generalisations of V that are also co-context-free, but have no obvious embeddings into V itself.

14.2.10 Groups that are not co-context-free In 2.6.22 we defined *linear* and *semilinear* subsets of $\mathbb{N}_0^n$, and proved Parikh's Theorem 2.6.23, which says that a context-free language over an alphabet $A := \{a_1, \ldots, a_n\}$ has a semilinear image under the map $\psi : A^* \to \mathbb{N}_0^n$ defined by $\psi(w) = (|w|_{a_1}, \ldots, |w|_{a_n})$, where $|w|_{a_i}$ is the total number of occurrences of a_i in w.

Holt, Rees, Röver and Thomas [154, Section 4] used this result to prove that, for certain classes of groups G, coWP(G) is not context-free. These classes include polycyclic groups that are not virtually abelian, and Baumslag–Solitar groups BS(m, n) with $m \neq \pm n$. The following result enables us to achieve this for finitely generated nilpotent groups, and for the Baumslag–Solitar groups. A stronger version [154, Theorem 14] is needed for general polycyclic groups.

14.2.11 Proposition   *Let $L \subseteq \mathbb{N}_0^{r+1}$ for some $r \in \mathbb{N}$, and suppose that L has the following property. For every $K \in \mathbb{N}$ there exists*

$$(a_1, \ldots, a_r) \in \mathbb{N}_0^r \setminus \{(0, \ldots, 0)\}$$

such that

(i) *there is a unique $b \in \mathbb{N}_0$ with $(a_1, \ldots, a_r, b) \notin L$; and*
(ii) *for $b, a_1, \ldots, a_r$ as in (i), we have $b \geq K(a_1 + \cdots + a_r)$.*

Then L is not semilinear.

Proof   Suppose, for a contradiction, that L is the union of the linear sets $L_1, L_2, \ldots, L_t$, where

$$L_i = \{c_i + n_1 \alpha_{i1} + \cdots + n_{m_i} \alpha_{i m_i} : n_j \in \mathbb{N}_0\},$$

and let $P_i := \{\alpha_{ij} : 1 \leq j \leq m_i\}$ for $1 \leq i \leq t$. We may clearly assume that all of the elements of each P_i are nonzero.

Order the L_i such that, for $1 \leq i \leq s$, P_i contains an element $(0, \ldots, 0, N_i)$ (where, by assumption, $N_i > 0$) and, for $s + 1 \leq i \leq t$, P_i contains no such element. It is possible that $s = 0$ or $s = t$. Let N be the least common multiple of the N_i ($1 \leq i \leq s$), with $N = 1$ if $s = 0$.

Suppose that i satisfies $s + 1 \leq i \leq t$. Then there exists $q_1 \in \mathbb{N}$ such that $x_{r+1} < q_1(x_1 + \cdots + x_r)$ for all elements $(x_1, \ldots, x_{r+1}) \in P_i$. We see then that there exists $q_2 \in \mathbb{N}_0$ such that $x_{r+1} < q_1(x_1 + \cdots + x_r) + q_2$ for all $(x_1, \ldots, x_{r+1}) \in L_i$. Let $C \in \mathbb{N}$ be twice the maximum of all of the constants q_1, q_2 that arise for all P_i with $s + 1 \leq i \leq t$. Then $x_{r+1} < C(x_1 + \cdots + x_r)$ whenever $(x_1, \ldots, x_{r+1}) \in L_i$ with $s + 1 \leq i \leq t$ and $x_1, x_2, \ldots, x_r$ are not all zero. We can take $C = 0$ if $s = t$.

Choose $K \geq 2 \max\{N, C\}$ and let $(a_1, \ldots, a_r)$ and b satisfy (i) and (ii) of the proposition for this K. The uniqueness of b implies that $(a_1, \ldots, a_r, b - N) \in L$ (note that $b > N$ by (ii)). In particular, $(a_1, \ldots, a_r, b - N) \in L_i$ for some i with $1 \leq i \leq t$.

Now, if $i \leq s$, then $(0, \ldots, 0, N_i) \in P_i$ for some $N_i > 0$, and we can add $(N/N_i)(0, \ldots, 0, N_i)$ to $(a_1, \ldots, a_r, b - N)$ to get $(a_1, \ldots, a_r, b) \in L_i$, contradicting (i). On the other hand, if $s + 1 \leq i \leq t$ then $b - N < C(a_1 + \cdots + a_r)$ and, since $a_1 + \cdots + a_r \geq 1$, we have

$$b < C(a_1 + \cdots + a_r) + N \leq 2 \max\{C, N\}(a_1 + \cdots + a_r) \leq K(a_1 + \cdots + a_r),$$

contradicting (ii). □

14.2.12 Theorem *A finitely generated nilpotent group has context-free co-word problem if and only if it is virtually abelian.*

Proof Since a finitely generated virtually abelian group is virtually a direct product of infinite cyclic groups, it follows from Propositions 14.2.1, 14.1.4 and Corollary 14.2.2 that $\mathrm{coWP}(G)$ is context-free for any such group.

Now assume that G is a finitely generated nilpotent but not virtually abelian group. Since the torsion subgroup of G is finite [84, Lemma 8.2.2] and G is residually finite [143, Theorem 3.25], G has a torsion-free subgroup of finite index, which cannot be virtually abelian. Furthermore, we claim that every non-abelian torsion-free nilpotent group has a subgroup isomorphic to the Heisenberg group $H = \langle x, y, z \mid [x, y] = z, [x, z] = [y, z] = 1 \rangle$. To see this, let x be a non-central element of the second term of the upper central series of G and let y be some element of G not commuting with x.

By Proposition 14.1.3, it suffices to show that $\mathrm{coWP}(H)$ is not context-free. Let $A := \{x^{\pm 1}, y^{\pm 1}, z^{\pm 1}\})$, and let L be the intersection of $\mathrm{coWP}(H, A)$ with the regular set

$$x^* y^* (x^{-1})^* (y^{-1})^* (z^{-1})^*.$$

Regard L as a language over $A' := A \setminus \{z\}$, and let $\psi \colon A' \to \mathbb{N}_0^5$ be the map defined above in 14.2.10. Then, since $[x^m, y^m] =_H z^{m^2}$ for all $m \in \mathbb{N}_0$, we see that L satisfies the hypothesis of Proposition 14.2.11. (Given K, consider (m, m, m, m) with $m \geq 4K$.) So L is not context-free and hence neither is $\mathrm{coWP}(H, A)$. □

The result below is stated incorrectly in [154, Theorem 13]: it is stated there that $\mathrm{BS}(p, q)$ is virtually abelian when $|p| = |q|$, whereas in fact this is only true when $|p| = |q| = 1$. The result that $\mathrm{coWP}(\mathrm{BS}(p, \pm p))$ is context-free is however correctly proved by Brough [55, Proposition 4.4], and we follow that proof here.

14.2.13 Theorem *Let $G = \mathrm{BS}(p, q) = \langle x, y \mid y^{-1} x^p y = x^q \rangle$. Then G is co-context-free if and only if $p = \pm q$.*

Proof Suppose first that $p = \pm q$, and assume without loss that $p > 0$. If $p = q$, then the quotient group $G / \langle x^p \rangle$ is isomorphic to the free product

$$\mathbb{Z} * C_p = \langle y, x \mid x^p = 1 \rangle,$$

which has the free group of rank p generated by $y, y^x, \dots, y^{x^{p-1}}$ as subgroup of index p. Since $\langle x^p \rangle \leq Z(G)$, it follows that G is virtually a direct product of free groups of ranks 1 and p and hence coWP(G) is context-free by Propositions 14.2.1, 14.1.4 and Corollary 14.2.2.

If $p = -q$, then by a straightforward Reidemeister–Schreier calculation, we find that the normal closure H of $\langle x, y^2 \rangle$ in G has index 2 in G and has the presentation $\langle a, b, c \mid a^p = b^p, c^{-1}a^p c = a^p \rangle$, where $a = x$, $b = y^{-1}x^{-1}y$, and $c = y^2$. The quotient H by its central subgroup $\langle a^p \rangle$ is isomorphic to the free product $\mathbb{Z} * C_p * C_p$, which is again virtually free, so coWP(G) is context-free.

Now suppose that $p \neq \pm q$, and assume without loss that $|p| < |q|$. Then, by considering the equation $y^{-m}x^{p^m}y^m =_G x^{q^n}$ for all $m \in \mathbb{N}_0$, we can apply Proposition 14.2.11 in a similar manner to the proof of Theorem 14.2.12 to derive a contradiction.   □

The following result, of which the proof is similar to that of Proposition 14.2.11, can be used to prove that the groups considered above are not poly-context-free. We could simplify the proof below by using the result that the intersection of finitely many semilinear sets is semilinear [102, Proof of Theorem 5.6.1].

14.2.14 Proposition *Let $L \subseteq \mathbb{N}_0^{r+1}$ for some $r \in \mathbb{N}$, and suppose that L has the following property. For every $K \in \mathbb{N}$ there exists*

$$(a_1, \ldots, a_r) \in \mathbb{N}_0^r \setminus \{(0, \ldots, 0)\},$$

such that

(i) *there is a unique $b \in \mathbb{N}_0$ with $(a_1, \ldots, a_r, b) \in L$; and*
(ii) *for $b, a_1, \ldots, a_r$ as in (i), we have $b \geq K(a_1 + \cdots + a_r)$.*

Then L is not an intersection of finitely many semilinear sets.

Proof Suppose, for a contradiction, that $L = L_1 \cap \cdots \cap L_t$, with each L_i semilinear. Then Parikh's Theorem implies that each L_i is a union of finitely many linear sets $L_{i1}, \ldots, L_{it_i}$. So, for $1 \leq i \leq t$ and $1 \leq j \leq t_i$, there exist $c_{ij}, \alpha_{ij1}, \ldots, \alpha_{ijm_{ij}} \in \mathbb{N}_0^{r+1}$ with

$$L_{ij} = \{c_{ij} + n_1\alpha_{ij1} + \cdots + n_{m_{ij}}\alpha_{ijm_{ij}} : n_k \in \mathbb{N}_0\},$$

where we may assume that the α_{ijk} are all nonzero.

For elements $v \in \mathbb{N}_0^{r+1}$, we denote the components of v by $v(1), \ldots, v(r+1)$. Let R be the maximum value of $\alpha_{ijk}(r+1)/\alpha_{ijk}(l)$ for any i, j, k, l with $1 \leq l \leq r$ and $\alpha_{ijk}(l) \neq 0$; if there is no such $\alpha_{ijk}(l)$ then put $R = 0$. Let S be the maximum value of $c_{ij}(r+1)$ for any i, j, and let $C = \max(R, S) + 1$. Let us call the vector α_{ijk} *simple* if its first r components $\alpha_{ijk}(l)$ ($1 \leq k \leq r$) are all 0, and *complex* otherwise. Then, if $(x_1, \ldots, x_{r+1}) = c_{ij} + n_1\alpha_{ij1} + \cdots + n_{m_{ij}}\alpha_{ijm_{ij}} \in L_{ij}$, and if some x_k with $1 \le k \le r$ is nonzero and each α_{ijk} with $1 \le k \le m_{ij}$ is complex, then $x_{r+1} \le S + R(x_1 + \cdots + x_r) \le C(x_1 + \cdots + x_r)$.

Choose $K > C$ and let $(a_1, \ldots, a_r, b) \in L$ be as in Hypothesis (i) of the proposition. So, for each i with $1 \le i \le t$, there exists j with $1 \le j \le t_i$ such that $(a_1, \ldots, a_r, b) \in L_{ij}$, and hence there exist $n_k \in \mathbb{N}_0$ for $1 \le k \le m_{ij}$ with $(a_1, \ldots, a_r, b) = c_{ij} + \sum_{k=1}^{m_{ij}} n_k \alpha_{ijk}$.

Then, since by Hypothesis (ii) $b > C(a_1 + \cdots a_r)$ and the a_i are not all zero, it follows that, for each i with $1 \le i \le t$, there must exist some k such that α_{ijk} is simple. So $\alpha_{ijk} = (0, \ldots, 0, n_i)$ for some $n_i > 0$. But then we have $(a_1, \ldots, a_r, b + ln_i) \in L_{ij}$ for all $l \in \mathbb{N}$, and so $(a_1, \ldots, a_r, b + n_1 n_2 \cdots n_t) \in L$, contradicting Hypothesis (i). $\square$

The following two results are due to Brough [55, Theorem 14.3, 14.5]. The first of these is proved for all virtually polycyclic groups in [55].

14.2.15 Theorem *A finitely generated nilpotent group has poly-context-free word problem if and only if it is virtually abelian.*

Proof We observed in 13.3 that the class of poly-context-free groups is closed under direct products and under finite index overgroups, so all virtually abelian groups are poly-context-free. Conversely, as in the proof of Theorem 14.2.12, it suffices to prove that the Heisenberg group H is not poly-context-free and this follows from Proposition 14.2.14 in the same way that the fact that coWP(H) is not context-free follows from Proposition 14.2.11. $\square$

14.2.16 Theorem *Let $G = \mathrm{BS}(p, q) = \langle x, y \mid y^{-1} x^p y = x^q \rangle$. Then G is poly-context-free if and only if $p = \pm q$.*

Proof If $p = \pm q$ then, as we saw in the proof of Theorem 14.2.13, $\mathrm{BS}(p, q)$ is virtually a direct product of free groups, so it is poly-context-free by the closure results in 13.3. When $p \ne \pm q$, the fact that WP($\mathrm{BS}(p, q)$) is not poly-context-free follows from Proposition 14.2.14 in the same way that the fact that coWP($\mathrm{BS}(p, q)$) is not context-free follows from Proposition 14.2.11. $\square$

14.3 Indexed co-word problems

As mentioned in Section 13.2, there are no known examples of groups G in which WP(G) is an indexed language but is not context-free.

In contrast to this, there are many examples of groups, including free products, such as $\mathbb{Z}^2 * \mathbb{Z}$ mentioned in Conjecture 14.2.3, which are known to be co-indexed, but not known to be co-context-free. Unfortunately, none of these examples has been proved not to be co-context-free.

Co-indexed groups are studied by Holt and Röver [157]. It is proved that the Higman–Thompson groups, and finitely generated groups of bounded permutational transducer mappings, which are described in Chapter 9 and include the much-studied Grigorchuk group, have this property. As mentioned earlier in 14.2.5, it was subsequently proved by Lehnert and Schweitzer [178] that the Higman–Thompson groups are co-context-free, but this is not known to be true for the bounded groups of permutational transducer mappings.

It is proved also by Holt and Röver [157] that, if G and H are co-indexed groups then, under a number of additional technical assumptions that hold in all known examples, the free product $G * H$ is co-indexed.

As is the case with many of the pda that accept co-word problems of groups, all of the nested stack automata that accept coWP(G) in the examples considered in [157] are *semi-deterministic*, in the sense that, before starting to read the input word, they write some arbitrary word non-deterministically onto the main stack, and operate deterministically thereafter.

14.4 The conjugacy problem

After our extensive considerations of the formal language properties of the word problem of a group, a natural question to ask is whether there are any corresponding properties of the conjugacy problem. This topic has received surprisingly little attention to date, but some basic results are developed by Holt, Rees and Röver [153].

For the conjugacy problem we have two input words, and there are various possibilities for how we could read them. We considered 2-variable fsa in Section 2.10, and we saw that the two words could be read either synchronously or asynchronously. We might also try reading one of the words in reversed order. It is easy to see that, with any of these options, the word problem arises as the special case in which one of the input words is empty and so, if the conjugacy problem of a group lies in some language class, then so does the word problem.

As might be expected, by analogy with Theorem 3.4.1, it turns out (and is proved in [153]) that, for any of these options, a finitely generated group has regular conjugacy problem if and only if it is finite.

Turning to context-free languages, by Theorem 11.1.1 any finitely generated group with context-free conjugacy problem must be virtually free. The main result of [153] is that, if we read the input words asynchronously and with one of the two words in reversed order, then any virtually free group has context-free conjugacy problem. With any of the other input options (read input words synchronously, or asynchronously with both words in correct order) the conjugacy problem is context-free if and only if the group is virtually cyclic.

Some results on groups with context-sensitive conjugacy problem are proved by Lakin in [174]. For example, the closure results for word problems outlined at the beginning of Chapter 12 all hold also for conjugacy problems of groups. But there are no such results on subgroups and extensions, and an example is constructed of groups $H \leq G$ with $|G : H| = 2$, where H has deterministic context-sensitive conjugacy problem but G has insoluble conjugacy problem. Examples of groups that do have deterministic context-sensitive conjugacy problem include certain one-relator groups, including all surface groups.

It could be interesting to consider conjugacy problems in other language classes, such as $\mathcal{RT}$.

References

[1] S.O. Aanderaa. On k-tape versus $(k-1)$-tape real time computation. In R.M. Karp, editor, *Complexity of Computation (Proc. Sympos., New York, 1973)*, pages 75–96. SIAM–AMS Proc., Vol. VII. Amer. Math. Soc., Providence, R.I., 1974.

[2] A. Akhavi, I. Klimann, S. Lombardy, J. Mairesse, and M. Picantin. On the finiteness problem for automaton (semi)groups. *Internat. J. Algebra Comput.*, 22(6):1250052-1 – 1250052-26, 2012.

[3] S.V. Alešin. Finite automata and the Burnside problem for periodic groups. *Mat. Zametki*, 11:319–328, 1972.

[4] J.M. Alonso. Inégalités isopérimétriques et quasi-isométries. *C. R. Acad. Sci. Paris Sér. I Math.*, 311(12):761–764, 1990.

[5] J.M. Alonso, T. Brady, D. Cooper, V. Ferlini, M. Lustig, M. Mihalik, M. Shapiro, and H. Short. Notes on word hyperbolic groups. In É. Ghys, A. Haefliger, and A. Verjovsky, editors, *Group Theory from a Geometrical Viewpoint* (Trieste, 1990), pages 3–63. World Sci. Publishing Co., Inc., River Edge, NJ, 1991.

[6] G. Amir, O. Angel, and B. Virág. Amenability of linear-activity automaton groups. *J. Eur. Math. Soc. (JEMS)*, 15(3):705–730, 2013.

[7] A.V. Anisimov. Certain algorithmic problems for groups and context-free languages. *Kibernetica*, 8:4–11, 1972.

[8] Y. Antolín and L. Ciobanu. Finite generating sets of relatively hyperbolic groups and applications to geodesic languages. *Trans. Amer. Math. Soc.*, 368(11):7965–8010, 2016.

[9] Y. Antolín and L. Ciobanu. Formal conjugacy growth in acylindrically hyperbolic groups. *Int. Math. Res. Notices*, pages 1–37, 2016. Available at arXiv:math/1508.06229.

[10] J. Avenhaus and K. Madlener. Subrekursive Komplexität bei Gruppen. I. Gruppen mit vorgeschriebener Komplexität. *Acta Informat.*, 9(1):87–104, 1977/78.

[11] P. Bahls. Some new biautomatic Coxeter groups. *J. Algebra*, 296(2):339–346, 2006.

[12] L. Bartholdi. Groups of intermediate growth, 2002. Available at arXiv:math/0201293.

[13] L. Bartholdi. Endomorphic presentations of branch groups. *J. Algebra*, 268(2):419–443, 2003.

[14] L. Bartholdi. A Wilson group of non-uniformly exponential growth. *C. R. Math. Acad. Sci. Paris*, 336(7):549–554, 2003.

[15] L. Bartholdi. FR functionally recursive groups – a GAP package, version 2.3.6, 2016. http://www.gap-system.org/Packages/fr.html.

[16] L. Bartholdi, R. Grigorchuk, and Z. Šuniḱ. Branch groups. In M. Hazewinkel, editor, *Handbook of Algebra, Vol. 3*, pages 989–1112. North-Holland, Amsterdam, 2003.

[17] L. Bartholdi, V.A. Kaimanovich, and V.V. Nekrashevych. On amenability of automata groups. *Duke Math. J.*, 154(3):575–598, 2010.

[18] L. Bartholdi and Z. Šuniḱ. On the word and period growth of some groups of tree automorphisms. *Comm. Algebra*, 29(11):4923–4964, 2001.

[19] G. Baumslag, S.M. Gersten, M. Shapiro, and H. Short. Automatic groups and amalgams. *J. Pure Appl. Algebra*, 76:229–316, 1991.

[20] G. Baumslag, M. Shapiro, and H. Short. Parallel poly-pushdown groups. *J. Pure Appl. Algebra*, 140:209–227, 1999.

[21] G. Baumslag and D. Solitar. Some two-generator one-relator non-Hopfian groups. *Bull. Amer. Math. Soc.*, 68:199–201, 1962.

[22] R. Berns-Zieve, D. Fry, J. Gillings, H. Hoganson, and H. Mathews. Groups with context-free co-word problem and embeddings into Thompson's group V, 2014. Available at arXiv:math/1407.7745.

[23] M. Bhattacharjee. The ubiquity of free subgroups in certain inverse limits of groups. *J. Algebra*, 172(1):134–146, 1995.

[24] C. Bleak, F. Matucci, and M. Neunhöffer. Embeddings into Thompson's group V and coCF groups, 2013. Available at arXiv:math/1312.1855.

[25] E. Bondarenko and V. Nekrashevych. Post-critically finite self-similar groups. *Algebra Discrete Math.*, 2(4):21–32, 2003.

[26] I. Bondarenko, N. Bondarenko, S. Sidki, and F. Zapata. On the conjugacy problem for finite-state automorphisms of regular rooted trees. *Groups Geom. Dyn.*, 7(2):323–355, 2013. With an appendix by Raphaël M. Jungers.

[27] I. Bondarenko, R. Grigorchuk, R. Kravchenko, Y. Muntyan, V. Nekrashevych, D. Savchuk, and Z. Šunić. Groups generated by 3-state automata over a 2-letter alphabet. I. *São Paulo J. Math. Sci.*, 1(1):1–39, 2007.

[28] I. Bondarenko, R. Grigorchuk, R. Kravchenko, Y. Muntyan, V. Nekrashevych, D. Savchuk, and Z. Šunić. On classification of groups generated by 3-state automata over a 2-letter alphabet. *Algebra Discrete Math.*, (1):1–163, 2008.

[29] I. Bondarenko, R. Grigorchuk, R. Kravchenko, Y. Muntyan, V. Nekrashevych, D. Savchuk, and Z. Šunić. Groups generated by 3-state automata over a 2-letter alphabet. II. *J. Math. Sci. (N. Y.)*, 156(1):187–208, 2009.

[30] R.V. Book and S.A. Greibach. Quasi-realtime languages. *Mathematical Systems Theory*, 4(1):97–111, 1970.

[31] W.W. Boone. Certain simple, unsolvable problems of group theory, I-VI. *Indag. Math.*, 16 (1954) 231–237,492–497; 17 (1955) 252–256; 19 (1957) 22–27,227–232, 1954–7.

[32] W.W. Boone. The word problem. *Proc. Nat. Acad. Sci. USA*, 44:1061–1065, 1958.

[33] W.W. Boone. The word problem. *Ann. of Math. (2)*, 70:207–265, 1959.

[34] Nicolas Bourbaki. *Groupes et Algèbres de Lie, Chapitres 4, 5 et 6*. Hermann, 1968.

[35] B.H. Bowditch. A short proof that a subquadratic isoperimetric inequality implies a linear one. *Michigan Math. J.*, 42(1):103–107, 1995.

[36] B.H. Bowditch. A course on geometric group theory, 2005. homepages.warwick.ac.uk/~masgak/papers/bhb-ggtcourse.pdf.

[37] N. Brady and M.R. Bridson. There is only one gap in the isoperimetric spectrum. *Geom. Funct. Anal.*, 10(5):1053–1070, 2000.

[38] T. Brady and J. McCammond. Three-generator Artin groups of large type are biautomatic. *J. Pure Appl. Algebra*, 151(1):295–310, 2000.

[39] Martin R. Bridson and André Haefliger. *Metric Spaces of Non-Positive Curvature*. Springer Verlag, 1999.

[40] M.R. Bridson. On the geometry of normal forms in discrete groups. *Proc. London Math. Soc. (3)*, 67:596–616, 1993.

[41] M.R. Bridson. Combings of groups and the grammar of reparameterization. *Comment. Math. Helv.*, 78(4):752–771, 2003.

[42] M.R. Bridson and R.H. Gilman. Formal language theory and the geometry of 3-manifolds. *Comment. Math. Helv.*, 71(4):525–555, 1996.

[43] M.R. Bridson and J. Howie. Conjugacy of finite subsets in hyperbolic groups. *Internat. J. Algebra Comput.*, 15(4):725–756, 2005.

[44] M.R. Bridson and K. Vogtmann. On the geometry of the automorphism group of a free group. *Bull. London Math. Soc.*, 27:544–552, 1995.

[45] M.R. Bridson and K. Vogtmann. Automorphism groups of free groups, surface groups and free abelian groups. In B. Farb, editor, *Problems on mapping class groups and related topics*, volume 74 of *Proc. Sympos. Pure. Math.*, pages 301–316, 2006.

[46] M.R. Bridson and K. Vogtmann. The Dehn functions of $\mathrm{Out}(F_n)$ and $\mathrm{Aut}(F_n)$. *Annales de l'Institut Fourier*, 62(5):1811–1817, 2012.

[47] E. Brieskorn and K. Saito. Artin-Gruppen und Coxeter-Gruppen. *Invent. Math.*, 17(4):245–271, 1972.

[48] J. Brieussel. Folner sets of alternate directed groups. *Ann. Inst. Fourier (Grenoble)*, 64(3):1109–1130, 2014.

[49] B. Brink and R.B. Howlett. A finiteness property and an automatic structure for Coxeter groups. *Math. Ann.*, 296:179–190, 1993.

[50] M. Brittenham and S. Hermiller. A uniform model for almost convexity and rewriting systems. *J. Group Theory*, 18(5):805–828, 2015.

[51] M. Brittenham, S. Hermiller, and D.F. Holt. Algorithms and topology of Cayley graphs for groups. *J. Algebra*, 415(1):112–136, 2014.

[52] M. Brittenham, S. Hermiller, and A. Johnson. Homology and closure properties of autostackable groups. *J. Algebra*, 452:596–617, 2016.

[53] M. Brittenham, S. Hermiller, and T. Susse. Geometry of the word problem for 3-manifold groups, 2016. Available at arXiv:math/1609.06253.

[54] T. Brough. Groups with poly-context-free word problem. PhD thesis, University of Warwick, 2010.

[55] T. Brough. Groups with poly-context-free word problem. *Groups–Complexity–Cryptology*, 6(1):9–29, 2014.

[56] A.M. Brunner and S. Sidki. The generation of GL(n, $\mathbf{Z}$) by finite state automata. *Internat. J. Algebra Comput.*, 8(1):127–139, 1998.

[57] B. Buchberger. The history and basic features of the critical-pair/completion procedure. *J. Symbolic Comput.*, 3:3–38, 1987.

[58] D.J. Buckley and D.F. Holt. The conjugacy problem in hyperbolic groups for finite lists of group elements. *Internat. J. Algebra Comput.*, 23:1127–1150, 2013.

[59] W. Burnside. On an unsettled question in the theory of discontinuous groups. *Quart. J. Math.*, 33:230–238, 1902.

[60] J.W. Cannon. The combinatorial structure of cocompact discrete hyperbolic group. *Geom. Dedicata*, 16:123–148, 1984.

[61] J.W. Cannon. Almost convex groups. *Geom. Dedicata*, 22:197–210, 1987.

[62] J.W. Cannon, W.J. Floyd, and W.R. Parry. Introductory notes on Richard Thompson's groups. *Enseign. Math. (2)*, 42(3-4):215–256, 1996.

[63] L. Caponi. *On the Classification of Groups Generated by Automata with 4 States over a 2-Letter Alphabet.* PhD thesis, University of South Florida, 2014. Available at http://scholarcommons.usf.edu/etd/4995.

[64] P.-E. Caprace and B. Mühlherr. Reflection triangles in Coxeter groups and biautomaticity. *J. Group Theory*, 8(4):467–489, 2005.

[65] R. Charney. Geodesic automation and growth functions for Artin groups of finite type. *Math. Ann.*, 301:307–324, 1995.

[66] R. Charney and J. Meier. The language of geodesics for Garside groups. *Math. Zeitschrift*, 248:495–509, 2004.

[67] R. Charney and L. Paris. Convexity of parabolic subgroups in Artin groups. *Bull. London Math. Soc.*, 46:1248–1255, 2014.

[68] Ian M. Chiswell. *A Course in Formal Languages, Automata and Groups.* Springer Verlag, 2009.

[69] L. Ciobanu and S. Hermiller. Conjugacy growth series and languages in groups. *Trans. Amer. Math. Soc.*, 366:2803–2825, 2014.

[70] L. Ciobanu, S. Hermiller, D.F. Holt, and S. Rees. Conjugacy languages in groups. *Israel J. Math*, 211:311–347, 2015.

[71] D.J. Collins. A simple presentation of a group with unsolvable word problem. *Illinois J. Math.*, 30 (2):230–234, 1986.

[72] F. Dahmani and V. Guirardel. The isomorphism problem for all hyperbolic groups. *Geom. Funct. Anal.*, 21:223300, 2011.

[73] M.M. Day. Amenable semigroups. *Illinois J. Math.*, 1:509–544, 1957.

[74] Pierre de la Harpe. *Topics in Geometric Group Theory.* Chicago Lectures in Mathematics. University of Chicago Press, Chicago, IL, 2000.

[75] M. Dehn. Über unendliche diskontinuierliche Gruppen. *Math. Ann.*, 71:116–144, 1911.

[76] M. Dehn. Transformationen der Kurve auf zweiseitigen Flächen. *Math. Ann.*, 72:413–420, 1912.

[77] P. Dehornoy. Groupes de Garside. *Ecole Norm. Sup.(4)*, 35:267–306, 2002.

[78] P. Dehornoy and L. Paris. Garside groups, two generalisations of Artin groups. *Proc. London Math. Soc.*, 73:569–604, 1999.

[79] M.J. Dunwoody. The accessibility of finitely presented groups. *Invent. Math.*, 81:449–457, 1985.

[80] M. Edjvet and A. Juhàsz. The groups $G^{m,n,p}$. *J. Algebra*, 319:248–266, 2008.

[81] M. Elder. Finiteness and the falsification by fellow traveler property. *Geom. Dedicata*, 95:103–113, 2002.

[82] M. Elder. Regular geodesic languages and the falsification by fellow traveler property. *Algebr. Geom. Topol. (electronic)*, 5:129–134, 2005.

[83] M. Elder and J. Taback. *C*-graph automatic groups. *J. Algebra*, 413:289–319, 2014.

[84] David B.A. Epstein, J.W. Cannon, D.F. Holt, S.V.F. Levy, M.S. Paterson, and W.P. Thurston. *Word Processing in Groups*. Jones and Bartlett, Boston, 1992.

[85] D.B.A. Epstein and D.F. Holt. The linearity of the conjugacy problem in word-hyperbolic groups. *Internat. J. Algebra Comput.*, 16:287–305, 2006.

[86] D.B.A. Epstein, D.F. Holt, and S.E. Rees. The use of Knuth-Bendix methods to solve the word problem in automatic groups. *J. Symbolic Comput.*, 12:397–414, 1991.

[87] D.B.A. Epstein, A.R. Iano-Fletcher, and U. Zwick. Growth functions and automatic groups. *Experiment. Math.*, 5(4):297–315, 1996.

[88] A. Erschler. Not residually finite groups of intermediate growth, commensurability and non-geometricity. *J. Algebra*, 272(1):154–172, 2004.

[89] S.M. Gersten. Isoperimetric and isodiametric functions of finite presentations. In G.A. Niblo and M.A. Roller, editors, *Geometric Group Theory, Volume 1*, volume 181 of *London Mathematical Society Lecture Note Series*, pages 79–96. Cambridge University Press, 1993.

[90] S.M. Gersten and T.R. Riley. Filling length in finitely presentable groups. *Geom. Dedicata*, 92:4158, 2002.

[91] S.M. Gersten and H. Short. Small cancellation theory and automatic groups. *Invent. Math.*, 102(2):305–334, 1990.

[92] S.M. Gersten and H. Short. Small cancellation theory and automatic groups, II. *Invent. Math.*, 105(3):641–662, 1991.

[93] S.M. Gersten and H.B. Short. Rational subgroups of biautomatic groups. *Ann. of Math.*, 134(1):125–158, 1991.

[94] É. Ghys and P. de la Harpe, editors. *Sur les groupes hyperboliques d'après Mikhael Gromov*, volume 83 of *Progress in Mathematics*. Birkhäuser Boston, Inc., Boston, MA, 1990.

[95] P. Gillibert. The finiteness problem for automaton semigroups is undecidable. *Internat. J. Algebra Comput.*, 24(1):1–9, 2014.

[96] R.H. Gilman. Presentations of groups and monoids. *J. Algebra*, 57:544–554, 1979.

[97] R.H. Gilman. Enumerating infinitely many cosets. In M.D. Atkinson, editor, *Computational Group Theory*, pages 51–55, London, New York, 1984. Academic Press.

[98] R.H. Gilman. On the definition of word hyperbolic groups. *Math. Z.*, 242:529–541, 2002.

[99] R.H. Gilman, D.F. Holt, S. Hermiller, and S. Rees. A characterisation of virtually free groups. *Arch. Math.*, 89:289–295, 2007.

[100] R.H. Gilman, D.F. Holt, and S. Rees. Combing nilpotent and polycyclic groups. *Internat. J. Algebra Comput.*, 9:135–155, 1999.

[101] S. Ginsburg and E. Spanier. Semigroups, Presburger formulas, and languages. *Pacific J. Math.*, 16(2):285–296, 1966.

[102] Seymour Ginsburg. *The Mathematical Theory of Context-free Languages.* McGraw-Hill, 1966.

[103] Y. Glasner and S. Mozes. Automata and square complexes. *Geom. Dedicata,* 111:43–64, 2005.

[104] E. Godelle and L. Paris. Basic questions on Artin-Tits groups. In *Configuration Spaces,* volume 14 of *CRM Series,* pages 299–311. Ed. Norm., Pisa, 2012.

[105] J. Goldstine. A simplified proof of Parikh's theorem. *Discrete Mathematics,* 19:235–239, 1977.

[106] E.S. Golod. On nil-algebras and finitely approximable p-groups. *Izv. Akad. Nauk SSSR Ser. Mat.,* 28:273–276, 1964. English translation: Amer. Math. Transl. (2), 48:103–106, 1965.

[107] E.S. Golod and I.R. Šafarevič. On the class field tower. *Izv. Akad. Nauk SSSR Ser. Mat.,* 28:261–272, 1964.

[108] O. Goodman and M. Shapiro. On a generalization of Dehn's algorithm. *Internat. J. Algebra Comput.,* 18:1137–1177, 2008.

[109] I. Gorun. A hierarchy of context-sensitive languages. *Lect. Notes Comput. Sc.,* 45:299–303, 1976.

[110] M. Greendlinger. Dehn's algorithm for the word problem. *Comm. Pure Appl. Math.,* 13:67–83, 1960.

[111] M. Greendlinger. On Dehn's algorithm for the conjugacy and word problem. *Comm. Pure Appl. Math.,* 13:641–677, 1960.

[112] M. Greendlinger. On the word problem and the conjugacy problem (Russian). *Akad. Nauk SSSR Ser. Mat.,* 29:245–268, 1965.

[113] R. Gregorac. On generalized free products of finite extensions of free groups. *J. London Math. Soc.,* 41:662–666, 1966.

[114] R.I. Grigorchuk. On Burnside's problem on periodic groups. *Funktsional. Anal. i Prilozhen.,* 14(1):53–54, 1980.

[115] R.I. Grigorchuk. On the Milnor problem of group growth. *Dokl. Akad. Nauk SSSR,* 271(1):30–33, 1983.

[116] R.I. Grigorchuk. Degrees of growth of finitely generated groups and the theory of invariant means. *Izv. Akad. Nauk SSSR Ser. Mat.,* 48(5):939–985, 1984.

[117] R.I. Grigorchuk. Degrees of growth of p-groups and torsion-free groups. *Mat. Sb. (N.S.),* 126(168)(2):194–214, 286, 1985.

[118] R.I. Grigorchuk. An example of a finitely presented amenable group that does not belong to the class EG. *Mat. Sb.,* 189(1):79–100, 1998.

[119] R.I. Grigorchuk. Just infinite branch groups. In M. du Sautoy, D. Segal, and A. Shalev, editors, *New horizons in pro-p groups,* volume 184 of *Progr. Math.,* pages 121–179. Birkhäuser Boston, Boston, MA, 2000.

[120] R.I. Grigorchuk. Solved and unsolved problems around one group. In L. Bartholdi, T. Ceccherini-Silberstein, T. Smirnova-Nagnibeda, and A. Żuk, editors, *Infinite groups: geometric, combinatorial and dynamical aspects,* volume 248 of *Progr. Math.,* pages 117–218. Birkhäuser, Basel, 2005.

[121] R.I. Grigorchuk, P. Linnell, T. Schick, and A. Żuk. On a question of Atiyah. *C. R. Acad. Sci. Paris Sér. I Math.,* 331(9):663–668, 2000.

[122] R.I. Grigorchuk, V.V. Nekrashevich, and V.I. Sushchanskiĭ. Automata, dynamical systems, and groups. *Tr. Mat. Inst. Steklova,* 231(Din. Sist., Avtom. i Beskon. Gruppy):134–214, 2000.

[123] R.I. Grigorchuk and A. Żuk. The lamplighter group as a group generated by a 2-state automaton, and its spectrum. *Geom. Dedicata*, 87(1-3):209–244, 2001.

[124] M. Gromov. Groups of polynomial growth and expanding maps. *Publications Mathématique d'IHÉS*, 53:53–78, 1981.

[125] M. Gromov. Hyperbolic groups. In S.M. Gersten, editor, *Essays in Group Theory*, volume 8 of *MSRI Publ.*, pages 75–263. Springer, 1987.

[126] M. Gromov. Asymptotic invariants of infinite groups. In G.A. Niblo and M.A. Roller, editors, *Geometric Group Theory, Volume 2*, volume 182 of *London Mathematical Society Lecture Note Series*, pages 1–295. Cambridge University Press, 1993.

[127] Mikhael Gromov. *Structures métriques pour les variétés riemanniennes*, volume 1 of *Textes Mathématiques [Mathematical Texts]*. CEDIC, Paris, 1981. Edited by J. Lafontaine and P. Pansu.

[128] C.K. Gupta, N.D. Gupta, and A.S. Oliynyk. Free products of finite groups acting on regular rooted trees. *Algebra Discrete Math.*, (2):91–103, 2007.

[129] N. Gupta and S. Sidki. Some infinite p-groups. *Algebra i Logika*, 22(5):584–589, 1983.

[130] U. Hamenstädt. Geometry of the mapping class group II: A biautomatic structure, 2009.

[131] A. Harkins. Combing lattices of soluble Lie groups. PhD thesis, University of Newcastle, 2001.

[132] J. Hartmanis and R.E. Stearns. On the computational complexity of algorithms. *Trans. Amer. Math. Soc.*, 117(5):285–306, 1965.

[133] G. Havas and D.F. Holt. On Coxeter's families of group presentations. *J. Algebra*, 324(5):1076–1082, 2010.

[134] T. Herbst. On a subclass of context-free groups. *RAIRO Inform. Théor. Appl.*, 25:255–272, 1991.

[135] T. Herbst and R.M. Thomas. Group presentations, formal languages and characterizations of one-counter groups. *Theoret. Comput. Sci.*, 112(2):187–213, 1993.

[136] S. Hermiller, D.F. Holt, and S. Rees. Star-free geodesic languages for groups. *Internat. J. Algebra Comput.*, 17:329–345, 2007.

[137] S. Hermiller, D.F. Holt, and S. Rees. Groups whose geodesics are locally testable. *Internat. J. Algebra Comput.*, 18:911–923, 2008.

[138] S. Hermiller and C. Martínez-Pérez. HNN-extensions and stackable groups, 2016. Available at arXiv:1605.06145.

[139] S. Hermiller and J. Meier. Algorithms and geometry for graph products of groups. *J. Algebra*, 171(1):230–257, 1995.

[140] Israel N. Herstein. *Noncommutative rings*, volume 15 of *Carus Mathematical Monographs*. Mathematical Association of America, Washington, DC, 1994. Reprint of the 1968 original, with an afterword by Lance W. Small.

[141] G. Higman, B.H. Neumann, and H. Neumann. Embedding theorems for groups. *J. London Math. Soc.*, 24:247–254, 1949.

[142] Graham Higman. *Finitely presented infinite simple groups*, volume 8 of *Notes on Pure Mathematics*. The Australian National University, 1974.

[143] K.A. Hirsch. On infinite soluble groups, III. *Proc. London Math. Soc. (2)*, 49:184–194, 1946.

[144] Derek F. Holt, Bettina Eick, and Eamonn A. O'Brien. *Handbook of Computational Group Theory*. Chapman & Hall/CRC, 2005.

[145] D.F. Holt. The Warwick automatic group software. In G. Baumslag et al., editors, *Geometric and Computational Perspectives on Infinite Groups*, volume 25 of *Amer. Math. Soc. DIMACS Series*, pages 69–82. (DIMACS, 1994), 1995.

[146] D.F. Holt. KBMAG (Knuth-Bendix in Monoids and Automatic groups), 2000. http://www.warwick.ac.uk/staff/D.F.Holt/download/kbmag2/.

[147] D.F. Holt. Word-hyperbolic groups have real-time word problem. *Internat. J. Algebra Comput.*, 10(2):221–227, 2000.

[148] D.F. Holt. Garside groups have the falsification by fellow-traveller property. *Groups Geom. Dyn.*, 4:777–784, 2010.

[149] D.F. Holt and D.F. Hurt. Computing automatic coset systems and subgroup presentations. *J. Symbolic Comput.*, 27:1–19, 1999.

[150] D.F. Holt, M.D. Owens, and R.M. Thomas. Groups and semigroups with a one-counter word problem. *J. Australian Math. Soc.*, 85:197–209, 2008.

[151] D.F. Holt and S. Rees. Solving the word problem in real time. *J. Lond. Math. Soc. (2)*, 63:623–639, 2001.

[152] D.F. Holt and S. Rees. Artin groups of large type are shortlex automatic with regular geodesics. *Proc. Lond. Math. Soc. (3)*, 104(3):486–512, 2012.

[153] D.F. Holt, S. Rees, and C.E. Röver. Groups with context-free conjugacy problems. *Internat. J. Algebra Comput.*, 21(1–2):193–216, 2011.

[154] D.F. Holt, S. Rees, C.E. Röver, and R.M. Thomas. Groups with context-free coword problem. *J. Lond. Math. Soc. (2)*, 71:643–657, 2005.

[155] D.F. Holt, S. Rees, and M. Shapiro. Groups that do and do not have growing context-sensitive word problem. *Internat. J. Algebra Comput.*, 18:1179–1191, 2008.

[156] D.F. Holt and C.E. Röver. On real-time word problems. *J. Lond. Math. Soc. (2)*, 67:289–301, 2003.

[157] D.F. Holt and C.E. Röver. Groups with indexed co-word problem. *Internat. J. Algebra Comput.*, 16(5):985–1014, 2006.

[158] John E. Hopcroft, Rajeev Motwani, and Jeffrey D. Ullman. *Introduction to Automata Theory, Languages, and Computation*. Addison-Wesley, 2006. Third edition.

[159] John E. Hopcroft and Jeffrey D. Ullman. *Introduction to Automata Theory, Languages, and Computation*. Addison-Wesley, 1979. First edition.

[160] H. Hopf. Enden offener Räume und unendliche diskontinuierliche Gruppen. *Comment. Math. Helv.*, 16:81–100, 1944.

[161] R.B. Howlett. Miscellaneous facts about Coxeter groups. Lectures given at the ANU Group Actions Workshop, October 1993, 1993. http://www.maths.usyd.edu.au:8000/res/Algebra/How/anucox.html.

[162] James E. Humphreys. *Reflection groups and Coxeter groups*, volume 29 of *Cambridge Studies in Advanced Mathematics*. Cambridge University Press, Cambridge, 1990.

[163] N. Immerman. Nondeterministic space is closed under complementation. *SIAM J. Comput.*, 17(5):935–938, 1988.

[164] R. Incitti. Regularities on the Cayley graphs of groups of linear growth. *Europ. J. Combinatorics*, 18:175–178, 1997.

278   *References*

[165] A. Karass, A. Pietrowski, and D. Solitar. Recursive predicates and quantifiers. *J. Austral. Math. Soc.*, 16:458–466, 1973.
[166] O. Kharlampovich, B. Khoussainov, and A. Miasnikov. From automatic structures to automatic groups. *Groups Geom. Dyn.*, 8(1):157–198, 2014.
[167] S.C. Kleene. Recursive predicates and quantifiers. *Trans. Amer. Math. Soc.*, 53:41–73, 1943.
[168] I. Klimann. The finiteness of a group generated by a 2-letter invertible-reversible Mealy automaton is decidable. In *30th International Symposium on Theoretical Aspects of Computer Science*, volume 20 of *LIPIcs. Leibniz Int. Proc. Inform.*, pages 502–513. Schloss Dagstuhl. Leibniz-Zent. Inform., Wadern, 2013.
[169] I. Klimann and M. Picantin. A characterization of those automata that structurally generate finite groups. In *LATIN 2014: theoretical informatics*, volume 8392 of *Lecture Notes in Comput. Sci.*, pages 180–189. Springer, Heidelberg, 2014.
[170] I. Klimann, M. Picantin, and D. Savchuk. A connected 3-state reversible Mealy automaton cannot generate an infinite Burnside group. In *Developments in language theory*, volume 9168 of *Lecture Notes in Comput. Sci.*, pages 313–325. Springer, Cham, 2015.
[171] I. Klimann, M. Picantin, and D. Savchuk. Orbit automata as a new tool to attack the order problem in automaton groups. *J. Algebra*, 445:433–457, 2016.
[172] D.E. Knuth and P.B. Bendix. Simple word problems in universal algebras. In J. Leech, editor, *Computational Problems in Abstract Algebra*, pages 263–297, Oxford, 1970. (Oxford, 1967), Pergamon Press.
[173] D. Krammer. The conjugacy problem for Coxeter groups. *Groups Geom. Dyn.*, 3(1):71–171, 2009.
[174] S. Lakin. Context-sensitive decision problems in groups. PhD thesis, University of Leicester, 2002.
[175] S.R. Lakin and R.M. Thomas. Context-sensitive decision problems in groups. In C.S. Calude, E. Calude, and M.J. Dinneen, editors, *Developments in Language Theory: 8th International Conference, DLT 2004, Auckland, New Zealand*, volume 3340 of *Lecture Notes in Computer Science*, pages 296–307. Springer-Verlag, 2004.
[176] S.R. Lakin and R.M. Thomas. Complexity classes and word problems of groups. *Groups Complex. Cryptol.*, 1(2):261–273, 2009.
[177] Y. Lavreniuk and V. Nekrashevych. Rigidity of branch groups acting on rooted trees. *Geom. Dedicata*, 89:159–179, 2002.
[178] J. Lehnert and P. Schweitzer. The co-word problem for the Higman-Thompson group is context-free. *Bull. Lond. Math. Soc.*, 39(2):235–241, 2007.
[179] Yu.G. Leonov. The conjugacy problem in a class of 2-groups. *Mat. Zametki*, 64(4):573–583, 1998.
[180] A. Lubotzky, A. Mann, and D. Segal. Finitely generated groups of polynomial subgroup growth. *Israel J. Math.*, 82(1-3):363–371, 1993.
[181] A. Lubotzky, L. Pyber, and A. Shalev. Discrete groups of slow subgroup growth. *Israel J. Math.*, 96(part B):399–418, 1996.
[182] R.C. Lyndon. On Dehn's algorithm. *Math. Ann.*, 166:208–228, 1966.
[183] Roger C. Lyndon and Paul E. Schupp. *Combinatorial Group Theory*. Classics in Mathematics. Springer-Verlag, Berlin, 2001. Reprint of the 1977 edition.

References 279

[184] I.G. Lysënok. A set of defining relations for the Grigorchuk group. *Mat. Zametki*, 38(4):503–516, 634, 1985. English translation: Math. Notes 38(3-4):784–792, 1985.

[185] O. Macedońska, V. Nekrashevych, and V. Sushchansky. Commensurators of groups and reversible automata. *Dopov. Nats. Akad. Nauk Ukr. Mat. Prirodozn. Tekh. Nauki*, (12):36–39, 2000.

[186] K. Madlener and F. Otto. Pseudonatural algorithms for the word problem for finitely presented groups and monoids. *J. Symbolic Comput.*, 1:383–418, 1985.

[187] Avinoam Mann. *How groups grow*, volume 395 of *London Mathematical Society Lecture Note Series*. Cambridge University Press, Cambridge, 2012.

[188] A.N. Maslov. The cyclic shift of languages. *Problemy Peredači Informacii*, 9(4):81–87, 1973.

[189] William S. Massey. *Algebraic Topology: An Introduction*. Springer-Verlag, New York-Heidelberg, 1977. Reprint of the 1967 edition, Graduate Texts in Mathematics, Vol. 56.

[190] J. Mennicke. Einige endliche Gruppen mit drei Erzeugenden und drei Relationen. *Arch. Math.*, 10:409–418, 1959.

[191] A. Miasnikov and Z. Šunić. Cayley graph automatic groups are not necessarily Cayley graph biautomatic. In A.-H. Dediu and C. Martín-Vide, editors, *Language and Automata Theory and Applications*, volume 7183 of *Lecture Notes in Computer Science*, pages 401–407. Springer Berlin Heidelberg, 2012.

[192] C.F. Miller III. Decision problems for groups – survey and reflections. In G. Baumslag and C.F. Miller III, editors, *Algorithms and Classification in Combinatorial Group Theory*, pages 1–59. Springer, New York, 1992.

[193] J. Milnor. A note on curvature and the fundamental group. *J. Differential Geom.*, 2:1–7, 1968.

[194] J. Milnor. Problem 5605. *Amer. Math. Monthly*, 75:685–686, 1968.

[195] L. Mosher. Mapping class groups are automatic. *Ann. of Math.*, 142(2):303–384, 1995.

[196] L. Mosher. Central quotients of biautomatic groups. *Comment. Math. Helv.*, 72(1):16–29, 1997.

[197] G. Moussong. Hyperbolic Coxeter groups. PhD thesis, Ohio State University, 1988.

[198] D.E. Muller and P.E. Schupp. Groups, the theory of ends, and context-free languages. *J. Comp. System Sci.*, 26:295–310, 1983.

[199] Y. Muntyan and D. Savchuk. automgrp automata groups – a GAP package, version 1.3, 2016. http://www.gap-system.org/Packages/automgrp.html.

[200] V. Nekrashevych and S. Sidki. Automorphisms of the binary tree: state-closed subgroups and dynamics of 1/2-endomorphisms. In *Groups: topological, combinatorial and arithmetic aspects*, volume 311 of *London Math. Soc. Lecture Note Ser.*, pages 375–404. Cambridge Univ. Press, Cambridge, 2004.

[201] Volodymyr Nekrashevych. *Self-similar groups*, volume 117 of *Mathematical Surveys and Monographs*. American Mathematical Society, Providence, RI, 2005.

[202] W.D. Neumann and L. Reeves. Central extensions of word hyperbolic groups. *Ann. of Math. (2)*, 145:183–287, 1997.

[203] W.D. Neumann and M. Shapiro. A short course in geometric group theory. Notes for the ANU Workshop January/February 1996. http://at.yorku.ca/i/a/a/i/13.htm.

[204] W.D. Neumann and M. Shapiro. Automatic structures, rational growth, and geometrically finite hyperbolic groups. *Invent. Math.*, 120:259–287, 1995.

[205] G.A. Niblo and L.D. Reeves. Coxeter groups act on CAT(0) cube complexes. *J. Group Theory*, 6:399–413, 2003.

[206] P.S. Novikov. On algorithmic unsolvability of the problem of the identity (Russian). *Doklady Akad. Nauk SSSR*, 85:709–712, 1952.

[207] A.Yu. Ol'shanskii. Hyperbolicity of groups with subquadratic isoperimetric inequalities. *Internat. J. Algebra Comput.*, 1:281–289, 1991.

[208] T. Oshiba. Closure property of the family of context-free languages under the cyclic shift operation. *Electron. Commun. Japan*, 55(4):119–122, 1972.

[209] P. Papasoglu. On the sub-quadratic isoperimetric inequality. In R. Charney, M. Davis, and M. Shapiro, editors, *Geometric Group Theory*, pages 149–158. de Gruyter-New York, 1995.

[210] P. Papasoglu. Strongly geodesically automatic groups are hyperbolic. *Invent. Math.*, 121(2):323–334, 1995.

[211] R.J. Parikh. Language generating devices. *MIT Res. Lab. Electron. Quart. Prog. Rep.*, 60:199–212, 1961.

[212] D.W. Parkes and R.M. Thomas. Groups with context-free reduced word problem. *Communications in Algebra*, 30:3143–3156, 2002.

[213] D. Peifer. Artin groups of extra-large type are biautomatic. *J. Pure Appl. Algebra*, 110:15–56, 1996.

[214] Jean-Éric Pin. *Varieties of Formal Languages*. Plenum Publishing Corp., New York, 1986.

[215] M.O. Rabin. Real time computation. *Israel J. Math.*, 1:203–211, 1963.

[216] D.Y. Rebecchi. Algorithmic properties of relatively hyperbolic groups. PhD thesis, Rutgers, Newark, 2003. Available at arXiv:math/0302245.

[217] I. Redfern. Automatic coset sytems. PhD thesis, University of Warwick, 1993.

[218] T.R. Riley. The geometry of groups satisfying weak almost-convexity or weak geodesic-combability conditions. *J. Group Theory*, 5:513–525, 2002.

[219] E. Rips. Subgroups of small cancellation groups. *Bull. London Math. Soc.*, 14:45–47, 1982.

[220] I. Rivin. Growth in free groups (and other stories)-twelve years later. *Illinois J. Math.*, 54:327–370, 2010.

[221] Derek J.S. Robinson. *A Course in the Theory of Groups*, volume 80 of *Graduate Texts in Mathematics*. Springer-Verlag, New York, second edition, 1996.

[222] A. Rosenberg. Real-time definable languages. *J. Assoc. Comput. Mach.*, 14:645–662, 1967.

[223] Joseph J. Rotman. *An Introduction to the Theory of Groups*. Springer-Verlag, Berlin and Heidelberg, 4th edition, 1995.

[224] C.E. Röver. Abstract commensurators of groups acting on rooted trees. *Geom. Dedicata*, 94:45–61, 2002. Proceedings of the Conference on Geometric and Combinatorial Group Theory, Part I (Haifa, 2000).

[225] A.V. Rozhkov. The conjugacy problem in an automorphism group of an infinite tree. *Mat. Zametki*, 64(4):592–597, 1998.

[226] M. Sapir. Asymptotic invariants, complexity of groups and related problems. *Bull. Math. Sci.*, 1(2):277–364, 2011.

[227] D. Savchuk and Y. Vorobets. Automata generating free products of groups of order 2. *J. Algebra*, 336:53–66, 2011.

[228] O. Schreier. Die Untergruppen der freien Gruppen. *Abh. Math. Sem. Univ. Hamburg*, 5:161–183, 1927.

[229] P.E. Schupp. On Dehn's algorithm and the conjugacy problem. *Math. Ann.*, 178:119–130, 1968.

[230] M.P. Schützenberger. On finite monoids having only trivial subgroups. *Information and Control*, 8:190–194, 1965.

[231] E.A. Scott. The embedding of certain linear and abelian groups in finitely presented simple groups. *J. Algebra*, 90(2):323–332, 1984.

[232] D. Segal. Decidability properties of polycyclic groups. *Proc. London Math. Soc. (3)*, 61:497–528, 1990.

[233] D. Segal. The finite images of finitely generated groups. *Proc. London Math. Soc. (3)*, 82(3):597–613, 2001.

[234] Z. Sela. The isomorphism problem for hyperbolic groups. *Ann. of Math. (2)*, 141:217–283, 1995.

[235] M. Shapiro. A note on context-sensitive languages and word problems. *Internat. J. Algebra Comput.*, 4:493–497, 1994.

[236] H. Short. Groups and combings, 1990. https://www.i2m.univ-amu.fr/~short/Papers/bicomball.pdf.

[237] S. Sidki. Automorphisms of one-rooted trees: growth, circuit structure, and acyclicity. *J. Math. Sci. (New York)*, 100(1):1925–1943, 2000. Algebra, 12.

[238] S. Sidki. Finite automata of polynomial growth do not generate a free group. *Geom. Dedicata*, 108:193–204, 2004.

[239] Said Sidki. *Regular trees and their automorphisms*, volume 56 of *Monografías de Matemática [Mathematical Monographs]*. Instituto de Matemática Pura e Aplicada (IMPA), Rio de Janeiro, 1998.

[240] Charles C. Sims. *Computation with Finitely Presented Groups*. Cambridge University Press, 1994.

[241] John Stallings. *Group Theory and Three-Dimensional Manifolds*, volume 4 of *Yale Mathematical Monographs*. Yale University Press, 1971.

[242] B. Steinberg, M. Vorobets, and Y. Vorobets. Automata over a binary alphabet generating free groups of even rank. *Internat. J. Algebra Comput.*, 21(1-2):329–354, 2011.

[243] R. Strebel. Small cancellation groups. In Ghys and de la Harpe [94], pages 215–259.

[244] Z. Šunić and E. Ventura. The conjugacy problem in automaton groups is not solvable. *J. Algebra*, 364:148–154, 2012.

[245] V.Ī. Suščans'kiĭ. Periodic *p*-groups of permutations and the unrestricted Burnside problem. *Dokl. Akad. Nauk SSSR*, 247(3):557–561, 1979. English translation: Soviet Math. Dokl. 20 (1979), no. 4, 766770.

[246] J. Tits. Le problème des mots dans les groupes de Coxeter. In *Symposia Matematica (INDAM Rome 1967/68)*, volume 1, pages 178–185. Academic Press, London, 1969.

[247] H. van der Lek. The homotopy type of complex hyperplane complements. PhD thesis, University of Nijmegen, 1983.

[248] M. Vorobets and Y. Vorobets. On a free group of transformations defined by an automaton. *Geom. Dedicata*, 124:237–249, 2007.

[249] M. Vorobets and Y. Vorobets. On a series of finite automata defining free transformation groups. *Groups Geom. Dyn.*, 4(2):377–405, 2010.

[250] Wikipedia. Geometrization conjecture. https://en.wikipedia.org/wiki/Geometrization_conjecture.

[251] A.J. Wilkie and L. van den Dries. An effective bound for groups of linear growth. *Arch. Math. (Basel)*, 42:391–396, 1984.

[252] A. Williams. MAF (Monoid Automata Factory), 2009. http://maffsa.sourceforge.net/.

[253] J.S. Wilson. Groups with every proper quotient finite. *Proc. Cambridge Philos. Soc.*, 69:373–391, 1971.

[254] J.S. Wilson. On just infinite abstract and profinite groups. In M. du Sautoy, D. Segal, and A. Shalev, editors, *New horizons in pro-p groups*, volume 184 of *Progr. Math.*, pages 181–203. Birkhäuser Boston, Boston, MA, 2000.

[255] J.S. Wilson. Further groups that do not have uniformly exponential growth. *J. Algebra*, 279(1):292–301, 2004.

[256] J.S. Wilson. On exponential growth and uniformly exponential growth for groups. *Invent. Math.*, 155(2):287–303, 2004.

[257] J.S. Wilson and P.A. Zalesskii. Conjugacy separability of certain torsion groups. *Arch. Math. (Basel)*, 68(6):441–449, 1997.

[258] H. Yamada. Real-time computation and recursive functions not real-time computable. *IRE Trans.*, EC-11:753–760, 1962.

[259] A. Żuk. Automata groups. In G. Cortiñas, editor, *Topics in noncommutative geometry*, volume 16 of *Clay Math. Proc.*, pages 165–196. Amer. Math. Soc., Providence, RI, 2012.

Index of Notation

283

Index of Names

Index of Topics and Terminology

surface group, 18, 28
symmetric closure (of group presentation), 26
symmetric closure (of relators), 110
synchronous
  2-variable fsa, 93
  automaton, 198
  fellow traveller property, 127
syntactic congruence, 51, 108
syntactic monoid, 52, 108
syntactic morphism, 52

terminal (of grammar), 37
terminating rewriting system, 120
thin triangle, 152
Thompson's group F, 35, 113, 196
Thurston's Geometrisation Conjecture, 146
tidy real-time language, 252
total function, 46
transducer
  deterministic, 195
  finite state, 194
  inverse, 197
  minimisation of, 214
  permutational, 197
transition function
  of fsa, 52
  of pda, 61
transition monoid, 54
transversal
  Schreier, 9
tree automorphism
  automaton of, 204
  bounded, 204, 239, 268
  circuitous, 206
  directed, 205
  rooted, 205
  section of, 204
triangle
  δ-slim, 151
  δ-thin, 152
  geodesic, 150
Turing machine, 36, 44, 82
  K-step real-time, 249
  deterministic, 44
  linearly bounded, 47
  non-deterministic, 44
  real-time, 47, 249
Type 0 grammar, 36

Type 1 grammar, 85
Type 2 grammar, 69
unique word-acceptor, 118
unrestricted grammar, 36
useful variable, 69
useless variable, 69, 229
van Kampen diagram, 27, 99
  diameter, 114, 243
  filling length, 114
  radius, 114
variable
  of grammar, 37
  useful, 69
  useless, 69, 229
virtual endomorphism, 207
virtually abelian group, 170, 182
weakly branch group, 208
Whitehead problem, 167
Wirtinger presentation, 33
word, 4
  cyclically reduced, 7, 26
  empty, 4
  geodesic, 4, 5, 21, 169
  positive, 30
  quasigeodesic, 24
  reduced, 7
  reduced (in rewriting system), 119
word metric, 153
word problem, 7, 39, 133, 166, 213
  generalised, 8, 167, 186
  of group, 97
  soluble, 7
word-acceptor, 117, 118, 127
  coset, 189, 190
  unique, 118
word-difference, 127
  automaton, 129
word-hyperbolic Coxeter group, 30, 141
word-hyperbolic group, 28, 30, 105, 111, 139, 145, 150, 170, 187, 191, 240, 252, 254
words over X, 4
wreath product, 12, 147
  restricted, 13
  restricted standard, 13
  standard, 13
wreath product ordering, 120

Printed in the United States
by Baker & Taylor Publisher Services